$\frac{2}{1}$ 11-2

elementary algebra
for college students

Mary P. Dolciani
  Hunter College of the City University of New York

Robert H. Sorgenfrey
  University of California, Los Angeles

Editorial Adviser
Albert E. Meder, Jr.
  Rutgers University, The State University of New Jersey

Houghton Mifflin Company · Boston
New York · Atlanta · Geneva, Illinois · Dallas · Palo Alto

COPYRIGHT © 1971 BY HOUGHTON MIFFLIN COMPANY

All rights reserved. No part of this work may be reproduced or transmitted in any form or by any means, electronic or mechanical, including photocopying and recording, or by any information storage or retrieval system, without permission in writing from the publisher. Printed in the U.S.A.

Library of Congress Catalog Card No. 78-146720

ISBN 0-395-12069-1

preface

This textbook is intended for the college student needing instruction in the fundamentals of elementary algebra. Completion of the course should lead to skill in performing algebraic processes and solving equations and inequalities, through quadratics. It should also lead to a realization that the subject rests on a firm structural basis.

The book is designed for a one-semester (or two-quarter) course but may be shortened by omission of topics marked Optional and by limiting the amount of exercise work required. If time is at a premium, the instructor may also wish to give less emphasis to some topics, for example, inequalities and graphing. For a class for which the course is largely review, the work should be easily covered in one quarter.

The notion of set is introduced early and used throughout to clarify concepts. The language is precise and explicit without being unduly technical. The sequence of topics makes use of a spiral development, by which most concepts appear repeatedly, first in simple terms and later with deeper significance. There is a summary and review at the end of each chapter.

Attention is called to the large number of illustrative examples as well as the abundance of exercises and problems. Answers to the odd-numbered exercises appear at the end of the book.

A second color has been used in the printing, not only to make the book more attractive, but especially to emphasize important properties and to point out parts of a solution or diagram. Special care has been taken to place diagrams close to the related part of the text.

Mary P. Dolciani
Robert H. Sorgenfrey

contents

1 Numbers and Sets

1-1	Naming Numbers; Equality and Inequality	1
1-2	Grouping Symbols	3
1-3	The Number Line	6
1-4	Comparing Numbers	9
1-5	Specifying Sets	12
1-6	Comparing Sets	15
	Chapter Summary	18
	Chapter Review	18

2 Variables and Open Sentences

2-1	Variables	20
2-2	Factors, Coefficients, and Exponents	24
2-3	Order of Operations	28
2-4	Variables and Open Sentences	30
2-5	Variables and Quantifiers	33
2-6	Applying Mathematical Expressions and Sentences	35
2-7	Using Open Sentences in Flow Charts (Optional)	36
	Chapter Summary	39
	Chapter Review	40

3 Addition and Multiplication of Real Numbers

3-1	Axioms of Closure and Equality	42
3-2	Commutative and Associative Axioms	44
3-3	Addition on the Number Line	47
3-4	The Opposite of a Real Number	51
3-5	Absolute Value	54
3-6	Rules for Addition	56
3-7	The Distributive Axiom	60
3-8	Rules for Multiplication	63
3-9	The Reciprocal of a Real Number	67
3-10	More About Flow Charts (Optional)	69
	Chapter Summary	73
	Chapter Review	74

4 Solving Equations and Problems

4–1	Transforming Equations by Addition	76
4–2	Subtracting Real Numbers	79
4–3	Transforming Equations by Multiplication	83
4–4	Dividing Real Numbers	85
4–5	Using Several Transformations	89
4–6	Using Equations to Solve Problems	91
4–7	Equations Having the Variable in Both Members	95
4–8	Equations and Functions	101
	Chapter Summary	104
	Chapter Review	105

5 Solving Inequalities; More Problems

5–1	Axioms of Order	106
5–2	Intersection and Union of Sets	110
5–3	Combining Inequalities	113
5–4	Absolute Value in Open Sentences	116
5–5	Problems about Integers	117
5–6	Uniform-Motion Problems	120
5–7	Mixture and Other Problems	125
	Chapter Summary	128
	Chapter Review	129

6 Operations with Polynomials

6–1	Adding Polynomials	130
6–2	Subtracting Polynomials	133
6–3	The Product of Powers	135
6–4	The Power of a Product	136
6–5	Multiplying a Polynomial by a Monomial	138
6–6	Multiplying Two Polynomials	140
6–7	Problems about Areas	142
6–8	Powers of Polynomials	145
6–9	The Quotient of Powers	147
6–10	Zero and Negative Exponents	150
6–11	Dividing a Polynomial by a Monomial	152
6–12	Dividing a Polynomial by a Polynomial	154
	Chapter Summary	157
	Chapter Review	158

7 Special Products and Factoring

7–1	Factoring in Algebra	160
7–2	Identifying Monomial Factors	163
7–3	Multiplying the Sum and Difference of Two Numbers	166
7–4	Factoring the Difference of Two Squares	167
7–5	Squaring a Binomial	169
7–6	Factoring a Trinomial Square	172
7–7	Multiplying Binomials at Sight	173
7–8	Factoring the Product of Binomial Sums or Differences	175
7–9	Factoring the Product of a Binomial Sum and a Binomial Difference	178
7–10	General Method of Factoring Quadratic Trinomials	179
7–11	Combining Several Types of Factoring	181
7–12	Working with Factors Whose Product is Zero	183
7–13	Solving Polynomial Equations by Factoring	185
7–14	Using Factoring in Problem Solving	188
	Chapter Summary	192
	Chapter Review	193

8 Operations with Fractions

8–1	Defining Fractions	195
8–2	Reducing Fractions to Lowest Terms	196
8–3	Ratio	199
8–4	Percent and Percentage Problems	203
8–5	Multiplying Fractions	206
8–6	Dividing Fractions	208
8–7	Expressions Involving Multiplication and Division	210
8–8	Sums and Differences of Fractions with Equal Denominators	212
8–9	Sums and Differences of Fractions with Unequal Denominators	214
8–10	Mixed Expressions	216
8–11	Complex Fractions	218
	Chapter Summary	220
	Chapter Review	220

9 Using Fractions

9–1	Solving Equations and Inequalities	222
9–2	Percent Mixture Problems	223
9–3	Investment Problems	225

9–4	Solving Fractional Equations	228
9–5	Rate-of-Work Problems	230
9–6	Motion Problems	232
9–7	Loops and Subscripts in Flow Charts (Optional)	235
	Chapter Summary	240
	Chapter Review	240

10 Functions, Relations, and Graphs

10–1	Functions Described by Tables	241
10–2	Coordinates in a Plane	245
10–3	Relations	247
10–4	Open Sentences in Two Variables	249
10–5	The Graph of a Linear Equation in Two Variables	252
10–6	Slope of a Line	255
10–7	The Slope-Intercept Form of a Linear Equation	258
10–8	Determining an Equation of a Line	260
10–9	Direct Variation and Proportion	262
10–10	Quadratic Functions	267
10–11	Inverse Variation	271
10–12	Joint Variation and Combined Variation	276
	Chapter Summary	280
	Chapter Review	281

11 Systems of Open Sentences in Two Variables

11–1	The Graphic Method	283
11–2	The Addition or Subtraction Method	286
11–3	Problems with Two Variables	288
11–4	Multiplication in the Addition or Subtraction Method	290
11–5	The Substitution Method	293
11–6	Digit Problems	295
11–7	Motion Problems	297
11–8	Age Problems	298
11–9	Problems about Fractions	300
11–10	Graph of an Inequality in Two Variables	302
11–11	Graphs of Systems of Linear Inequalities	304
	Chapter Summary	306
	Chapter Review	307

12 Rational and Irrational Numbers

12–1	The Nature of Rational Numbers	308
12–2	Decimal Forms for Rational Numbers	312
12–3	Roots of Numbers	315
12–4	Properties of Irrational Numbers	318
12–5	Geometric Interpretation of Square Roots	322
12–6	Multiplication, Division, and Simplification of Square-Root Radicals	325
12–7	Addition and Subtraction of Square-Root Radicals	328
12–8	Multiplication of Binomials Containing Square-Root Radicals	330
12–9	Radical Equations	331
	Chapter Summary	335
	Chapter Review	336

13 Quadratic Equations and Inequalities

13–1	The Square-Root Property	337
13–2	Sum and Product of the Roots of a Quadratic Equation	339
13–3	Solution by Completing a Trinomial Square	341
13–4	Solution by Using the Quadratic Formula	344
13–5	The Nature of the Roots of a Quadratic Equation	348
13–6	Solving Quadratic Inequalities	351
13–7	Using Graphs of Equations to Solve Inequalities	353
	Chapter Summary	355
	Chapter Review	356

Tables 357

Answers for Odd-Numbered Exercises 360

Index 391

list of symbols

		PAGE			PAGE
$=$	is equal to	1, 15	$\therefore$	therefore	30
$\neq$	is not equal to	2, 15	$\stackrel{?}{=}$	Is this statement true?	50
$\overline{YF}$	line segment Y, F	8	$|\ |$	absolute value	54
$<$	is less than	9	$\geq$	is greater than or equal to	55
$>$	is greater than	9	$\leq$	is less than or equal to	55
$\{\ \}$	set	12	$\mathcal{R}$	{the real numbers}	78
$\in$	is an element of	12	$\cap$	intersection of sets	110
$\notin$	is not an element of	12	$\cup$	union of sets	111
$\emptyset$	empty set	13	$\%$	percent	203
$\ldots$	continue unendingly	15	$\sqrt{a}$	the positive square root of a	315
	or and so on through	17	$\pm$	positive or negative	315
$\doteq$	is approximately equal to	23		*or* plus or minus	340

1 numbers and sets

1-1 Naming Numbers; Equality and Inequality

In algebra, numbers and the ways to name them play a very great part. The number of dots in Figure 1-1 can be named by the Arabic numeral "4" or by the Roman numeral "IV." The English word "four," the Spanish word "cuatro" (pronounced kwah-tro), and the Swahili word "nne" (pronounced en-nay) also name that number. In Morse code, $\cdots -$ represents four. Each of the expressions

$$3 + 1, \quad 8 \div 2, \quad 9 - 5$$

FIGURE 1-1

designates four. Names, or symbols, for numbers are called **numerical expressions** or **numerals**. The number named by a numerical expression is called the **value** of the expression. Thus, 4 is the value of "$3 + 1$," of "$8 \div 2$," and of "$9 - 5$."

To show that the numerical expressions "$3 + 1$" and "4" have the same value, we use the important symbol $=$, which stands for the word **"equals"** or for the words **"is equal to."** Thus, we write the *statement*

$$3 + 1 = 4$$

and say, "Three plus one equals four." Any statement of equality is called an **equation**. The following equations are true statements.

$$8 \div 2 = 9 - 5; \quad 3 + 1 = 8 \div 2.$$

The statement

$$3 \times 1 = 4$$

is false because "3×1" names the number three, not four. One way to change this statement into a true one is to use the symbol $\neq$, read "**is not equal to**" or "**does not equal**,"

$$3 \times 1 \neq 4.$$

Any statement asserting that two numerical expressions do not have the same value is called an **inequality**.

The statement $10 - 1 \neq 3 \times 3$ is false.

The statement $6 \div 1 \neq 1 \div 6$ is true.

EXERCISES

Replace the __?__ with one of the symbols $=$ or $\neq$ to make a true statement.

1. 8×24 __?__ 48×4
2. $56 \div 7$ __?__ 4×2
3. 8×0 __?__ $8 + 0$
4. $6 \div 3$ __?__ $\frac{3}{6}$
5. $\frac{1}{2} + \frac{5}{10}$ __?__ 1
6. $\frac{1}{3} + \frac{0}{3}$ __?__ $\frac{10}{3}$
7. $0 + 0$ __?__ 0×0
8. $\frac{1}{5} \times 6$ __?__ $\frac{5}{6}$
9. $1 + 1\frac{2}{4}$ __?__ $2\frac{1}{2}$
10. $\frac{3}{7} - \frac{2}{7}$ __?__ $3 \times \frac{1}{21}$
11. $0.7 + 0.2$ __?__ 3×0.3
12. $0.45 - 0.25$ __?__ $2 \div 10$

Make each sentence a true statement by replacing each __?__ with a numeral. In any sentence containing more than one __?__, the same replacement should be used for each __?__.

Example. $63 \neq 7 \times$ __?__.

Solution: $63 \neq 7 \times 4$. (In place of "4," any numeral that does not name 9 may be used.)

13. $7 + 2 = 2 +$ __?__
14. $8 \times 5 =$ __?__ $\times 8$
15. $14 -$ __?__ $\neq 10$
16. __?__ $+ 57 \neq 57$
17. $15 +$ __?__ $= 5 \times 3$
18. $6 \times$ __?__ $= 3 + 3$
19. __?__ $\div 9 = 0$
20. __?__ $\div 5 \neq 1$
21. $7 \div$ __?__ $= 1$
22. $1 \times$ __?__ $= 1$

23. $4 \div \underline{?} = \frac{1}{2}$
24. $\underline{?} \times \frac{1}{3} \neq 2$
25. $7 + 12 + \underline{?} = 42 - 9$
26. $33 + 16 + \underline{?} = 94 - 21$
27. $3\frac{1}{2} - \underline{?} = 2\frac{1}{4} + \frac{3}{4}$
28. $\frac{15}{6} - \frac{3}{2} = \underline{?} \times 3$
29. $0.24 \div 0.6 = 1 - \underline{?}$
30. $4.5 \div 0.9 = 1 \div \underline{?}$
31. $\underline{?} \times \underline{?} = 1$
32. $\underline{?} \div \underline{?} = 1$
33. $\underline{?} + \underline{?} = \underline{?}$
34. $\underline{?} - \underline{?} = \underline{?}$

1–2 Grouping Symbols

The numerical expression "$2 \times 5 + 6$" is ambiguous. Does it mean, "The sum of twice five and six, that is, 16"? Or does it mean, "Twice the sum of five and six, that is, 22"? To make the meaning of this expression clear, we use parentheses as punctuation marks. When we write

$$(2 \times 5) + 6,$$

we name the number 16. When we write

$$2 \times (5 + 6),$$

we name 22.

A pair of parentheses is called a **symbol of inclusion** or a **grouping symbol** because it is used to enclose, or include, an expression for a particular number. In the expression "$2 \times (5 + 6)$," the parentheses group the numerals "5" and "6" together with the symbol $+$, and thus show that the *sum* of 5 and 6 is to be multiplied by 2.

In writing "$2 \times (5 + 6)$," we usually omit the symbol $\times$ and write simply

$$2(5 + 6).$$

Similarly, the product 2×11 may be expressed in any of the forms

$$2(11), \quad (2)11, \quad \text{or} \quad (2)(11).$$

Brackets and braces are also used as grouping symbols:

Parentheses	Brackets	Braces
$2(5 + 6)$	$2[5 + 6]$	$2\{5 + 6\}$

The horizontal bar in a fraction symbol often acts as a grouping symbol, as well as a division sign. For example, in the expression below, the bar groups the "17" and "3"; it also groups the "8" and "1." The bar indicates that the number (17 − 3) is to be divided by the number (8 − 1).

$$\frac{17-3}{8-1} = \frac{14}{7} = 2.$$

Notice that each of the expressions $\frac{\text{"}17-3\text{"}}{8-1}$, $\frac{\text{"}14\text{"}}{7}$, and "2" names the number two. But the last numeral, "2," is a *simpler name* for two than the other expressions. Whenever we replace a numerical expression by the simplest or most common numeral for the number named, we have **simplified the expression**.

The statement

$$\frac{17-3}{8-1} = \frac{14}{7}$$

suggests that substituting "14" for "17 − 3" and "7" for "8 − 1" in the fraction $\frac{\text{"}17-3\text{"}}{8-1}$ does not change the value of the fraction. Whenever we simplify a numerical expression, we use the following:

Substitution Principle

Changing the numeral by which a number is named in an expression does not change the value of the expression.

In the expression "4[70 − (8 × 7)]," a pair of parentheses appears inside a pair of brackets. Using different grouping symbols (such as pairs of parentheses and brackets) helps keep track of the mates in each pair. To simplify an expression which shows one grouping inside another grouping, first simplify the numeral in the innermost grouping symbol and then work toward the outermost grouping symbol until all symbols of inclusion have been removed. Thus:

$$\begin{aligned} 4[70 - (8 \times 7)] &= 4[70 - 56] \\ &= 4[14] \\ &= 56 \end{aligned}$$

Fractions may also involve more than one grouping. For example,

$$4 + \frac{3 + (15 \div 5)}{7 - 1} = 4 + \frac{3 + 3}{6} = 4 + \frac{6}{6} = 4 + 1 = 5.$$

EXERCISES

Simplify each of the following expressions.

1. $(40 + 7) - 18$
2. $93 - (6 + 22)$
3. $0 \times [3 - (1 \times 2)]$
4. $4 \times [0 + (3 \div 1)]$
5. $(21 \times 3) + 2$
6. $22 + (16 \times 2)$
7. $\dfrac{8 \times 4}{5 - 1} + 3$
8. $7 - \dfrac{6 \times 18}{22 + 5}$
9. $[7 \times (8 + 4)] \div 70$
10. $144 \div [3 \times (8 + 4)]$
11. $(12 \times 4) \div (3 \times 2)$
12. $[8 - (15 \div 3)] \times 8$

Make a true statement by replacing each __?__ with one of the symbols = or $\neq$.

13. $\dfrac{100 - 64}{10 - 8}$ __?__ $10 + 8$
14. $\dfrac{100 + 64}{10 + 8}$ __?__ $10 - 8$
15. $\dfrac{15 \times 3}{8 + 1}$ __?__ $40 \div 8$
16. $\dfrac{7 + 28}{7 \times 5}$ __?__ $(27 \div 3) - 8$
17. $6 \times \{4 + 7\}$ __?__ $\{5 \times 5\} \times 5$
18. $28 \div (8 - 4)$ __?__ $(10 \div 5) \div 2$
19. $2 \times [3 \times (4 + 1)]$ __?__ $(6 \times 6) - 6$
20. $3 \times [24 \div (5 + 3)]$ __?__ $36 - (6 \times 6)$
21. $2 \times [2 \times (3 \times 4)]$ __?__ $64 \div [32 \div (2 \div 1)]$
22. $4 \times [3 \times (5 - 2)]$ __?__ $96 \div [32 \div (5 - 3)]$
23. $(5 + 7) \times (3 + 8)$ __?__ $8 + [(17 \times 3) + 5]$
24. $3 \times [(5 + 1) \times (6 \div 2)]$ __?__ $[18 + (144 \div 8)] + 3$
25. $\dfrac{[8 \times 6] \div 12}{4 - [6 \div 2]} + 1$ __?__ $27 - \dfrac{40 + 4}{8 - 6}$
26. $\dfrac{(6 \times 8) - 8}{(36 \div 2) + 2} + 2$ __?__ $1 + \dfrac{30 - 3}{1 + 4}$
27. $\dfrac{(4 \times 3) + 22}{8 + (27 \div 3)} - \dfrac{8 - 4}{2 + 2}$ __?__ $\dfrac{(7 \times 6) + 6}{30 - (2 \times 3)} - 1$
28. $\dfrac{\{[(2 \times 2) \times 2] + (2 \times 2)\} + 40}{\{[(3 \times 3) \times 3] - (3 \times 3)\} - 5}$ __?__ $\dfrac{[2 \times (3 \times 3)] + (2 \times 3)}{[3 \times (2 \times 2)] - (2 \times 3)}$
★29. $2 \left\{ \dfrac{[5 - (2 \times 2)] + (8 \times 3)}{[(24 \div 6) + 4] - 3} \right\}$ __?__ $[2 + 5 - (30 \div 6)] \times \dfrac{12 - 2}{5 - 3}$
★30. $\dfrac{90 - [(13 \times 18) \div 3]}{[7 + (6 \times 4)] - 19}$ __?__ $\{[(18 \times 3) \div 9] - 5\} \times \dfrac{18 \div 2}{3 \times 3}$

1-3 The Number Line

Figure 1-2 shows how to picture numbers as points on a line.

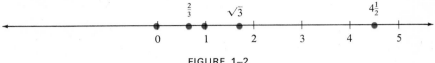

FIGURE 1-2

To construct such a **number line** (*number scale*), choose a starting point **(origin)** on a line and label it "0." At some convenient distance on one side of the origin, select a second point and mark it "1." The side containing the point paired with the number 1 is called the **positive side** of the line, and the direction from the origin to the point labeled "1" is called the **positive direction** on the number line. (On a horizontal number line, the side to the right of the origin is usually chosen to be the positive side of the line.)

Using the distance between the points labeled "0" and "1" as the *unit length* on the line, we can pair each point on the positive side of the line with the number which measures the distance along the line from the origin to the point. We call the numbers paired with points on the *positive* side of the line **positive numbers**. Thus, 1, $\frac{2}{3}$, $\sqrt{3}$, and $4\frac{1}{2}$ are all positive numbers.

On the number line, the point paired with a number is called the **graph** of that number. The number paired with a point is called the **coordinate** of that point.

Example 1. State the coordinate of each of the points *A*, *B*, and *C* on the given number line.

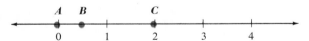

Solution: *A*: 0; *B*: $\frac{1}{2}$; *C*: 2.

Example 2. Draw a number line and on it show the graphs of $\frac{3}{4}$ and 2.5.

Solution:

To assign coordinates to points to the left of the origin on the number line, we introduce **negative numbers**. In Figure 1-3, the coordinate of *R* is ⁻3 (read "negative three") because *R* is 3 units **to the left** of the origin. Similarly, the coordinate of *S* is ⁻1 ("negative one"), and the coordinate of *T* is ⁻$\frac{1}{2}$ ("negative one-half"). Notice that a negative number is named by a numeral with a negative (minus) sign ⁻ written above and to the left of an ordinary number symbol.

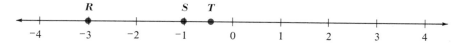

FIGURE 1-3

The side of the number line that contains the points with negative coordinates is called the **negative side** of the line. The direction from the origin to any point on the negative side of the line is called the **negative direction** on the line.

Any number which is either a positive number, a negative number, or 0 is known as a **real number**. In working with a number line, the following facts are taken for granted.

1. **There is exactly one point on the number line paired with any real number.**
2. **There is exactly one real number paired with any given point on the number line.**

Sometimes to emphasize that a number like 1 is a *positive* number, we call it "positive 1" and denote it by the symbol "⁺1." Notice that the small signs ⁺ and ⁻ in the symbols "⁺1" and "⁻1" indicate the directions of the corresponding points from the origin on a number line. They do *not* indicate addition and subtraction.

Because positive and negative numbers suggest opposite directions, they are sometimes called **directed numbers**. They are very useful in describing measurements that involve *direction* as well as *size*. Here are a few examples:

1. Let 10 refer to a *profit* of $10. Then ⁻10 refers to a *loss* of $10.
2. Let $3\frac{1}{2}$ refer to $3\frac{1}{2}$ miles *east*. Then ⁻$3\frac{1}{2}$ refers to $3\frac{1}{2}$ miles *west*.
3. Let 5 refer to 5 seconds *after* launch time of a rocket. Then ⁻5 refers to 5 seconds *before* launch time.

EXERCISES

Draw a horizontal number line and on it show the graphs of the given numbers.

1. 4
2. 2
3. ⁻5
4. ⁻3
5. ⁻1.5
6. ⁻2.5
7. ⁻$4\frac{1}{3}$
8. ⁻$5\frac{1}{4}$

State the coordinate of the point at which we would arrive on a number line if we were to start at the origin and:

9. Move five units in the positive direction, then three more units in that direction.
10. Move four units in the negative direction, then seven more units in that direction.
11. Move six units in the positive direction, then six units in the negative direction.
12. Move three units in the negative direction, then three units in the positive direction.
13. Move five units in the negative direction, then two units in the positive direction.
14. Move eight units in the positive direction, then nine units in the negative direction.

Name the coordinate of the point described. Refer to the number line below.

Example. The point that is one-fourth of the distance from A to R.

Solution: The length of $\overline{AR}$ is 6 units. Let us call the desired point X. Then X is $\frac{1}{4} \times 6 = \frac{3}{2}$, or $1\frac{1}{2}$, units from A toward R. As shown in the diagram below, its coordinate is $-\frac{1}{2}$.

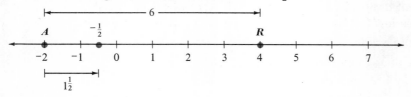

15. The midpoint of $\overline{DK}$
16. The midpoint of $\overline{SR}$
17. The point 5 units to the right of N
18. The point 3 units to the left of I
19. The point $2\frac{1}{2}$ units to the right of E
20. The point $3\frac{1}{2}$ units to the left of U
21. The point 1.5 units to the left of K
22. The point 5.5 units to the right of B
23. The point one-fourth of the way from B to E
24. The point one-third of the way from N to E

25. The point one-sixth of the distance from *S* to *I*
26. The point one-fifth of the distance from *A* to *L*
27. The point two-thirds of the distance from *C* to *A*
28. The point three-fourths of the distance from *H* to *S*
★29. The point between *D* and *H* that is twice as far from *H* as it is from *D*
★30. The point to the right of *U* that is half as far from *U* as it is from *I*

1-4 Comparing Numbers

The true statement "⁻3 ≠ 2" is pictured on the horizontal number line in Figure 1–4 by the fact that the graphs of ⁻3 and 2 are different points. The positive direction in this figure is toward the right, and the graph of ⁻3 lies *to the left* of the graph of 2. Thus, in moving *from* **the graph of ⁻3** *to* **the graph of 2**, we go in the positive direction on the line.

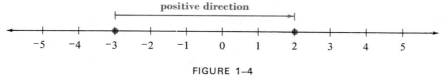

FIGURE 1–4

Because numbers increase as one moves in the positive direction, we say that

⁻3 is less than 2,

and we use the *directed inequality symbol* < (read "**is less than**") to write

⁻3 < 2.

The *directed inequality symbol* > stands for the words "**is greater than**";

2 > ⁻3

means

2 is greater than ⁻3.

On the number line in Figure 1–4, this means that the graph of 2 lies *to the right* of the graph of ⁻3. Therefore, a move *from* **the graph of 2** *to* **the graph of ⁻3** is in the negative direction on the line (Figure 1–5).

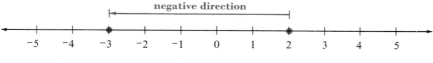

FIGURE 1–5

Note that the inequalities "$^-3 < 2$" and "$2 > {}^-3$" give the same information. Notice also that both inequality symbols point toward the numeral naming the smaller number.

The statement "2 is *between* $^-5$ and 3" is true because $^-5$ is less than 2 *and* 2 is less than 3 (Figure 1–6). This latter statement may be expressed as

$$^-5 < 2 \quad \text{and} \quad 2 < 3.$$

However, a more compact way to state this fact is

$$^-5 < 2 < 3$$

(read "$^-5$ is less than 2, which is less than 3").

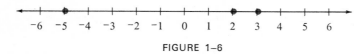

FIGURE 1–6

The relationship among $^-5$, 2, and 3 can also be stated

$$3 > 2 > {}^-5$$

(read "3 is greater than 2, which is greater than $^-5$").

Example. Write each statement in compact form and then tell whether the statement is true or is false.

 a. $^-5 < {}^-2$ and $^-2 < 4$. **b.** $^-5 < 3$ and $3 < 1$.

Solution: **a.** $^-5 < {}^-2 < 4$; true. **b.** $^-5 < 3 < 1$; false.

A statement, such as "$^-5 < {}^-2$ and $^-2 < 4$" or "$^-5 < 3$ and $3 < 1$," which is formed by joining two statements by the word *and* is called a **conjunction** of statements. For a conjunction to be true, *both* of the joined statements must be true. Thus, "$^-5 < 3$ and $3 < 1$" is false because "$3 < 1$" is false.

EXERCISES

Copy each sentence and replace the __?__ with one of the symbols =, <, or > to make a true statement.

1. 4 __?__ 7
2. 0 __?__ 0
3. $\frac{15}{3}$ __?__ $3 + 1$
4. 12×0 __?__ $12 + 0$

5. $^-1$ __?__ 1
6. 5 __?__ $^-5$
7. $\dfrac{8-3}{5-2}$ __?__ $1 + 0.6$
8. $^-2$ __?__ $\dfrac{24-6}{9}$
9. $6 \div 3$ __?__ $12 \div 6$
10. 72×1 __?__ 73
11. $^-3$ __?__ $^-3 \times 0$
12. $^-2$ __?__ $^-2 \times 1$
13. $\tfrac{1}{2}$ __?__ $\tfrac{1}{3} + \tfrac{0}{2}$
14. $0.75 + 0.25$ __?__ $\tfrac{3}{4} + \tfrac{1}{4}$
15. 1.2431 __?__ 1.2432
16. $^-1.2431$ __?__ $^-1.2432$
17. $\tfrac{1}{5} \div 4$ __?__ $\tfrac{4}{5}$
18. $\tfrac{1}{5} \times 4$ __?__ $\tfrac{5}{4}$
19. $0.9 \div 3$ __?__ $0.1 + 0.4$
20. $0.7 - 0.3$ __?__ 0.2×2

Copy each sentence and make it a true statement by replacing each __?__ with a numeral. In any sentence containing more than one __?__, the same replacement should be used for each __?__.

Example. $25 \times 3 > 70 + $ __?__

Solution: $25 \times 3 > 70 + 1$. (In place of "1," the numeral for any number less than 5 may be written.)

21. $26 +$ __?__ $= 2 \times 13$
22. $6 + 2 = 2 +$ __?__
23. __?__ $\div 5 = 0$
24. $124 +$ __?__ $= 124$
25. $13 -$ __?__ > 0
26. __?__ $\times 4 > 0$
27. __?__ $\div 2 = 1$
28. $5 \div$ __?__ $= 5$
29. $16 - 6 <$ __?__ $\times 2$
30. $9 \times 4 < 4 \times$ __?__
31. $0 \times$ __?__ $> ^-3$
32. $1 \times$ __?__ $> ^-1$
33. $\tfrac{3}{4} \times \tfrac{8}{15} =$ __?__ $\times \tfrac{3}{4}$
34. $6 \div$ __?__ $= \tfrac{1}{2}$
35. $9 \div$ __?__ > 2
36. __?__ $\times \tfrac{1}{2} < 3$
37. $\dfrac{48 + \text{?}}{12} = 4 + 1$
38. $\dfrac{42 - \text{?}}{6} = 7 - 2$
39. $5 - 1 <$ __?__ $< 5 + 1$
40. $4 + 1 >$ __?__ $> 4 - 1$
41. __?__ $\times$ __?__ $= 1$
42. __?__ $\div$ __?__ $= 1$
43. $\dfrac{32 - 5}{8} >$ __?__ $> \dfrac{32 - 7}{8}$
44. $\dfrac{56 + 16}{14} <$ __?__ $< \dfrac{56 + 18}{14}$
45. $\tfrac{2}{5} <$ __?__ $< \tfrac{3}{5}$
46. $\tfrac{5}{7} >$ __?__ $> \tfrac{3}{7}$
47. $3.2 - 2.2 <$ __?__ $< 3.4 - 2.2$
48. $7.5 + 3.1 >$ __?__ $> 7.3 + 3.1$
49. $3.14159 <$ __?__ < 3.15159
50. $\sqrt{7} >$ __?__ $> \sqrt{6}$
★51. __?__ $\times$ __?__ $<$ __?__
★52. __?__ $\times$ __?__ $>$ __?__

1–5 Specifying Sets

A useful concept in mathematics is that of **set**, that is, a collection of objects. Each object in a set is called a **member** or **element** of the set.

For example, all the Presidents of the United States form a set, and Abraham Lincoln is a member, or element, belonging to that set. However, objects such as the letter r, William Jennings Bryan, and the number 9 are not elements in the set of all the Presidents. Thus, a set is any collection of objects so well described that one can always tell whether or not an object belongs to the set.

Suppose that a set is composed of the five real numbers 0, 3, $^-7$, 8, $^-14$. Use a capital letter, say F, to name or refer to the set. If we specify the set by listing the objects forming the set within braces { }, then we have

$$F = \{0, 3, {}^-7, 8, {}^-14\}.$$

This is read, "F is the set of numbers 0, 3, $^-7$, 8, $^-14$." It is easily seen that the number 3 is a member of this set and that the number 4 is not. We use a special symbol, $\in$, to mean "**is an element of**," and $\notin$ to mean "**is not an element of**." Thus, $3 \in F$ and $4 \notin F$.

Specifying a set by listing its elements in braces gives a **roster**, or **list**, of the set. The elements of a set need have no relation with one another other than being grouped together. Furthermore, the order of listing the elements is unimportant. What is important is that each element be named in the listing and **no element is listed more than once**.

Often it is inconvenient or impossible to specify a set by roster. For example, we cannot list the names of all members of the **set of real numbers**. So we simply write

{the real numbers}

(read "**the set whose members are** the real numbers" or "**the set of** the real numbers"). Thus, we have specified the set by writing within braces a *description*, or *rule*, that identifies the members of the set.

Example 1. Specify the following set by roster: {the days of the week whose names begin with T}

Solution: {Tuesday, Thursday}

Example 2. Specify by rule: {Washington, John Adams, Jefferson}

Solution: {the first three Presidents of the U.S.A.}

A set of numbers can also be specified by graphing the numbers on a number line. The set of points corresponding to a set of numbers is called

the **graph** of the set. For example,

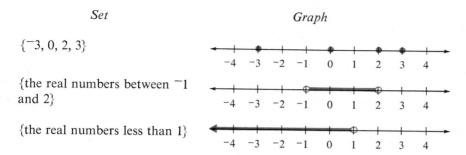

Set	Graph
$\{^-3, 0, 2, 3\}$	
{the real numbers between $^-1$ and 2}	
{the real numbers less than 1}	

A heavy shading is used to show that all points on the indicated portion of the line belong to the graph. Open dots or portions of the line not shaded show points not belonging to the graph. A heavy arrowhead implies that the graph continues without end in the indicated direction.

To specify a set, its elements can be identified by:

1. A roster, listing the names of the elements, or
2. A rule, describing the elements, or (if the elements are real numbers)
3. A graph, locating the elements as points on a number line.

Consider the set of those positive numbers that are less than 0. This set certainly has no members because every positive number is greater than 0. We call the set with no members the **empty set** or the **null set**, and we denote it by the symbol ∅. Note that there is only one empty set. Thus,

$$\emptyset = \{\text{the positive numbers less than } 0\},$$

and

$$\emptyset = \{\text{the New England states west of the Mississippi River}\}.$$

EXERCISES

Specify each nonempty set by roster.

1. {the living ex-Presidents of the U.S.A.}
2. {the Governor and U.S. Senators of your state}
3. {the first five months of the year}
4. {the last six months of the year}
5. {the authors of this textbook}
6. {the years between 1850 and 1860}

Specify each set by rule.

7. {Washington, California, Florida, Maine}
8. {Lake Superior, Lake Michigan, Lake Huron, Lake Erie, Lake Ontario}
9. {Houston, Dallas}
10. {New York, Chicago, Los Angeles}
11. ∅
12. {30, 31, 32, 33, 34, 35, 36, 37, 38, 39}

Graph the given set of numbers.

13. {⁻4, 6, 1}
14. {⁻1, 1, 2, ⁻2}
15. {0}
16. ∅
17. {$\frac{1}{2}$, $\frac{3}{2}$, $\frac{5}{2}$}
18. {3, 5, 7, 11}
19. {the real numbers between ⁻2 and 1}
20. {the real numbers between ⁻3 and 2}
21. {the real numbers greater than ⁻2}
22. {the real numbers less than ⁻1}
23. {the positive real numbers less than 4}
24. {the negative real numbers greater than ⁻3}
25. {the negative real numbers less than or equal to ⁻2}
26. {the positive real numbers greater than or equal to 1}
27. {the real numbers less than $\frac{3}{2}$ and greater than ⁻4}
28. {the real numbers greater than ⁻1$\frac{1}{2}$ and less than 5}
29. {the positive and negative real numbers}
30. {the real numbers that are neither positive nor negative}

Copy each sentence and make it a true statement by replacing each __?__ with a numeral or with one of the symbols ∈ or ∉.

31. $\frac{8 \times \underline{}}{4} \in \{8\}$

32. $5 \times \underline{} \notin$ {the positive and negative numbers}

33. $\frac{27 - 3}{6}$ __?__ {the nonnegative numbers}

34. $5(16 - 6)$ __?__ {the nonpositive numbers}

35. {[(3 × 5) + 7] − (3 × 5)} __?__ {the numbers greater than or equal to 2}
 (Notice that the braces on the left side of the sentence in Exercise 35 form a grouping symbol and do not mean "the set whose members are.")

36. [(7 × 5) − {75 ÷ (5 × 5)}] __?__ {the numbers less than ⁻1}
37. [(4 · 15) − (18 ÷ 9)] − (2 · 29) ∈ {the numbers between ⁻3 and __?__}
38. [5(3 + 6) − 7(2 + 4)] − (48 ÷ __?__) ∉ {the positive numbers}

1–6 Comparing Sets

The sets $\{^-1, 0, 6\}$ and $\{0, 6, ^-1\}$ have the same members. They both name the set whose members are $^-1$, 0, and 6. Therefore, we write

$$\{^-1, 0, 6\} = \{0, 6, ^-1\}$$

which is read, "The set whose members are negative one, zero, and six **equals** (or **is equal to**) the set whose members are zero, six, and negative one."

On the other hand, $\{^-1, 0, 6\}$ and $\{^-1, 0, 2\}$ do *not* have the same members. Hence, we write

$$\{^-1, 0, 6\} \neq \{^-1, 0, 2\},$$

which is read, "The set whose members are negative one, zero, and six **does not equal** (or **is not equal to**) the set whose members are negative one, zero, and two."

Although the sets $\{^-1, 0, 6\}$ and $\{^-1, 0, 2\}$ are not equal, Figure 1–7 suggests an important relationship between them. It shows a pairing which assigns to each member of each set *one and only one* member of the other set. Such a pairing of the elements of two sets is called a **one-to-one correspondence**. The number line is a one-to-one correspondence between the set of real numbers and the set of the points on a line.

"Counting" is another one-to-one correspondence, which exists between the set being counted and a set of numbers. Thus, in counting the sides of a triangle (Figure 1–8), the members of the set of sides are paired with the members of $\{1, 2, 3\}$.

$\{^-1, 0, 6\}$
$\updownarrow \updownarrow \updownarrow$
$\{^-1, 0, 2\}$

FIGURE 1–7   FIGURE 1–8

The set of numbers used in counting is called the set of **counting numbers** or **natural numbers**; we name it N.

$$N = \{1, 2, 3, 4, 5, \ldots\}$$

The three dots after the "5" are read "and so on," and indicate that the pattern shown in the list continues without end. We say that N is an **infinite set** because the process of counting its members would *never* come to an end. We say that it is specified above by an *incomplete roster*. The set N may also be specified by the graph

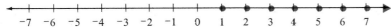

where the heavy arrowhead implies that the graph continues without end.

Here are some other infinite sets, each specified by rule and by incomplete roster and by graph.

W = {the whole numbers} = {0, 1, 2, 3, 4, 5, 6, ...}

J = {the integers} = {0, 1, ⁻1, 2, ⁻2, 3, ⁻3, ...}
     = {..., ⁻3, ⁻2, ⁻1, 0, 1, 2, 3, ...}

E = {the even integers} = {0, 2, ⁻2, 4, ⁻4, 6, ⁻6, ...}
     = {..., ⁻6, ⁻4, ⁻2, 0, 2, 4, 6, ...}

D = {the positive odd integers} = {1, 3, 5, 7, 9, ...}

Notice that when an infinite set is specified by incomplete roster, *enough elements must be listed to show the pattern in the list.*

The **set of real numbers** is another infinite set. It cannot be represented by roster. Its graph is the entire number line:

If there is a counting number which is the number of members in a set, or if the set is the empty set, then the set is a **finite set**. Thus, the sides of a triangle form a finite set. Each of the following sets is also finite:

A = {2, 4, 6, 8, 10}
B = {the negative numbers greater than 0}
C = {the whole numbers named by two-digit decimal numerals}

C can also be specified by writing

$$C = \{10, 11, 12, 13, \ldots, 99\}.$$

The three dots between "13" and "99" are read "and so on through" and mean that the missing numerals follow the pattern indicated by the first four items in the list.

EXERCISES

Give a (complete or incomplete) roster for each set, and state whether or not the set is finite.

1. {the natural numbers less than 7}
2. {the whole numbers between 7 and 19}
3. {the whole numbers greater than 12}
4. {the even integers less than $^-2$}
5. {the odd integers between 7 and 39, inclusive (that is, including 7 and 39)}
6. {the even integers between $^-4$ and 4, inclusive}
7. {the even integers between 3 and 4}
8. {the odd integers between $^-2$ and $^-3$}

Specify each set by rule.

9. $\{21, 22, 23, 24, \ldots, 29\}$
10. $\{0, ^-1, ^-2, \ldots, ^-12\}$
11. $\{1, 7, 5, 3, 9\}$
12. $\{^-5, ^-3, ^-1, ^-2, ^-4\}$
13. $\{0, 5, 10, 15, \ldots\}$
14. $\{2, 4, 6, 8, \ldots\}$

Draw the graph of each set.

15. {the positive integers less than $5\frac{3}{4}$}
16. {the natural numbers less than 7 but greater than 5}
17. {the integers between 3 and 7, inclusive}
18. {the negative integers greater than or equal to $^-4$}

Let $A = \{1, 3, 5, 7\}$ and $B = \{2, 3, 5, 7\}$. Specify each of the following sets.

19. {the integers belonging to A and also to B}
20. {the integers belonging to A, to B, or to both A and B}
21. {the integers in A but not in B}
22. {the integers in B but not in A}
★23. {the positive integers that are neither in A nor in B}
★24. {the positive integers not in both A and B}

chapter summary

1. Names, or symbols, for numbers are called **numerical expressions** or **numerals**. Numerical expressions that name the same number are said to have the same **value** and to be **equal**. A statement which asserts that two numerical expressions are **equal** is called an **equation**. A statement asserting that two numerical expressions do not have the same value is called an **inequality**.

2. **Symbols** of **inclusion**, or **grouping symbols**, are used to make clear the meaning of *ambiguous* numerical expressions. The **Substitution Principle** is used in **simplifying** numerical expressions.

3. On the number line the **point** paired with a number is called the **graph** of that number. The number paired with a point is called the **coordinate** of the point. *Positive* and *negative* numbers are sometimes called **directed numbers**, since, on the number line, their graphs are in opposite directions from the **origin**.

4. The **directed inequality symbol** $<$ is read "is less than." The **directed inequality symbol** $>$ is read "is greater than."

5. A statement which is formed by joining two statements by the word *and* is called a **conjunction** of statements.

6. To specify a **set**, identify its **members**, or **elements**, by **roster**, **rule**, or **graph**. The set with no members is called the **empty set** or the **null set**.

7. Two sets are **equal** when they have the same members. Given two sets, a pairing which assigns to each member of each set *one and only one* member of the other set is called a **one-to-one correspondence**. A set is an **infinite set** if the process of counting its members would never end. If the number of members in a set can be named by a **counting number**, or if the set is the empty set, the set is a **finite set**.

CHAPTER REVIEW

1–1
1. Replace each __?__ with one of the symbols $=$ or $\neq$ to make a true statement.
    a. $0 + 2$ __?__ 0×2
    b. $0.37 - 0.02$ __?__ 5×0.07

2. Replace each __?__ with a numeral which makes the statement true.
    a. $6 +$ __?__ $= 3 \times 3$
    b. __?__ $+ 29 \neq 29$

1–2 3. Simplify: $[6 - (18 \div 3)] + (4 \times 3)$

1–3 4. Draw a number line, and on it show the graphs of the given points.
   a. 2
   b. The point 5 units to the right of 2.
   c. The point 3 units to the left of the origin.
   d. The point halfway between $^-1$ and 3.

1–4 5. Replace each __?__ with one of the symbols =, >, or < to make a true statement.
   a. $^-4$ __?__ $^-4 \times 0$ b. $\frac{16}{4}$ __?__ 3   c. $\frac{9-3}{4-2}$ __?__ $9-3$

1–5 6. Specify the given set by roster:

   {months of the year whose names begin with A}

   7. Specify the given set by rule:

   {Washington, Oregon, California}

   8. Graph the given set:

   {even counting numbers less than ten}

1–6 9. Tell whether the two given sets are equal.
   a. {1, 2, 3} and {3, 1, 2}
   b. {0, 1, 2, 3, 4, 5} and {counting numbers less than 5}

   10. Tell whether the following sets are finite or infinite.
   a. {even integers less than 7}
   b. {people living on the earth}

2 variables and open sentences

2–1 Variables

Each of the numerical expressions

$$7 \times 2, \quad 7 \times 5, \quad 7 \times 8$$

fits the pattern

$$7 \times n$$

where the letter n may stand for "2," for "5," or for "8"; n is called a *variable*. A **variable** is a symbol which may represent any of the members of a specified set, called the **domain** or **replacement set** of the variable. Thus, the domain of n in this example is $\{2, 5, 8\}$. An individual member of the replacement set, such as 2, 5, or 8, is called a **value** of the variable. A variable with just one value is called a **constant**.

An expression, such as "$7 \times n$," which contains a variable is called a **variable expression** or an **open expression**. A variable expression or a numerical expression is sometimes called a **mathematical expression**.

To read "$7 \times n$," say "seven times n," or "seven multiplied by n," or "the product of seven and n." Of course, in referring to "$7 \times n$" as a "product," we mean that when n is replaced by the name of a number, the resulting expression names a *product of numbers*. For example, if we write "2" in place of n, we have the expression "7×2," which names the product of 7 and 2, that is, 14.

In writing a product that contains a variable, the multiplication sign, $\times$, is usually omitted. Thus, "$7 \times n$" is written "$7n$." But notice that in the expressions "7 × 2" and "7(2)" we do not omit the multiplication symbol, for "72" names seventy-two, not 7 times 2.

Furthermore, in any expression involving "7n," the "7" and "n" are considered to be grouped. For example, "7n + 5" means "(7 × n) + 5." Similarly, "ab ÷ 7n" means "(a × b) ÷ (7 × n)."

Given a particular value for the variable in an open expression, the process of replacing the variable by the numeral for the given value and simplifying the result is known as **evaluating the expression** or **finding its value**.

Example. If the value of a is 4 and the value of d is 1, find the value of

$$\frac{4a + 11d}{a - d}.$$

Solution:
1. Replace the letter a by "4" and d by "1," and insert the necessary multiplication and grouping symbols.

$$\frac{4a + 11d}{a - d}$$

$$\frac{(4 \times 4) + (11 \times 1)}{4 - 1}$$

2. Simplify the expression obtained in Step 1.

$$\frac{16 + 11}{3} = \frac{27}{3} = 9$$

A mathematical expression using numerals or variables or both to indicate a product or a quotient is called a **term**. The expression "3n" is a term, and so are "8," "5xy," and $\frac{x}{a-b}$. In the expression "3ab + 3(a − b) − $\frac{b - 2a}{5}$," there are three terms: "3ab," "3(a − b)," and "$\frac{b - 2a}{5}$." Note that "a + b" has two terms, but "(a + b)" is a single term.

EXERCISES

Evaluate each of the following expressions, given that the value of a is 1, b is 2, c is 3, x is 12, y is 0, and z is $\frac{1}{2}$.

1. bc
2. cx
3. $x - bc$
4. $b - cy$
5. $xz - bc$
6. $ac - yz$
7. $b(x - c)$
8. $c(a - y)$
9. $(bb)(cc)$
10. $(zz)(bb)$

22 CHAPTER TWO

11. $bc(x - y)$
12. $bz(x - c)$
13. $abcxy$
14. $abcxz$
15. $(x - 2c)(2c - b)$
16. $(c - 2y)(bc - a)$
17. $\dfrac{x - 2c}{b}$
18. $\dfrac{bx - c}{2c + a}$
19. $b(c - a) + \dfrac{x}{bc}$
20. $5(b + c) - \dfrac{bx}{c}$
21. $\dfrac{4(bcc - xz)}{b}$
22. $\dfrac{6(bbc - 2zc)}{2x + c}$
23. $\dfrac{16cz - 2x}{a - 3xy} + \dfrac{x - 2b}{bz}$
24. $\dfrac{x(2c - b)}{b(c - b)} - \dfrac{x(bz + 1)}{2c + b}$

PROBLEMS

Evaluate each of the following expressions which are taken from practical situations. In each problem assign the indicated values to the variables. Refer to the diagrams shown.

1. Distance traveled at constant rate r in time t: rt
   Let r be 280 (miles per hour) and t be 18 (hours). Distance is then in miles.

2. Area of a rectangle of width w and length l: lw
   Let l be 84 (feet) and w be 27 (feet). Area is then in square feet.

EX. 2

3. Perimeter of a rectangle: $2l + 2w$
   Let l be 210 (feet) and w be 162 (feet). Perimeter is then in feet.

4. Amount (in dollars) of an investment of P dollars at simple interest: $P + Prt$
   Let P be 2500 (dollars), r be 0.04 (4% per year), and t be 10 (years).

5. Perimeter of a triangle: $a + b + c$
   Let a be 6 (inches), b be 8 (inches), and c be 10 (inches). Perimeter is then in inches.

6. Area of a right triangle: $\tfrac{1}{2}ab$
   Let a be 12 (inches) and b be 16 (inches). Area is then in square inches.

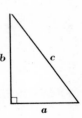

EX. 5–6

7. Area of a parallelogram: bh
   Let b be 12.5 (centimeters) and h be 7.5 (centimeters). Area is then in square centimeters.

8. Perimeter of a parallelogram: $2(a + b)$
   Let a be 5.5 (meters) and b be 25.2 (meters). Perimeter is then in meters.

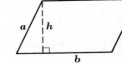

EX. 7–8

9. Perimeter of an isosceles trapezoid: $2a + b + c$
   Let a be 54 (yards), b be 98 (yards), and c be 48 (yards).
   Perimeter is then in yards.

10. Area of a trapezoid: $\frac{1}{2}h(b + c)$
    Let h be 25 (yards), b be 50 (yards), and c be 25 (yards). Area is then in square yards.

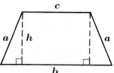
EX. 9–10

11. Area of a circle: $(\pi r)r$
    Let r be 14 (inches). $\pi \doteq \frac{22}{7}$ ($\doteq$ means **"is approximately equal to"**).
    Area is then in square inches.

12. Circumference of a circle: πd
    Let d be $3\frac{1}{2}$ (feet). $\pi \doteq \frac{22}{7}$. Circumference is then in feet.

13. Perimeter of a Norman window: $2(r + h) + \pi r$
    Let r be 2.00 (feet), h be 6.00 (feet). $\pi \doteq 3.14$. Perimeter is then in feet.

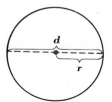
EX. 11–12

14. Perimeter of a trefoil: πr
    Let r be 3.75 (feet). $\pi \doteq 3.14$. Perimeter is then in feet.

15. Volume of a rectangular solid: lwh
    Let l be 12.5 (inches), w be 5.5 (inches), and h be 7.5 (inches). Volume is then in cubic inches.

16. Surface area of a rectangular solid: $2(lw + wh + lh)$
    Let l be 10 (inches), w be 6 (inches), and h be 8 (inches). Surface area is then in square inches.

17. Cutting speed of a tool on a lathe: $\dfrac{\pi dn}{12}$
    Let d be 1.80 (inches) and n be 200 (rpm, that is, revolutions per minute). $\pi \doteq 3.14$. Speed is then in feet per minute.

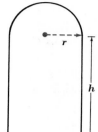
EX. 13

18. Velocity of a satellite orbiting the earth at the equator: $\dfrac{2\pi(3960 + h)}{P}$

    Let h be 22,300 (miles) and P be 24 (hours). The velocity is then in miles per hour. $\pi \doteq 3.14$.

19. A value that is $\frac{4}{5}$ of the way from l to r: $l + \frac{4}{5}(r - l)$
    Let l be 3502 and r be 3522.

EX. 14

20. Temperature (in degrees) on Kelvin (absolute) scale: $\frac{5}{9}(F - 32) + 273$
    Let F be 77 (degrees Fahrenheit).

★21. Focal length of a thin lens: $\dfrac{rs}{(n - 1)(r + s)}$

    Let r be 11.6 (centimeters), s be 9.4 (centimeters), and n be 1.6 (index of refraction).

★22. Total resistance of three resistances in parallel: $\dfrac{rst}{st + rt + rs}$

    Let r be 19 (ohms), s be 90 (ohms), and t be 318 (ohms).

EX. 15

2–2 Factors, Coefficients, and Exponents

When two or more numbers are multiplied, each of the numbers is called a **factor** of the product. Thus, 5 and 7 are factors of 35; two other factors are 1 and 35.

In such an expression as $\frac{1}{4}yz$, each factor is the **coefficient** of the product of the other factors. Thus, in the product $\frac{1}{4}yz$, $\frac{1}{4}$ is the coefficient of yz, y is the coefficient of $\frac{1}{4}z$, $\frac{1}{4}z$ is the coefficient of y, and yz is the coefficient of $\frac{1}{4}$. Usually, we refer to the numerical part of a term as its (numerical) coefficient. For example, the coefficient of $25yz$ is 25. Also, the coefficient of x is 1, since $x = 1 \cdot x$.

In writing a product, a raised dot $\cdot$ is often used as a symbol of multiplication. Thus,

$$8 \cdot 9 = 8 \times 9$$

and

$$7 \cdot 7 = 7 \times 7.$$

"7×7" can also be written "7^2," read "the *square* of 7" or "the *second power* of 7." The powers of 7 are defined as follows:

First power: $\quad 7^1 = 7$

Second power: $\quad 7^2 = 7 \cdot 7 \quad$ (also read "seven squared" or "seven-square")

Third power: $\quad 7^3 = 7^2 \cdot 7 = (7 \cdot 7) \cdot 7 \quad$ (read "the cube of seven" or "seven cubed" or "seven-cube")

Fourth power: $\quad 7^4 = 7^3 \cdot 7 = [(7 \cdot 7) \cdot 7] \cdot 7 \quad$ (read "seven to the fourth" or "seven-fourth")

and so on.

In general, if x denotes any real number and n denotes any positive integer, then:

$$x^n = \underbrace{x \cdot x \cdots \cdot x}_{n \text{ factors}}$$

We call x^n the **nth power** of x. In x^n, the number represented by the small raised symbol n is called an **exponent**. The exponent tells the number of times the **base** x occurs as a factor in the product. Thus:

$$x^n = \text{the } n\text{th power of } x$$

(Exponent — n; Base — x)

In writing a power such as "x^3" in the form "$x \cdot x \cdot x$," we say that the power has been written in **factored form** or **expanded form**. The expression "x^3" is called the **exponential form**.

Example. If the value of m is 5, find the value of:
    a. m^3 **b.** $3m$

Solution: In each case, replace m by 5 and simplify the resulting expression.

    a. $m^3 = (m \cdot m) \cdot m$ b. $3m = 3 \cdot m$
                $= (5 \cdot 5) \cdot 5$ $= 3 \cdot 5$
                $= 25 \cdot 5$ $= 15$
                $= 125$

As the Example above shows, the expressions "m^3" and "$3m$" may have very different values for a given value of m. In "m^3," 3 is an exponent and shows that m is a factor three times. In "$3m$," 3 is the coefficient and thus is itself a factor.

In an expression such as "$5y^2$," 2 is the exponent of the base y. On the other hand, "$(5y)^2$" stands for "$5y \cdot 5y$"; in this case, the parentheses show that the base is $5y$. Consider the following examples:

$4 \cdot 3^2 = 4(3 \cdot 3) = 36$, but $(4 \cdot 3)^2 = (4 \cdot 3)(4 \cdot 3) = 144$
$yz^3 = y(z \cdot z \cdot z)$,   but $(yz)^3 = (yz)(yz)(yz)$
$6 + x^2 = 6 + (x \cdot x)$, but $(6 + x)^2 = (6 + x)(6 + x)$

EXERCISES

Write each expression in exponential form.

1. $x \cdot x \cdot x$
2. $b \cdot b \cdot b \cdot b$
3. z squared
4. y cubed
5. $10 \cdot x \cdot y \cdot y$
6. $11 \cdot c \cdot c \cdot c \cdot d \cdot d$
7. $23u \cdot u \cdot u(w + 3)$
8. $(r + s)(r + s)(r + t)$
9. $(x - 2)(x - 2)(x - 2)$
10. $3 \cdot y \cdot y \cdot (x + 1)(x + 1)$
11. The cube of $x + 8$
12. The fourth power of $z - 3$
13. The product of 5 and the cube of $y + z$
14. The sum of 7 and the square of x
15. The fourth power of the sum of a and b
16. The fifth power of the square of $x + 1$

Evaluate each expression for the given value of the variable.

17. y^2; 6
18. z^2; 11
19. $2x^3$; 3
20. $7y^3$; 2
21. $2x^2$; $\frac{1}{3}$
22. $3a^4$; $\frac{1}{2}$
23. $(3x)^2$; 2
24. $(4p)^2$; 3
25. $(6m)^3$; $\frac{1}{2}$
26. $(12n)^4$; $\frac{1}{12}$
27. $(z + 2)^2$; 3
28. $2(t + 1)^3$; 4

If the values of x, y, and z are 2, 3, and 5, respectively, evaluate the given expression.

29. $\dfrac{x^2 + y}{z}$ 30. $\dfrac{z^2 - y}{x}$ 31. $\dfrac{y^2 + 3x}{z}$ 32. $\dfrac{2z + 4x}{y^2}$

33. $(\tfrac{1}{5}z)^2 + (2y)^2$ 34. $(3x)^3 - (2y)^3$ 35. $\dfrac{y^3 - 27}{xy}$ 36. $\dfrac{x^4 + y^2}{z^2}$

PROBLEMS

Evaluate each of the following expressions, replacing the variables as indicated.

1. Volume of a square prism: lw^2
   Let l be 6 (feet) and w be 3 (feet). Volume is then in cubic feet.
2. Surface area of a square prism: $2w^2 + 4lw$
   Let l be 4 (yards) and w be 2 (yards). Surface area is then in square yards.

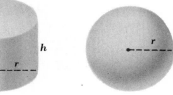

EX. 1–2   EX. 3–4   EX. 5–6   EX. 7–8

3. Surface area of a right circular cylinder: $2\pi r(r + h)$
   Let r be 14 (inches) and h be 15 (inches). $\pi \doteq \tfrac{22}{7}$. Surface area is then in square inches.
4. Volume of a right circular cylinder: $\pi r^2 h$
   Let r be 2 (meters) and h be $3\tfrac{1}{2}$ (meters). $\pi \doteq \tfrac{22}{7}$. Volume is then in cubic meters.
5. Surface area of a sphere: $4\pi r^2$
   Let r be 7 (centimeters). $\pi \doteq \tfrac{22}{7}$. Surface area is then in square centimeters.
6. Volume of a sphere: $\tfrac{4}{3}\pi r^3$
   Let r be 21 (feet). $\pi \doteq \tfrac{22}{7}$. Volume is then in cubic feet.
7. Volume of a right circular cone: $\tfrac{1}{3}\pi r^2 h$
   Let r be 6 (inches) and h be 20 (inches). $\pi \doteq 3.14$. Volume is then in cubic inches.
8. Surface area of a right circular cone: $\pi r(s + r)$
   Let r be 3 (feet) and s be 10 (feet). $\pi \doteq 3.14$. Surface area is then in square feet.

9. Area of a ring: $\pi(R^2 - r^2)$
   Let R be 12 (feet) and r be 5 (feet). $\pi \doteq 3.14$. Area is then in square feet.

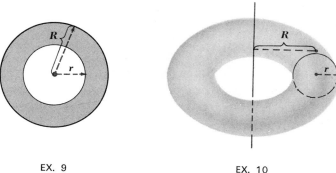

EX. 9   EX. 10

10. Volume of a torus: $2\pi^2 r^2 R$
    Let r be 4 (inches) and R be 12 (inches). $\pi \doteq 3.14$. Volume is then in cubic inches.

11. Power in an electrical circuit: $I^2 R$
    Let I be 8 (amperes) and R be 0.2 (ohms). Power is then in watts.

12. Illumination on a surface: $\dfrac{I}{s^2}$
    Let I be 28 (candles) and s be 2 (feet). Illumination is then in foot-candles (or lumens per square foot).

13. Kinetic energy of a moving body: $\dfrac{Wv^2}{2g}$
    Let W be 30,000 (pounds) and v be 44 (feet per second). $g \doteq 32$ (feet per second). Energy is then in foot-pounds.

14. Height of a body thrown vertically upward with velocity k: $kt - \dfrac{gt^2}{2}$
    Let k be 48 (feet per second) and t be 10.2 (seconds). $g \doteq 32$ (feet per second per second). Height is then in feet.

★15. Length of a pendulum: $\dfrac{gP^2}{4\pi^2}$
    Let P be 2 (seconds). $g \doteq 32.2$ (feet per second per second). $\pi \doteq 3.14$. Length is then in feet.

★16. Inductance of circuit: $\dfrac{1}{4\pi^2 f^2 C}$
    Let f be 1000 (cycles per second) and C be 0.0000004 farads. $\pi \doteq 3.14$. Inductance is then in henries.

2–3 Order of Operations

In order to avoid questions about the meaning of an expression such as "$6 \times 11 - 2^3$," in which grouping symbols have been omitted, mathematicians have agreed on the following steps to simplify such expressions. However, it is best to use enough grouping symbols to avoid ambiguity.

1. **Simplify the names of powers.**
2. **Then simplify the names of products and quotients in order from the left to right.**
3. **Then simplify the names of sums and differences in order from left to right.**

Using these rules we find that

$$6 \times 11 - 2^3 = (6 \times 11) - 2^3 = 66 - 8 = 58.$$

When grouping symbols are used, apply these rules within each grouping symbol, beginning with the innermost pair.

Example 1. $392 \div (12 - 4) \cdot 3 = 392 \div 8 \cdot 3$
$= (392 \div 8) \cdot 3 = 49 \cdot 3 = 147$

Example 2. $[(3 \times 5)5 + 1]2 - 6 = [15 \cdot 5 + 1]2 - 6$
$= [76]2 - 6 = 146$

Example 3. If the value of x is 2 and the value of y is 1, find the value of $5x^3 - 4x^2y + 3y^2 - 6y^4$.

Solution: In the given expression, replace x with "2" and y with "1," insert necessary multiplication signs, and simplify the resulting numerical expression.

$5x^3 - 4x^2y + 3y^2 - 6y^4$

$= 5 \cdot 2^3 - 4 \cdot 2^2 \cdot 1 + 3 \cdot 1^2 - 6 \cdot 1^4$
$= 5 \cdot 8 \quad - 4 \cdot 4 \cdot 1 \quad + 3 \cdot 1 \quad - 6 \cdot 1$
$= 40 \quad\quad - 16 \quad\quad\quad + 3 \quad\quad - 6$
$= \quad\quad\quad 24 \quad\quad\quad\quad\quad + 3 \quad\quad - 6$
$= \quad\quad\quad\quad\quad 27 \quad\quad\quad\quad\quad\quad - 6$
$= 21.$

EXERCISES

Simplify each expression.

1. $(8 + 4 + 3) \div 3 - 2$
2. $(8 + 4 + 5) \div (3 - 2)$
3. $3(11 + 5) \div 4 + 6$
4. $19 + 7(2 + 8) - 49$
5. $7 + 4 - (3 - 1)$
6. $7 + 4 - 3 + 1$
7. $2 \cdot 3^2 - 5 \cdot 3 - 1$
8. $7 \cdot 2 - 2^2 + 3^2$
9. $\dfrac{2^5 \div 8 + 3}{4 + 3}$
10. $\dfrac{5 \cdot 3 + 5 \cdot 2^2}{9 - 2}$
11. $\dfrac{3^3}{2^3 - 2^2 - 1}$
12. $\dfrac{17 \cdot 5 - 3 \cdot 5}{3^2 + 1}$
13. $(5^2 - 5)(5^2 + 5)$
14. $\dfrac{7^2 - 3^2}{7 - 3}$
15. $7(8 - 5) \div 3 - 1$
16. $4(9 + 2) - 21 \div 7$
17. $12 \times 6 \div 3 \times 2 \div 48$
18. $18 \div 2 \times 3 - 5 - 20$
19. $2^6 \div 2^2 \div 2^3 \div 2$
20. $3^4 - 3^2 \div 3^2 - 3$
21. $\dfrac{29 - 5(3 - 2)}{2 - 1 + 7}$
22. $\dfrac{6(2 + 4) - 1}{2 \cdot 3 + 1}$
23. $(24 - 3 + 9 \div 3) \div (8 \times \tfrac{1}{2})$
24. $9(7 - 3) - 30 \div (6 - 1)$

Given that the values of *x, y,* and *z* are 2, 4, and 3, evaluate each of the following expressions.

25. $y^2 - 3y + 2$
26. $2x^2 - x - 1$
27. $z^2 - 3z - 2$
28. $3z^2 + 2z - 10$
29. $3x^3 + 2x^2 - x - 5$
30. $y^3 - 3y^2 + 7y - 10$
31. $y^2 - z^2 + x^2$
32. $z^2 - 2x^2 + 3y^2$
33. $xy + z - x^3$
34. $2yz - x + z^3$
35. $\dfrac{z^2 + 2z + 1}{y^2}$
36. $\dfrac{3x^2 + 5x + 3}{(z + 2)^2}$
37. $\dfrac{2y^2 - 3y + 16}{y^2 - x^2}$
38. $\dfrac{4z^2 - 10z + 15}{y^2 - z^2}$

Evaluate each expression for the given values of the variables.

★39. $\dfrac{3(u - 2)^2 + 6v}{u + (v - 2)^2}$; $u: 4, v: 5$

★40. $\dfrac{2[r + 2(s + r)^2]}{r^2 + 2rs + 5s^2}$; $r: 8, s: 2$

★41. $\dfrac{5(m - 2n)(m^2 + n^2)}{5m^2 - 2mn + n^2}$; $m: 6, n: 3$

★42. $\dfrac{k(l + 3n)^2(l + k)}{2l^2 - 4k^2 + n^2 - 6}$; $k: 3, l: 5, n: 4$

2-4 Variables and Open Sentences

A remark that is true about some members of a set may be false about the other members of the set. For example, if $S = \{0, 1, 2, 3\}$, the sentence

$$x \in S \quad \text{and} \quad x + 4 = 6$$

is true when x is replaced by "2" but false when the replacement is "1" or any other numeral that does not name two.

An equation, such as "$x + 4 = 6$," or an inequality, such as "$y - 1 > 5$," which contains one or more variables is called an **open sentence**. An open sentence is a pattern for the different statements — some true, some false — which are obtained by replacing each variable by the names for the different values of the variable. Consider the equation

$$3r + 8 = 20.$$

If the replacement set of r is $\{0, 4\}$, we can find the values of r, if any, which make the given equation a true statement as follows:

$$3r + 8 = 20 \qquad\qquad 3r + 8 = 20$$
$$3 \cdot 0 + 8 = 20 \qquad\qquad 3 \cdot 4 + 8 = 20$$
$$0 + 8 = 20 \qquad\qquad 12 + 8 = 20$$
$$8 = 20, \text{ which is false} \qquad 20 = 20, \text{ which is true}$$

The set that consists of the members of the domain of the variable for which an open sentence is true is called the **truth set** or the **solution set** of the open sentence *over that domain*. Thus, the solution set of "$3r + 8 = 20$" over $\{0, 4\}$ is $\{4\}$.

Each member of the solution set is said to **satisfy** and to be a **solution** of the open sentence. A solution of an *equation* is also known as a **root** of the equation; thus, 4 is a solution, or root, of "$3r + 8 = 20$." To **solve** an open sentence *over a given domain* means to determine its solution set, or truth set, in that domain. The solution set of an open sentence may be specified by roster, by rule, or by graph. The graph of the solution set is called the **graph of the open sentence**.

Example 1. If $z \in \{\text{the integers}\}$, specify the solution set of $z - 2 = 3$ by roster and by graph.

Solution: The one and only integer that satisfies $z - 2 = 3$ is 5.

∴ (read "therefore") the solution set is $\{5\}$.

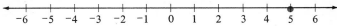

Example 2. Solve $^-1 < x + 3 < 6$ if $x \in \{0, 1, 2, 3\}$, and draw the graph of the sentence.

Solution: Replace x in turn by "0," "1," "2," and "3."
$^-1 < x + 3 < 6$
$^-1 < 0 + 3 < 6$, or $^-1 < 3 < 6$; true.
$^-1 < 1 + 3 < 6$, or $^-1 < 4 < 6$; true.
$^-1 < 2 + 3 < 6$, or $^-1 < 5 < 6$; true.
$^-1 < 3 + 3 < 6$, or $^-1 < 6 < 6$; false.
The solution set is $\{0, 1, 2\}$.

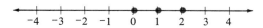

EXERCISES

Substitute members of the given replacement set in the open sentence, tell whether the resulting statements are true, and specify each solution set by roster.

1. $x + 3 = 7$; $\{3, 4, 5\}$
2. $x + 2 = 3$; $\{0, 1, 2\}$
3. $y > 4$; $\{3, 4, 5\}$
4. $z < 2$; $\{0, 1, 2\}$
5. $x \neq 4$; $\{2, 3, 4\}$
6. $y \neq 0$; $\{0, 1, 2\}$
7. $2x = 6$; $\{1, 2, 3\}$
8. $3y = 12$; $\{3, 4, 5\}$
9. $2z < 8$; $\{1, 2, 3\}$
10. $3x > 8$; $\{3, 4, 5\}$
11. $x^2 < 9$; $\{1, 2, 3\}$
12. $y^2 > 2$; $\{0, 1, 2\}$

Solve each sentence over the given set. Specify the solution set of each equation in roster form. Graph the solution set of each inequality.

Example 1. $2x + 1 = 5$; {the positive numbers}

Solution: By inspection, 4 is the only arithmetic number whose sum with 1 is 5. This means that $2x$ must represent 4. The one and only positive number replacement for x for which $2x = 4$ is 2. $\{2\}$.

Example 2. $3x < 6$; {the nonnegative numbers}

Solution: Since $3x$ is to be less than 6, x must be less than 2.

Solve each sentence over the given set. Specify the solution set of each equation in roster form. Graph the solution set of each inequality.

13. $y + 3 = 9$ {the whole numbers}
14. $z - 2 = 10$ {the whole numbers}
15. $4t = 12$ {the whole numbers}
16. $7r = 28$ {the whole numbers}
17. $2 = \frac{1}{3}y$ {the whole numbers}
18. $8 = \frac{1}{2}n$ {the whole numbers}
19. $2x + 1 = 11$ {the positive numbers}
20. $3y + 2 = 11$ {the positive numbers}
21. $11 = 6x - 1$ {the real numbers}
22. $14 = 8z - 2$ {the real numbers}
23. $x + 1 < 7$ {the positive integers}
24. $y + 2 < 10$ {the positive integers}
25. $2x < 13$ {the whole numbers}
26. $5x < 11$ {the whole numbers}
27. $x + 1 > 2$ {the nonnegative numbers}
28. $t + 3 > 5$ {the nonnegative numbers}
29. $10 < 5l$ {the positive numbers}
30. $21 < 7l$ {the positive numbers}
31. $x + 3 = 3 + x$ {the real numbers}
32. $2x = x + x$ {the real numbers}
33. $x^2 < 4$ {the nonnegative numbers}
34. $y^2 > 9$ {the nonnegative numbers}
35. $2x = x + 2$ {the integers}
36. $3x = 6 + x$ {the integers}

Substitute the members of the given replacement sets in the open sentence, and tell whether the resulting statements are true or false.

Example 3. $3l + m = 4; l \in \{1, 2\}, m \in \{0, 1\}$

Solution: $3(1) + 0 = 4;$ false $\qquad 3(2) + 0 = 4;$ false
$3(1) + 1 = 4;$ true $\qquad 3(2) + 1 = 4;$ false

37. $r + s = 7$ $\qquad r \in \{3, 4\}, s \in \{3, 4\}$
38. $p - q = 5$ $\qquad p \in \{9, 10\}, q \in \{4, 5\}$
39. $x + 2y = 11$ $\qquad x \in \{5, 6\}, y \in \{3, 4\}$
40. $2w + 3z = 12$ $\qquad w \in \{1, 2\}, z \in \{3, 4\}$

41. $x < 2y$ $x \in \{3, 4\}, y \in \{1, 2\}$
42. $3x > y + 2$ $x \in \{2, 3\}, y \in \{4, 5\}$
43. $x + yz = x(y + z)$ $x \in \{1\}, y \in \{2\}, z \in \{0, 1\}$
44. $l + mn = lm + n$ $l \in \{0\}, m \in \{1\}, n \in \{2, 3\}$
45. $3x - 2y = 9$ $x \in \{5, 6, 7\}, y \in \{3, 4, 5\}$
46. $2m + 4n = 20$ $m \in \{3, 4, 5\}, n \in \{2, 3, 4\}$
47. $3x + 2y > 26$ $x \in \{4, 5, 6\}, y \in \{4, 5, 6\}$
48. $2x < 4z + 2$ $x \in \{4, 5, 6\}, z \in \{0, 1, 2\}$
49. $x + 3y \neq 2x + y$ $x \in \{0, 1, 2\}, y \in \{0, 2, 4\}$
50. $2t - r \neq 3t - 2r$ $t \in \{1, 2, 3\}, r \in \{1, 2, 3\}$

2–5 Variables and Quantifiers

Whatever numeral is used in place of y in the sentence

$$y + 9 = 9 + y$$

a true statement results. To assert this fact, we may write:

For all real numbers y, $y + 9 = 9 + y$.

Other forms of this statement are:

For each real number y, $y + 9 = 9 + y$.
For every real number y, $y + 9 = 9 + y$.
For any real number y, $y + 9 = 9 + y$.
If y is any real number, then $y + 9 = 9 + y$.

Consider the assertion,

For each positive integer x, $x + 1 > 7$.

This statement is surely false, because when x is replaced by "2" we obtain the false statement

$$2 + 1 > 7.$$

On the other hand, if x is replaced by "8," then the open sentence "$x + 1 > 7$" becomes the true statement

$$8 + 1 > 7.$$

Because there is a positive integer x for which "$x + 1 > 7$" is a true statement, we can make the following true assertion:

There is a positive integer x such that $x + 1 > 7$.

Of course, there are many positive integers for which "$x + 1 > 7$" is a true statement. But, the existence of *at least one* is enough to guarantee the truth of the given assertion. Other ways to state this fact are:

There exists a positive integer x such that $x + 1 > 7$.
For at least one positive integer x, $x + 1 > 7$.
For some positive integer x, $x + 1 > 7$.

Such key words as *all, each, every, any, there is, there exists, there are, there exist, some,* and *at least one* involve the idea of "how many" or of "quantity." For this reason such an expression is called a **quantifier** when it is used in combination with a variable in an open sentence.

Example. Use a quantifier to change each open sentence into a true statement. Assume that the domain of each variable is the set of real numbers.
    a. $2n - 5 = 11$ **b.** $x \neq x + 1$

Solution: **a.** There exists a real number n such that $2n - 5 = 11$.
    b. For all real numbers x, $x \neq x + 1$.

EXERCISES

Show that each of the following statements is true by finding a value of the variable for which the statement is true.

1. For some natural number t, $2t = 6$.
2. There is a real number s such that $s \div 4 \neq 3$.
3. There exists an integer y such that $3y - 1 > 5$.
4. At least one positive number satisfies the equation "$x^2 = 1$."

Show that each of the following statements is false by finding a value of the variable for which the statement is false.

5. For every integer x, $x + 2 = 3$.
6. All real numbers y are greater than 0.
7. Any whole number m is less than $2m$.
8. The square of each natural number k is greater than k.

Show that each of the following statements is true by finding an expression for *b* for which it is true.

Example. For each whole number *a*, there is a whole number *b* that exceeds *a* by 5.

Solution: $b = a + 5$

9. For each real number *a*, there is a real number *b* that is 5 times as great as *a*.
10. For each real number *a*, there is a real number *b* that is half as great as *a*.

2–6 Applying Mathematical Expressions and Sentences

Mathematical expressions and sentences are often applied in real life situations to describe numerical relationships.

For example, the expression

$$x + 3x$$

can represent the sum of the weights of two books if one book weighs *x* pounds and the other weighs three times as much, that is, $3x$ pounds.

The equation

$$x + 3x = 6$$

describes mathematically the situation in which the sum of the weights of the two books is 6 pounds.

The same mathematical expression or sentence may represent the numerical facts in more than one situation. For instance, the equation "$x + 3x = 6$" also describes the following sentence in symbols.

A certain number when added to 3 times itself yields a sum of 6.

Below are two ways of interpreting the inequality "$2y > y + 7$."

Twice a given number is greater than 7 more than the number.

Here *y* represents the given number.

Carlos is twice as old as his sister, Maria. In fact, Carlos is more than 7 years older than Maria.

Here *y* represents Maria's age in years, and $2y$ represents Carlos' age in years.

EXERCISES

Find an open sentence to represent each of the following numerical relationships described in words.

Example. A rectangle whose perimeter is 152 centimeters is 8 centimeters longer than it is wide.

Solution: $w + (w + 8) + w + (w + 8) = 152.$

1. After depositing $8, I had $53 in my savings-bank account.
2. The population of Bertown is 6000 less today than it was ten years ago when 24,500 people lived there.
3. In a certain class of 19 students, the number of boys exceeds the number of girls by 1.
4. Nineteen-year-old Fred is 1 year more than twice as old as his brother.
5. In an isosceles triangle whose perimeter is 35 inches, the base is 8 inches shorter than each of the two congruent sides.
6. The sum of three numbers is 22. The second of the numbers is two less than the first, and the third is three more than the first.
7. Two years from now, Marge will be twice as old as she was 7 years ago.
8. In a certain office, a keypunch operator earns seventy-five cents more per hour than a typist. In 39 hours, the keypunch operator earns as much as a typist does in 48 hours.
★9. A newsboy has more than $2.50 in dimes and nickels. He has 11 more nickels than dimes.
★10. Mr. Hoyer spent less than $3.00 to buy some six-cent and some ten-cent postage stamps. He bought 15 more ten-cent stamps than six-cent ones.

2–7 Using Open Sentences in Flow Charts (Optional)

A program is any list of steps used to carry out a particular job. Consider the following example.

On each of the cards pictured in Figure 2–1 is a numeral printed in color. Suppose that we begin with Card 1 and take the following steps:

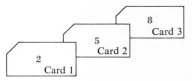

FIGURE 2–1

1. Read the color numeral on the card.
2. Multiply the number whose name we read in Step 1 by 7.
3. Write the simplest numeral for the product obtained in Step 2.
4. If we have worked with all the cards, then write the word "Stop"; otherwise, return to Step 1 and work with the next card.

These four steps describe a computation to be carried out, and they form a program.

Figure 2–2 is a diagram of the program described above. To picture the steps in the program, the diagram uses boxes with directions of what to do written in them. The order in which to carry out these directions is shown by arrows. Such a diagram is called a **flow chart** or a **flow diagram**. Flow charts are often used in analyzing lengthy computations to be carried out by an electronic computer.

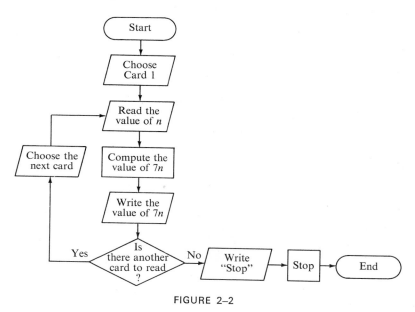

FIGURE 2–2

Notice that the shape of each box shows the kind of operation described in the box. Flow charts in this book will use boxes of the four different shapes to indicate four different kinds of operation.

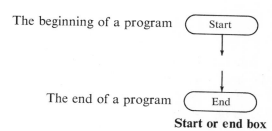

The beginning of a program

The end of a program

Start or end box

In a parallelogram, describe a step to begin or to complete an *input* operation (reading) or an *output* operation (writing). Variables or expressions whose values are to be read or written may be named in the box. Input or output word messages may also be stated in the box. Such messages are shown in quotation marks.

Input-output box

In a rectangular box, describe computations to be carried out, or assign a value to a variable.

Computation or assignment box

In a diamond-shaped box, ask a question to be answered "Yes" or "No." The box has two exits, labeled to show how the program continues, depending on the answer to the question.

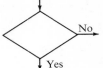

Decision box

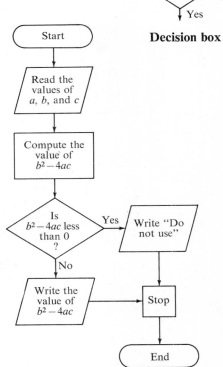

Example. Given the flow chart at the right, what value will be written by the program if the values read for *a*, *b*, and *c* are 2, 5, and 3 ?

Solution: The values of *a*, *b*, and *c* are 2, 5, and 3.

$$\therefore b^2 - 4ac = 5^2 - 4 \cdot 2 \cdot 3$$
$$= 25 - 24$$
$$= 1.$$

Since $b^2 - 4ac = 1$ and $1 > 0$, the answer to the question "Is $b^2 - 4ac$ less than 0?" is "No." Therefore, the "No" path is followed in the flow chart, and the value of $b^2 - 4ac$ is written, that is, 1.

Often the question asked in a decision box is written as an open sentence. When an open sentence is used in a decision box, the exits from the box may be labeled T (for True) and F (for False). For instance, in the Example above, the decision box shown as

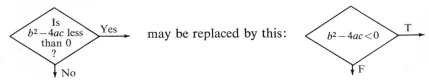

EXERCISES

1. If the box shown at the right describes a step in a program and if the value of x is 3, under what condition will the T path be taken in the program? the F path?

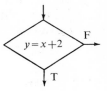

Exercises 2 and 3 refer to the flow chart at the right.

2. What will be written if the value of a is -3 and the value of b is 4?

3. What will be written if the values of a and b are both 6?

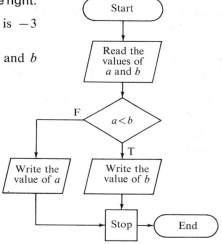

chapter summary

1. A **variable** is a symbol which may represent any of the members of a specified set, called the **domain** or **replacement set**. A variable with just one value is called a **constant**. An expression containing a variable is called an **open expression**. **Evaluating an expression** is the process of replacing the variable in an open expression with the numeral for the given value and simplifying the result. A **term** is a mathematical expression using numerals or variables or both to indicate a product or a quotient.

2. When two or more numbers are multiplied, each is called a **factor** of the product. In a term containing numerical and variable factors, the numerical part of the term is called the **numerical coefficient**. The expression $x \cdot x$ can be written in **exponential form** as x^2, where the **exponent** 2 indicates the number of times the **base** x occurs as a factor in the product.

3. To eliminate ambiguity about the meaning of an expression in which grouping symbols have been omitted, follow the rules for the **order of operations**:

   a. Simplify the names of powers.

   b. Simplify the names of products and quotients in order from left to right.

   c. Simplify the names of sums and differences in order from left to right.

4. An **open sentence** is a sentence that contains one or more variables. The set consisting of the members of the domain of the variable for which an open sentence is true is called the **truth set** or the **solution set** of the open sentence over a given domain. Each member of this set is known as a **solution**. A solution of an equation is also called a **root**.

5. A **quantifier** is a word denoting the idea of "how many" used in combination with a variable in an open sentence.

6. Mathematical expressions and sentences can be applied to real life situations to describe numerical relationships.

7. A **program** is a list of steps describing how to perform a certain task. A program is often shown by a diagram called a **flow chart**.

CHAPTER REVIEW

2–1 Evaluate the following expressions if x has the value 4 and y has the value $\frac{1}{4}$.

    1. $3x$ **2.** $4y + x$ **3.** $3(x - 6y) + x$ **4.** $(x + y)(x - y)$

2–2 Give the set of factors of each of the expressions.

    5. t **6.** $3ab$ **7.** $\frac{xy}{2}$

Give the missing coefficients, as indicated.

    8. $7xyz = (?)yz$ **9.** $\frac{pq}{8} = (?)pq$

Evaluate the following expressions if a, b, and c have the values 1, 2, and 3.

    10. ab^2c **11.** $\frac{c^2}{ab}$ **12.** $2a^2b + c^3$

2–3 Simplify each of the expressions.

13. $28 \div 7 \times 2^2$
14. $5(6 + 2) \div 4$
15. $14 \div 2 + 2 - 1$
16. $\dfrac{7 \cdot 3 + 7 \cdot 2^2}{9 - 2}$

2–4 The replacement set for x is $\{2, 4, 6, 8\}$. Which of the elements make each of the following open sentences true?

17. $x - 2 = 2$
18. $x - 2 > 2$

From $\{1, 2, 3, 4 \ldots, 10\}$ determine the solution set of each sentence.

19. $3x + 4 = 10$
20. $3x + 4 > 10$

2–5 Tell which of the following statements are true and which are false.

21. For all real numbers r, $r^3 \neq r$.
22. For some real numbers t, $t = \frac{1}{2}t$.
23. If y is any real number, then $y - 2 = y \div 2$.

2–6
24. Write an algebraic expression for the profit realized on the sale of 12 portable radios if the cost to the dealer for each radio is c, and the selling price for each radio to the customer is s.

25. Write an open sentence stating that the length of a hallway is 2 feet more than four times its width.

3 addition and multiplication of real numbers

3–1 Axioms of Closure and Equality

When two whole numbers are added, is the sum *always* a whole number? To try every example would be an endless task. After checking a large number of varied examples,

$$138 + 51 = 189, \quad 174 + 236 = 410, \quad \text{and so on,}$$

we would probably *assume* that the answer is, "Yes, the sum of two whole numbers is always a whole number."

Similarly, we would assume that the product of two whole numbers is always a whole number.

Any set S is said to be **closed under an operation** performed on its elements, provided that each result of the operation is an element of S. Thus, the set of whole numbers is closed under both addition and multiplication. In general, calculations with real numbers are based on the often unstated assumption that the set of real numbers is closed under addition and multiplication.

In mathematics, assumptions (statements accepted as true without proof) are called **axioms** or **postulates**. The **axioms of closure** follow.

Axiom of Closure for Addition

For all real numbers a and b, the sum $a + b$ is a unique (one and only one) real number.

Axiom of Closure for Multiplication

For all real numbers a and b, the product ab is a unique real number.

Notice that closure under any operation depends on *both* the particular *operation* and the *set* of numbers used. For example, the set of odd numbers is closed under multiplication. But it is *not* closed under addition because there are odd numbers whose sum is not odd, for instance, $3 + 5 = 8$. On the other hand, under division the set of whole numbers is not closed, but the set of positive real numbers is.

Can a finite set be closed under an operation? Looking at the addition and multiplication tables at the right, we see that $\{0, 1\}$ is closed under multiplication, but *not* under addition (since $2 \notin \{0, 1\}$).

The way that we have agreed to use the equals symbol $=$ gives equality the following properties.

+	0	1
0	0	1
1	1	2

×	0	1
0	0	0
1	0	1

Axioms of Equality

For all real numbers *a*, *b*, and *c*:

Reflexive Property $a = a$
Symmetric Property   If $a = b$, then $b = a$.
Transitive Property   If $a = b$ and $b = c$, then $a = c$.

Example 1. Name the property of equality which is illustrated.
  a. If $5 + 4 = 9$, then $9 = 5 + 4$.
  b. If $1 + 6 = 7$ and $7 = 5 + 2$, then $1 + 6 = 5 + 2$.

Solution: a. The symmetric property
  b. The transitive property

Example 2. Which property of equality asserts that
  a. every equation is reversible?
  b. every real number is equal to itself?
  c. if each of two given numbers equals a third number, then the given numbers are equal?

Solution: a. The symmetric property
  b. The reflexive property
  c. The symmetric and transitive properties

EXERCISES

Which of the following sets are closed under each of the operations of addition, multiplication, subtraction, and division (excluding division by zero)? When the set is not closed, give an example which shows this.

1. $\{1\}$
2. $\{2\}$
3. $\{0\}$
4. $\{\frac{1}{2}\}$
5. $\{0, 1\}$
6. $\{1, 2\}$
7. $\{1, \frac{1}{2}, 2\}$
8. $\{0, 5, 10, 15, \ldots\}$

9. {the natural numbers}
10. {the whole numbers}
11. {the positive real numbers}
12. $\{1, \frac{1}{3}, \frac{1}{9}, \frac{1}{27}, \frac{1}{81}, \ldots\}$
13. {the real numbers greater than 1}
14. {the positive real numbers that are not whole numbers}

Name the closure axiom or the property of equality illustrated by each of the following true statements.

15. If $15 - 3 = 12$, then $12 = 15 - 3$.
16. $7 \cdot 3$ is a real number.
17. $8 + (^-9)$ is a real number.
18. Given that $2 = 8 - 6$; therefore, $8 - 6 = 2$.
19. For all real numbers x and y, $x + y = x + y$.
20. If r denotes a real number and $r + 2 = 0$, then $0 = r + 2$.
21. Given that $x + 4 = 13$ and that $13 = 9 + 4$; therefore, $x + 4 = 9 + 4$.
22. Given that $7y = 35$ and that $35 = 7 \cdot 5$; therefore, $7y = 7 \cdot 5$.
23. Given that $2(1 + 0) = 2$ and that $2 = 2 + 0$; therefore, $2(1 + 0) = 2 + 0$.
24. Given that $\frac{3}{5} \div \frac{2}{7} = \frac{3}{5} \times \frac{7}{2}$ and that $\frac{3}{5} \times \frac{7}{2} = \frac{21}{10}$; therefore, $\frac{3}{5} \div \frac{2}{7} = \frac{21}{10}$.

3–2 Commutative and Associative Axioms

We assume that when we add two numbers, we get the same sum no matter what order we use in adding them. Thus,

$$8 + 2 = 2 + 8 \quad \text{and} \quad 10 + 3 = 3 + 10.$$

This assumption is called the **commutative axiom of addition**.

Commutative Axiom of Addition

For all real numbers *a* and *b*, *a* + *b* = *b* + *a*.

Similarly, $8 \times 2 = 2 \times 8$ and $10 \times 3 = 3 \times 10$. When we multiply numbers, we obtain the same product regardless of the order of the factors.

Commutative Axiom of Multiplication

For all real numbers *a* and *b*, *ab* = *ba*.

Addition and multiplication of real numbers are *commutative operations*. Notice that subtraction and division are *not* commutative operations. For example, $8 - 2 \neq 2 - 8$ and $8 \div 2 \neq 2 \div 8$.

Addition and multiplication of real numbers are **binary operations** because either operation is performed on only *two* numbers at a time. Therefore, to find a sum or a product of three or more numbers, these numbers must somehow be grouped in pairs.

The example

$$(93 + 7) + 78 = 93 + (7 + 78)$$

suggests that the way that we group, or *associate*, real numbers in a sum of three (or more) numbers makes no difference in the result. We say that addition in the set of real numbers is an *associative operation* or that addition satisfies the **associative axiom**.

Associative Axiom of Addition

For all real numbers a, b, and c, $(a + b) + c = a + (b + c)$.

Similarly, products of real numbers do not depend on the way we group the factors. For example,

$$(5 \cdot 2) \cdot 6 = 5 \cdot (2 \cdot 6) \quad \text{and} \quad (14 \cdot \tfrac{1}{3}) \cdot 9 = 14 \cdot (\tfrac{1}{3} \cdot 9).$$

Associative Axiom of Multiplication

For all real numbers a, b, and c, $(ab)c = a(bc)$.

However, subtraction and division are not associative operations as the following examples show.

Example 1. $(24 - 12) - 3 = 12 - 3 = 9;$
$24 - (12 - 3) = 24 - 9 = 15.$
$\therefore (24 - 12) - 3 \neq 24 - (12 - 3).$

Example 2. $(24 \div 12) \div 3 = 2 \div 3 = \tfrac{2}{3};$
$24 \div (12 \div 3) = 24 \div 4 = 6.$
$\therefore (24 \div 12) \div 3 \neq (24 \div 12) \div 3.$

Wise use of the associative and commutative axioms of addition and multiplication can often make computations easier.

Example 3. $17 + 2\frac{1}{3} + 63 + 7\frac{2}{3} = (17 + 63) + (2\frac{1}{3} + 7\frac{2}{3})$
$= 80 + 10$
$= 90$

Example 4. $\frac{7}{9} \times 4 \times 27 \times 25 = (\frac{7}{9} \times 27)(4 \times 25)$
$= 21 \times 100$
$= 2100$

EXERCISES

Simplify each of the following expressions.

1. $396 + 134 + 6 + 4$
2. $530 + 28 + 70 + 32$
3. $25 \times 17 \times 2 \times 4$
4. $2 \times 16 \times 19 \times 5$
5. $\frac{1}{2} \cdot 21 \cdot 10 \cdot \frac{1}{7}$
6. $\frac{1}{3} \cdot \frac{5}{11} \cdot 44 \cdot 12$
7. $6\frac{1}{2} + 2\frac{1}{3} + 1\frac{1}{2} + \frac{2}{3}$
8. $99\frac{3}{5} + 1\frac{3}{7} + \frac{2}{5} + 8\frac{4}{7}$
9. $\frac{7}{5} \cdot \frac{4}{3} \cdot \frac{5}{7} \cdot 18$
10. $2 + 31 + 8 + 17\frac{1}{2} + 9 + 2\frac{1}{2}$

Given that the replacement set of the variable in each of the following sentences is the set of real numbers, specify the solution set of the sentence.

Example. $(5 + z) + 6 = 6 + (5 + z)$

Solution: The commutative axiom of addition guarantees that the equation is true for every real value of z. The solution set is the set of all real numbers.

11. $4 \cdot 5a = 20a$
12. $2 + (b + 4) = b + 6$
13. $(1 + n) + 5 = 5 + (6 + 1)$
14. $9d \cdot 7 = 63 \cdot 2$
15. $4 + (x + 2) < (x + 4) + 2$
16. $3 \cdot 4m > 4 \cdot 3m$

In each exercise, an operation $*$ is defined over the set of natural numbers. In each case:

a. Find $2 * 5$.
b. Determine whether or not the set of natural numbers is closed under $*$.
c. State whether or not $*$ is (1) commutative, (2) associative.

★17. $a * b = a + (b + 1)$
★18. $a * b = 2a + b$
★19. $a * b = a - b$
★20. $a * b = ab^2$

3-3 Addition on the Number Line

On the number line, we can illustrate some of the axioms and other facts about sums of real numbers by representing numbers by *displacements* (changes of position) on the line. A displacement in the positive direction represents a positive number, and a displacement in the negative direction represents a negative number.

For instance, to picture adding 3 and 4 on the number line in Figure 3–1, start at the origin and move 3 units to the right. The short black arrow in the diagram shows this displacement and represents the number 3. Then, starting at the graph of 3, move 4 units to the right (the arrow in color).

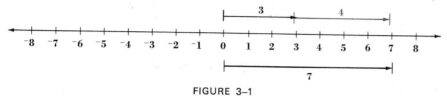

FIGURE 3–1

Together, the two displacements amount to a displacement of 7 units to the right from the origin. Thus, the diagram pictures the fact that

$$3 + 4 = 7.$$

Figure 3–2 pictures the sum $4 + 3 = 7$.

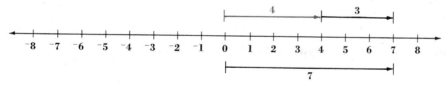

FIGURE 3–2

Together, Figures 3–1 and 3–2 illustrate the fact that

$$3 + 4 = 4 + 3.$$

Figure 3–3 shows a way to determine the sum of $^-4$ and 3. Notice that $^-4$ has been represented by a displacement from the origin of 4 units to the left. When this displacement is followed by a move of 3 units to the right,

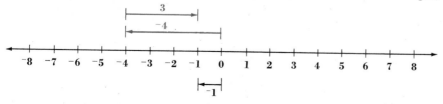

FIGURE 3–3

the net effect is a displacement from the origin of 1 unit to the left (black arrow). Thus, the diagram suggests that

$$^-4 + 3 = {}^-1.$$

A slightly different way to picture this fact is simply to show the displacement of 3 units, starting at the graph of ⁻4 and ending at the graph of ⁻1 (Figure 3–4).

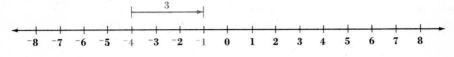

FIGURE 3–4

To add ⁻4 to 3, start at 3 and move 4 units to the left, again arriving at the graph of ⁻1 (Figure 3–5).

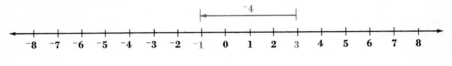

FIGURE 3–5

Thus, $3 + {}^-4 = {}^-1$, the same as $^-4 + 3$, illustrating again the commutative axiom.

Example. State the addition fact suggested by each diagram.

a.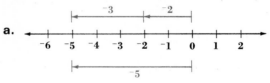

b.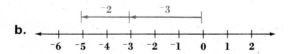

c.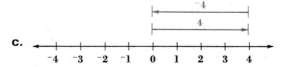

Solution: a. $^-2 + {}^-3 = {}^-5$
b. $^-3 + {}^-2 = {}^-5$
c. $4 + {}^-4 = 0$

ADDITION AND MULTIPLICATION OF REAL NUMBERS

On the number line, the sum of any two real numbers can be found by following either one of the following rules.

To add two real numbers:
1. **Start at the origin and draw an arrow representing the first number. Then, from the head of that arrow, draw an arrow representing the second number. The arrow from the origin to the head of the second arrow represents the sum of the two numbers; or**
2. **Start at the graph of the first number, and draw an arrow representing the second number. The head of this arrow points to the graph of the sum of the two numbers.**

To visualize $^-4 + 0$ on the number line, interpret "add 0" to mean "take no displacement."

$$^-4 + 0 = {^-4} \quad \text{and} \quad 0 + {^-4} = {^-4}$$

These equations illustrate the special role that zero plays for addition in the set of real numbers: When 0 is added to any given number, the sum is identical with the given number. We call 0 the **identity element for addition** and accept the following statement as true.

Additive Axiom of Zero

The set of real numbers contains a unique element 0 having the property that for every real number a,

$$a + 0 = a \quad \text{and} \quad 0 + a = a.$$

EXERCISES

In Exercises 1–14:
a. Simplify each expression using the number line if necessary.
b. Use the associative axiom of addition to regroup the terms in the given expression, and then simplify the resulting expression.

1. $(^-4 + 3) + {^-7}$
2. $(6 + {^-42}) + {^-8}$
3. $(12 + {^-24}) + 6$
4. $(^-15 + {^-17}) + 27$
5. $38 + (^-8 + 30)$
6. $48 + (8 + {^-22})$
7. $^-2.3 + (1.3 + 1.6)$
8. $^-0.12 + (^-0.28 + 0.08)$
9. $(\frac{7}{2} + {^-\frac{1}{2}}) + 0$
10. $(^-\frac{3}{5} + {^-\frac{7}{5}}) + 2$
11. $(^-2 + 5) + {^-3\frac{2}{3}}$
12. $(^-6 + 6) + {^-7\frac{3}{4}}$
13. $[(^-2 + 3) + {^-5}] + {^-8}$
14. $[2 + (^-12 + {^-8})] + {^-22}$

Solve each of the following equations given that the replacement set of the variable is the set of real numbers. Use the number line as needed.

Example. $^-4 + x = 2$

Solution: To arrive at 2 from $^-4$, take a displacement of 6 units to the right.

Check: $^-4 + 6 \stackrel{?}{=} 2$ ($\stackrel{?}{=}$ means "Is this statement true?")
$2 = 2$

∴ the solution set is $\{6\}$.

15. $y + {}^-3 = {}^-5$
16. $6 + x = {}^-1$
17. $^-8 + z = {}^-6$
18. $t + {}^-2 = {}^-5$
19. $a + {}^-9 = {}^-9$
20. $s + 6 = 0$
21. $0 = {}^-7 + b$
22. $12 = {}^-4 + c$
23. $t + 8 = 20$
24. $r + {}^-4 = {}^-3$
25. $w + 24 = 0$
26. $^-30 + x = {}^-30$
27. $x + 5 = {}^-1$
28. $4 + y = {}^-4$
29. $n + n = 0$
30. $m + m = 8$
31. If $A = \{{}^-1, 0, 1\}$, find the set of all sums of pairs of elements of A. Be sure to include the sum of each element with itself. Is A closed under addition? Justify your answer.

PROBLEMS

Example. An elevator starts at the ground floor and goes up to the 7th floor. It then goes down 5 floors and up 3 floors. At what floor is the elevator then located?

Solution: $7 + {}^-5 + 3 = 5$.
The elevator is at the 5th floor.

1. A ship sails directly north from a point A for 42 miles and then sails directly south for 57 miles. Where is the ship then located with reference to point A?
2. A salesman drives from his office on a turnpike 90 miles west to see a customer. He then drives 110 miles east to see a second customer. Where is he then located with respect to his office?

ADDITION AND MULTIPLICATION OF REAL NUMBERS   51

3. The pilot of a jet traveling at 27,000 feet above sea level is ordered by ground control to descend 8000 feet. Later, the pilot is ordered to climb 3500 feet to a new altitude. At what altitude will the jet then be flying?

4. A submarine dives to a level 730 feet below the surface of the ocean. Later it climbs 200 feet and then dives another 80 feet. What is then the depth of the submarine?

5. Jack brought four objects for $2.15, $3.05, $3.40, and $2.85. He later sold them for $2.25, $2.85, $3.15, and $2.75, respectively. What was the net financial result of the transactions?

6. A stock selling for $30 per share rose 2 dollars per share each of two days and then fell $1.75 per share for each of three days. What was the selling price per share of the stock after these events?

7. Mr. Allen owned 650 shares of stock in the McNiff Corporation. On Monday he sold 75 shares, on Tuesday he sold 50 shares, on Wednesday he purchased 350 shares, and on Thursday sold 125 shares. How many shares of the stock did he then own?

8. On a revolving charge account, Mrs. Dallins purchased $27.50 worth of clothing, and $120.60 worth of furniture. She then made two monthly payments of $32.00 each. If the interest charges for the period of two months were $3.25, what did Mrs. Dallins then owe the account?

9. Mr. Gordon's normal blood pressure was 142. During a particularly trying period his blood pressure rose 17 points, fell 22 points, and then rose 8 points. How did his blood pressure then compare with his normal blood pressure?

10. During and after the passage of a cold front, the temperature at Pokesville fell 45°, rose 8°, fell 3°, rose 12°, and then rose 5°. How did the temperature then compare to the temperature prior to the passage of the cold front?

3–4 The Opposite of a Real Number

Figure 3–6 suggests a useful way to pair points on the number line. The paired points are at the same distance from the origin, but on opposite sides of the origin. Notice that Figure 3–6 shows the origin paired with itself.

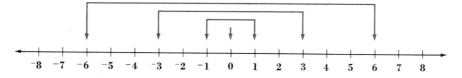

FIGURE 3–6

This pairing of points suggests that we also pair the coordinates of the points; for example, ⁻1 with 1, ⁻3 with 3, ⁻6 with 6, 0 with 0, and so on. We can check that adding two such paired numbers on the number line gives 0. For example, as shown in Figure 3–7,

$$^-6 + 6 = 0.$$

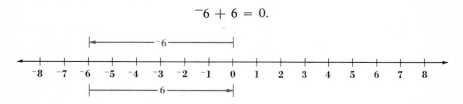

FIGURE 3–7

Each number in such a pair is called the **opposite** or the **additive inverse** or the **negative** of the other number. The symbol

$-a$ (note the lowered position of the minus sign)

denotes **the opposite of *a*** or **the additive inverse of *a*** or **the negative of *a***. For example:

$-6 = {}^-6$, read "the opposite of six equals negative six";
$-({}^-5) = 5$, read "the opposite of negative five equals five";
$-0 = 0$, read "the opposite of zero equals zero."

The following axiom is a formal way of saying that every real number has a unique opposite and that the sum of a number and its opposite is always zero.

Axiom of Opposites

For every real number *a* there is a unique real number −*a*, such that

$$a + (-a) = 0 \quad \text{and} \quad (-a) + a = 0.$$

The equation "$-6 = {}^-6$" indicates that the numerals "-6" and "${}^-6$" name the same number. This means that we can always use the numeral "-6" (lowered minus sign) in place of the numeral "${}^-6$" (raised minus sign). It also means that we can read "-6" either as "negative 6" or as "the opposite of 6." *Throughout the rest of this book, lowered minus signs will be used in the numerals for negative numbers.*

Caution! Be careful about reading *variable* expressions like $-a$. This should be read "the opposite of *a*" or "the additive inverse of *a*" or "the negative *of a*." Never call it "negative *a*," because $-a$ may denote either a negative number, a positive number, or zero.

By looking at the number line in Figure 3–6, we can see that the following statements are true:

1. **If *a* is a positive number, then $-a$ is a negative number; if *a* is a negative number, then $-a$ is a positive number; if *a* is 0, then $-a$ is 0.**
2. **The opposite of $-a$ is *a*; that is, $-(-a) = a$.**

Example 1. Simplify: a. $-(-3)$;   b. -0;   c. $-(3+4)$;

Solution: a. 3   b. 0   c. -7

Example 2. $-(-3) + (-5) = 3 + (-5) = -2$

EXERCISES

Simplify each of the following expressions.

1. $-(-4) + 6$
2. $-(-8) + 3$
3. $-8 + [-(-3)]$
4. $7 + [-(-4)]$
5. $-(6 + 8)$
6. $-(-3 + 10)$
7. $-(-\frac{1}{2}) + \frac{1}{2}$
8. $-\frac{1}{3} + [-(-\frac{1}{3})]$
9. $-(2.1 + 5.2)$
10. $-[6.75 + (-0.75)]$
11. $-[-2 + (-1)]$
12. $-[8 + (-6)]$
13. $-(-2 + 3) + (-1)$
14. $[4 + (-7)] + [-(-2)]$
15. $2\frac{1}{2} + (-3\frac{1}{2}) + \frac{1}{2}$
16. $-4\frac{2}{3} + (-\frac{1}{3}) + 2\frac{2}{3}$

In each of the following exercises, replace the variable in turn by the name of each member of $\{-2, -1, 0, 1\}$, and then:
a. Tell whether each resulting statement is true or false.
b. Give the solution set of the open sentences over that replacement set.

17. $-x = 2$
18. $-1 = -t$
19. $-y = 0$
20. $-s = 1$
21. $-4 = -a$
22. $-b = -2$
23. $d + 1 = -1$
24. $c + (-2) = 1$
25. $-z > 1$
26. $-x < 1$
27. $-m < -2$
28. $-y > -1$
29. $1 + [-(-x)] = -1$
30. $-(-r) + 3 = 0$
31. $-1 < -y < 2$
32. $-2 < -x < 0$
33. $0 > -m > -3$
34. $2 > -q > -2$

Given the values $\frac{5}{2}$ for x, $-\frac{1}{2}$ for y, 0.25 for z, and 1.75 for w, evaluate each expression.

35. $x + (-y) + z$
36. $-x + y + (-z)$
37. $-[x + y + (-z)]$
38. $-[-y + z + (-w)]$
39. $x + [-(w + z)]$
40. $(x + y) + [-(w + z)]$

3–5 Absolute Value

In any pair of nonzero opposites, like 5 and -5, one number is a positive number, while the other is a negative number. We call the positive number of any pair of opposite real numbers the **absolute value** of each of the numbers. Thus, 5 is the absolute value of 5; 5 is also the absolute value of -5.

The absolute value of a number is denoted by writing a name of the number between a pair of vertical bars $|\ |$. For example,

$$|5| = 5 \quad \text{and} \quad |-5| = 5.$$

The **absolute value of 0** is defined to be 0 itself: $|0| = 0$.

In terms of displacement on the number line, the absolute value of a number is the length of the arrow representing the number, without regard to the direction of the arrow (Figure 3–8). We can also think of the absolute value of a number as the *distance* between the origin and the graph of the number.

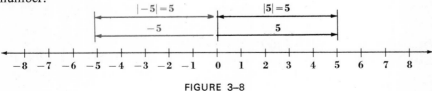

FIGURE 3–8

Example 1. Draw the graph of $|a| = 3$ if $a \in \{\text{the real numbers}\}$.

Solution: The graph of $|a| = 3$ consists of the two points which are 3 units from the origin, that is, the points with coordinates -3 and 3.

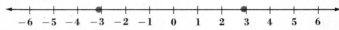

Example 2. Draw the graph of $|x| < 2$ if $x \in \{\text{the real numbers}\}$.

Solution: The graph of $|x| < 2$ consists of all the points which are less than 2 units from the origin in either direction.

Another inequality with this graph is $-2 < x < 2$.

The following example uses the inequality symbol ≥, which stands for "**is greater than or equal to.**" The symbol ≤ stands for "**is less than or equal to.**"

Example 3. Draw the graph of the set of all real numbers y such that $|y| \geq 1$.

Solution: The graph of $|y| \geq 1$ consists of all the points which are *at least* 1 unit from the origin on either side of the origin.

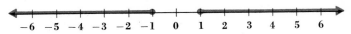

EXERCISES

Determine the value of each expression.

1. $6|-5|$
2. $-|3|$
3. $2|-5| + |2|$
4. $3|7| + |-7|$
5. $|5 + (-2)| \div 3$
6. $|-5 + 8| - 3$
7. $-|-8 + 10|$
8. $-|15 + (-9)|$
9. $2|-5| + |3|$
10. $3|-2| + |-6|$
11. $-[3|2|] + [-|2|]$
12. $-[4|-3| + (-6)]$

Solve each equation over the set of real numbers.

Example. $|x| + 1 = 5$

Solution: Since the sum of 4 and 1 is 5, $|x|$ must be equal to 4. That is, $|x| = 4$. The only two real numbers whose absolute value is 4 are -4 and 4. The solution set is $\{-4, 4\}$.

13. $|x| = 2$
14. $|y| = 7$
15. $|y| + 3 = 8$
16. $|z| + 5 = 20$
17. $|t| + (-2) = 3$
18. $|x| + (-4) = 10$

Graph the solution set of each open sentence over the set of real numbers.

19. $|x| < 3$
20. $|y| > 2$
21. $|t| \geq 4$
22. $|s| \leq 4$
23. $|-a| = 5$
24. $|-b| = 1$
25. $|y| \leq 0$
26. $|w| > 0$

★27. Explain why the following statement is true or why it is false.
If a is a real number, then

$$|a| = a \quad \text{if } a \geq 0;$$
$$|a| = -a \quad \text{if } a < 0.$$

★28. For what real values of y is it true that $|y| < -y$?

3–6 Rules for Addition

Numerical work often suggests general properties of real numbers. Consider the following example.

Example 1. Show on the number line that
 a. $-(3 + 5) = -8$ and
 b. $-3 + (-5) = -8$; and, therefore, that
 c. $-(3 + 5) = -3 + (-5)$.

Solution: a. Add 3 and 5, and then find the opposite of this sum, -8.

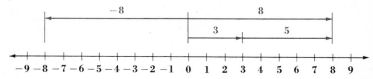

 b. Add the opposite of 3 and the opposite of 5; that is, add -3 and -5. The sum is -8.

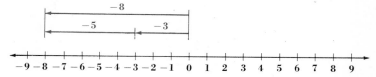

 c. By part a, $-(3 + 5) = -8$.
 By part b, $-3 + (-5) = -8$, so that by the symmetric property of equality, $-8 = -3 + (-5)$.
 Therefore, by the transitive property of equality,
$$-(3 + 5) = -3 + (-5).$$

We can use the method of Example 1 to show that the equations below are also true:

$$-[3 + (-5)] = -3 + 5, \quad -(-3 + 5) = 3 + (-5),$$
$$-[-3 + (-5)] = 3 + 5.$$

All these examples suggest the following general statement.

Property of the Opposite of a Sum

The opposite of a sum of real numbers is the sum of the opposites of the numbers; that is, for all real numbers *a* and *b*,

$$-(a + b) = (-a) + (-b).$$

ADDITION AND MULTIPLICATION OF REAL NUMBERS

The property of the opposite of a sum is very useful in making such substitutions for negative numbers as these:

$$-9 = -(6 + 3) = -6 + (-3), \qquad -4\tfrac{1}{2} = -(4 + \tfrac{1}{2}) = -4 + (-\tfrac{1}{2}).$$

Also, by using this property along with the familiar addition facts for positive numbers and the axioms, we can compute sums of any real numbers without having to think of the number line.

Example 2. Simplify: $-8 + (-5)$

Solution:
$-8 + (-5) = -(8 + 5)$ Property of the opposite of a sum
$ = -13 $ Substitution principle
$\therefore -8 + (-5) = -13.$ Transitive property of equality

Example 3. Simplify: $14 + (-6)$

Solution:
$14 + (-6) = (8 + 6) + (-6)$
$ = 8 + [6 + (-6)]$
$ = 8 + 0$
$ = 8$
$\therefore 14 + (-6) = 8.$

After computing many sums by using either the number line or the methods of the examples above, the following rules would become evident:

Rules for Addition

1. **If a and b are each positive numbers or zero, then $a + b = |a| + |b|$.**
   Example. $6 + 8 = 14$

2. **If a and b are each negative numbers, then $a + b = -(|a| + |b|)$.**
   Example. $(-6) + (-8) = -(6 + 8) = -14$

3. **If a is a positive number and b is a negative number and $|a| \geq |b|$, then $a + b = |a| - |b|$.**
   Example. $14 + (-6) = 14 - 6 = 8$

4. **If a is a positive number or zero and b is a negative number and $|b| \geq |a|$, then $a + b = -(|b| - |a|)$.**
   Example. $8 + (-14) = -(14 - 8) = -6$

Example 4. Simplify: $6 + (-11) + 13 + (-5)$

Solution 1:

Step 1	*Step 2*	*Step 3*
$6 + (-11) = -5;$	$-5 + 13 = 8;$	$8 + (-5) = 3$

Solution 2:

Step 1	*Step 2*	*Step 3*
6	-11	19
13	$-\ 5$	-16
$\overline{19}$	$\overline{-16}$	$\overline{\ \ 3}$

Example 5. Add:
-214
$\ \ 132$
$\ \ 211$
-142
-100

Solution:

Step 1	*Step 2*	*Step 3*
-214	132	-456
-142	211	$\ \ 343$
-100	$\overline{343}$	$\overline{-113}$
$\overline{-456}$		

EXERCISES

Simplify each expression for a sum.

1.	2.	3.	4.	5.	6.
8	-8	$2\frac{1}{2}$	4.3	123	412
3	-2	$3\frac{1}{4}$	-2.6	-148	-213
-5	6	$-\frac{3}{4}$	-8.1	215	-309
$\ \ 2$	$\ \ 5$	-4	10.0	-300	156

7. $-8 + 7 + (-9) + 5$
8. $-12 + (-5) + 8 + 20$
9. $28 + (-17) + (-48) + 30$
10. $(-62) + (-18) + 40 + 3$
11. $122 + (-47) + (-83) + 28$
12. $-210 + (-80) + 250 + 65$
13. $-[25 + (-3)] + [-(-2 + 5)]$
14. $[-3 + (-5)] + [-(3 + 5)]$
15. $6.5 + (-2.3) + 0 + (-5.4) + (-7.2) + 15.1$
16. $-0.7 + 1.38 + (-4.4) + (-12.9) + 2$
17. $\frac{1}{2} + 3\frac{1}{2} + (-2\frac{1}{2}) + 4\frac{1}{2} + 2\frac{1}{3} + (-\frac{1}{3}) + (-1\frac{2}{3}) + 4\frac{2}{3}$
18. $\frac{4}{5} + 3\frac{2}{5} + (-2\frac{1}{5}) + 0 + (-7)$

Replace each __?__ with a numeral to make a true statement.

19. __?__ $+ 5 = 2$
20. $11 +$ __?__ $= 7$

21. $5 + \underline{?} = -5$
22. $\underline{?} + (-6) = 3$
23. $-2 + \underline{?} = -5$
24. $-3 + \underline{?} = 7$
25. $\frac{2}{5} + \underline{?} = -1$
26. $\underline{?} + \frac{3}{4} = -\frac{1}{4}$

Write a chain of equations leading to the stated equation. Justify each step.

Example. $(a + b) + [(-a) + (-b)] = 0$

Solution: $(a + b) + [(-a) + (-b)]$
$= [a + (-a)] + [b + (-b)]$ Commutative and associative axioms of addition
$= 0 + 0$ Axiom of opposites
$= 0$ Additive axiom of zero
$\therefore (a + b) + [(-a) + (-b)] = 0.$*   Transitive property of equality

★27. $(-5) + [(-s) + (s + 5)] = 0$ ★28. $-[(-a) + b] + b = a$
★29. $(a + b) + [-(a + b + c)] = -c$ ★30. $t + [-(t + r)] = -r$

PROBLEMS

1. The Appleton Arms apartments has two levels of garage below the ground level. Starting at the lowest garage level, an elevator went to the fourth floor. From there it rose 2 floors and then dropped 4 floors. Where was the elevator then located?

2. Julie was on a diet. Her weight over a 4-week period fell 12 pounds, rose $7\frac{1}{2}$ pounds, fell 8 pounds, and rose $3\frac{1}{2}$ pounds. How did her weight then compare with her weight at the start of her diet?

3. McCoy, the left halfback on a football team, made the following yardage on successive carries of the ball: 12, -3, 5, -11, 2, -3. What was his net yardage on these six plays?

4. The ferryboat that constituted the only connection between Argyle Island and the mainland made 3 round trips. It carried 83 persons to the island on its first trip and returned 114 to the mainland. It then delivered 109 and returned 121, delivered 114 and returned 98. How did the population of the island then compare with the population before the first trip?

---

* This statement means that the sum of $a + b$ and $(-a) + (-b)$ is zero. But the one and only number whose sum with $a + b$ is zero is the opposite of $a + b$, that is, $-(a + b)$. Hence, $-(a + b) = (-a) + (-b)$, that is, the property of the opposite of a sum (page 56).

5. During 4 days, Mercy Hospital received 12 new patients and discharged 9, received 14 and discharged 21, received 5 and discharged 12, and received 11 and discharged 10. How did the number of patients in the hospital then compare with the number at the start of the 4-day period?

6. Mr. Thompkins had a balance of $121.50 in his checking account. During the next week, he wrote checks for $47.20, $18.55, and $32.40. On Friday, he made a deposit of $52.00. What was his balance after making the deposit?

7. A helicopter was flying in Death Valley at an altitude of 47 feet below sea level. If it climbed 125 feet and then dropped 117 feet, at what altitude was it then flying?

8. A book dealer had 7200 copies of a novel in its warehouse. During a one-week period, it shipped out 3140 books and received 2700 return copies. What was then the number of copies of this book in the warehouse?

★9. The following values of n are shown on successive cards: 1, 0, 3, −5, 6. Given the flow chart at the right, what numbers will be written by the program? (*Note:* It is not always necessary to have a "Stop" box in a flow chart.)

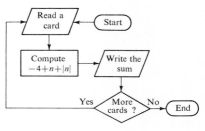

3–7 The Distributive Axiom

A student works Friday evenings and Saturdays at a local branch of the public library and is paid $1.60 an hour. He works 3 hours on Friday and 7 hours on Saturday. Since he works $3 + 7$ hours, his weekly earnings in dollars are

$$1.60(3 + 7) = 1.60 \times 10 = 16.00.$$

His weekly earnings are also the sum of his Friday earnings and his Saturday earnings:

$$(1.60 \times 3) + (1.60 \times 7) = 4.80 + 11.20 = 16.00.$$

Either way we compute them, his weekly earnings are the same; that is,

$$1.60(3 + 7) = (1.60 \times 3) + (1.60 \times 7).$$

Note that 1.60, the coefficient (multiplier) of the sum $3 + 7$, is *distributed* as a multiplier of each term of $3 + 7$. This example illustrates a fact that we will use in working with real numbers: multiplication is *distributive* with respect to addition.

Distributive Axiom of Multiplication with Respect to Addition

For all real numbers a, b, and c,

$$a(b + c) = ab + ac \quad \text{and} \quad (b + c)a = ba + ca.$$

By applying the symmetric property of equality, we can also state the distributive axiom in the following form.

For all real numbers a, b, and c:

$$ab + ac = a(b + c) \quad \text{and} \quad ba + ca = (b + c)a.$$

The following examples show some uses of the distributive axiom.

Example 1. a. $36(\frac{1}{2} + \frac{1}{9}) = 36 \times \frac{1}{2} + 36 \times \frac{1}{9} = 18 + 4 = 22$
b. $7(5\frac{2}{7}) = 7(5 + \frac{2}{7}) = 7 \times 5 + 7 \times \frac{2}{7} = 35 + 2 = 37$
c. $\frac{1}{8} \cdot 23 + \frac{1}{8} \cdot 9 = \frac{1}{8}(23 + 9) = \frac{1}{8} \cdot 32 = 4$

Example 2. Show that for every real number x,

$$4x + 3x = 7x.$$

Solution: $\quad 4x + 3x = (4 + 3)x \quad$ Distributive axiom
$\qquad\qquad\qquad\quad = 7x \qquad\qquad$ Substitution principle
$\quad \therefore 4x + 3x = 7x. \qquad\quad$ Transitive property of equality

Because properties of real numbers guarantee that for *all* values of the variable, each of the expressions

$$4x + 3x \quad \text{and} \quad 7x$$

represents the same number as the other expression, they are called **equivalent expressions**. When a given expression is replaced by an equivalent expression with as few terms as possible, the given expression is said to be **simplified**.

Example 3. Simplify: $7y^2 + 6 + (-5)y^2$
Solution: $\quad 7y^2 + 6 + (-5)y^2 = 7y^2 + (-5)y^2 + 6$
$\qquad\qquad\qquad\qquad\qquad\quad = [7 + (-5)]y^2 + 6$
$\qquad\qquad\qquad\qquad\qquad\quad = 2y^2 + 6$
$\quad \therefore 7y^2 + 6 + (-5)y^2 = 2y^2 + 6.$

Terms such as "$7y^2$" and "$(-5)y^2$" are called *similar terms* or *like terms*. Two terms are **similar** if they are exactly alike or if they differ only in their numerical coefficients. Examples 2 and 3 show how a sum of similar terms can be replaced by a single term by use of the distributive axiom. Notice that an expression like "$2y^2 + 6$" in which the terms are unlike cannot be replaced by a single term.

Example 4. Simplify: $3(2k + 5m) + 9(7m + 4k)$

Solution: $3(2k + 5m) + 9(7m + 4k) = 6k + 15m + 63m + 36k$
$= (6 + 36)k + (15 + 63)m$
$= 42k + 78m$

∴ $3(2k + 5m) + 9(7m + 4k) = 42k + 78m.$

EXERCISES

Simplify each expression.

1. $31x + 16x$
2. $45y + 32y$
3. $100b + (-18)b$
4. $(-16)c + 37c$
5. $12z + 3z + 37z$
6. $15t + (-2)t + (-8)t$
7. $15a^2 + 14 + 9a^2$
8. $17d^3 + 9 + 73d^3$
9. $2m + 3k + 5m + 6k$
10. $4g + 2h + 9h + 45g$
11. $12m + 19 + (-8)m + 5$
12. $16e + 38 + 13e + (-17)$
13. $2s + 5r + 3s + (-1)r$
14. $6s + (-7)t + 12s + 9t$
15. $5c + 7d + (-1)c + 2d + 3c$
16. $4y + 3y + 5 + 28 + (-11)$
17. $12 + 9h^2 + 5h + (-6) + (-3)h$
18. $23 + 5w^2 + 8w + (-2)w + (-8)$
19. $4ab + 7a + 5ab + (-2)a$
20. $7xy + 3x + (-2)xy + 8x$
21. $5(x + y) + 12(x + y) + 3x$
22. $2(a + b) + 7(a + b) + 3b$
23. $6(a + 2) + 9(3 + a)$
24. $4(r + 3) + 13(r + 5)$
25. $3(x^2 + x) + 6(2x^2 + 4)$
26. $12(4y + 5y^2) + 2(7 + 8y^2)$
27. $3(a + 2b + 4) + 5(2a + 3)$
28. $4(r + 3s + 6) + 7(s + 2)$
29. $5(2t^2 + 3t + 4) + 6(7t^2 + 5t)$
30. $2(3x^2 + 5x + 9) + 9(8x^2 + 2x)$

31. Five times the sum of a and b, increased by twice the sum of $2a$ and $3b$.
32. Twice the sum of 5 and the square of x, increased by three times the sum of $6x^2$ and 11.
33. Ten more than the sum of negative seven, y, and the square of y.

34. The product of eight and the cube of d, increased by the sum of nine and negative six times the cube of d.
35. $3[7y + 4(5 + 2y)] + (-16)$
36. $21 + 2[3z + 5(3z + 8)]$
37. $8(2p + q) + 13[2p + 3(4q + 2p + 6)]$
38. $9[2(4x + 3y + 4) + 7(x + 14)] + 8(5x + 3y)$

Determine the value of each numerical expression. Whenever possible, use properties of numbers to simplify the calculation.

39. $98 \times 54 + 2 \times 54$
40. $66 \times 23 + 34 \times 23$
41. $50(\frac{4}{5} + \frac{1}{2})$
42. $827 \cdot 11 + 827 \cdot 9$
43. $3\frac{1}{2} \cdot 9 + 3\frac{1}{2} \cdot 1$
44. $\frac{1}{2} \cdot 83 + 99\frac{1}{2} \cdot 83$
45. 628×1001
46. $40 \times 7\frac{3}{10}$
47. $2 \times 37 + 5 \times 37 + 3 \times 37$
48. $7 \times 59 + 1 \times 59 + 2 \times 59$

★49. Draw a flow chart of a program to compute and write the area of the shaded region bounded by the rectangles shown in the diagram. The values of a, b, and c are to be read from a card. Assume that more than one card is to be read and processed.

3–8 Rules for Multiplication

The following equations suggest the reason that 1 is called the *identity element for multiplication:*

$$6 \times 1 = 6 \quad \text{and} \quad 1 \times 6 = 6$$

One is the **identity element for multiplication** because whenever we multiply a given real number and 1, the product is the given real number.

Multiplicative Axiom of One

The set of real numbers has a unique element 1 having the property that for every real number a,

$$a \cdot 1 = a \quad \text{and} \quad 1 \cdot a = a.$$

The equations

$$6 \times 0 = 0 \quad \text{and} \quad 0 \times 6 = 0$$

illustrate the *multiplicative property of zero:* when one of the factors of a product is 0, the product itself is 0.

Multiplicative Property of Zero

For each real number a,

$$a \cdot 0 = 0 \quad \text{and} \quad 0 \cdot a = 0.$$

The following examples suggest that $a(-1) = -a$.

$$2 \times (-1) = -1 + (-1) = -2$$
$$3 \times (-1) = -1 + (-1) + (-1) = -3, \text{ and so on.}$$

To verify that $a(-1)$ is the opposite of a for every real number a, we must show that the sum of $a(-1)$ and a is zero:

$$\begin{aligned}
a(-1) + a &= a(-1) + a(1) & &\text{Multiplicative axiom of one} \\
&= a[(-1) + 1] & &\text{Distributive axiom} \\
&= a(0) & &\text{Axiom of opposites} \\
&= 0 & &\text{Multiplicative property of zero}
\end{aligned}$$

Thus, multiplying any real number by -1 produces the opposite of the number.

Multiplicative Property of -1

For all real numbers a,

$$a(-1) = -a \quad \text{and} \quad (-1)a = -a.$$

A special case of this property occurs when the value of a is -1; we have $(-1)(-1) = 1$.

The multiplicative property of -1 together with the multiplication facts for positive numbers can be used to compute the product of any two real numbers. For example:

1. $3 \cdot 4 = 12$
2. $(-3) \cdot 4 = (-1 \cdot 3)4 = -1(3 \cdot 4) = -1(12) = -12$
3. $3(-4) = 3[4(-1)] = (3 \cdot 4)(-1) = 12(-1) = -12$
4. $(-3)(-4) = (-1 \cdot 3)(-1 \cdot 4) = [-1(-1)](3 \cdot 4) = 1 \cdot 12 = 12$

Similarly, for all real numbers a and b:

$$(-a)b = (-1 \cdot a)b = -1(ab) = -ab$$
$$a(-b) = a[b(-1)] = (ab)(-1) = -ab$$
$$(-a)(-b) = (-1 \cdot a)(-1 \cdot b) = [-1(-1)](ab) = 1(ab) = ab$$

Property of Opposites in Products

For all real numbers a and b,

$$(-a)b = -ab, \qquad a(-b) = -ab, \qquad (-a)(-b) = ab.$$

Practice in computing products leads to the following rules.

Rules for Multiplication

1. The product of a positive and negative number is a negative number.
2. The product of two positive numbers or of two negative numbers is a positive number.
3. The absolute value of the product of two real numbers is the product of the absolute values of the numbers:

$$|ab| = |a| \times |b|$$

Of course, by pairing the negative numbers in a product, we can extend these rules to any number of factors.

A product of nonzero real numbers of which an even number are negative is a positive number. A product of nonzero real numbers of which an odd number are negative is a negative number.

The absolute value of the product of real numbers is the product of the absolute values of the numbers.

Example 1. State whether the given expression names a positive number, a negative number, or zero. Then simplify the expression.
    a. $-7(-2)(-5)$ b. $3^2 \cdot (-2)^3 \cdot 0$
    c. $3^2 \cdot (-2)^3 \cdot (-1)^5$

Solution: a. A negative number; -70 b. 0
    c. A positive number; $3^2 \cdot (-2)^3 \cdot (-1)^5 = 9 \cdot (-8)(-1) = 9 \cdot 8 = 72$

Example 2. Simplify each expression.
    a. $(-4x)(-6x)$ b. $3y + (-4y)$

Solution: a. $(-4x)(-6x) = [-4(-6)]x \cdot x = 24x^2$
    b. $3y + (-4y) = 3y + (-4)y = [3 + (-4)]y = (-1)y = -y$

EXERCISES

Evaluate each of the following expressions.

1. $(-3)(7 + 5)$
2. $(-2)[(-1) + (-8)]$
3. $[15 + (-6)](-1)$
4. $[-4 + 18](-2)$
5. $0[-6 + (-15)]$
6. $12[3 + (-3)]$
7. $5 \cdot 1 + 5(-11)$
8. $-6 \cdot 13 + (-6) \cdot 1$
9. $-7(-4) + (-7)$
10. $15 + 15(-21)$
11. $23(99) + 23(-99)$
12. $(-11)(-4) + (-4)(11)$
13. $-5 + 5(-69)$
14. $-99(7) + (-7)$
15. $-(-8 + 25)$
16. $-[-6 + (-19)]$
17. $(-12)(\frac{5}{6} + \frac{2}{3})$
18. $42(-\frac{2}{7} + \frac{16}{21})$
19. $0.7 - [2.8 + (-0.4)]$
20. $-0.5 - [-0.3 + (-5.6)]$

Simplify each expression.

21. $2x + 3y + (-x)$
22. $3a + 3c + (-2a)$
23. $4m + (-2n) + m + (-3n)$
24. $2z + (-5w) + (-3z) + w$
25. $-4t + 3 + 6t + (-7)$
26. $-7p + 14 + 3p + (-21)$
27. $3xy + (-2yz) + 7yz + (-6xy)$
28. $5rs + 2rt + (-8rs) + 10rt$
29. $-x + 2 + 3x + (-4) + (-6x)$
30. $3 + (-y) + 3y + (-7) + 2y$
31. $10t^2 + (-4t) + 7t + (-8t^2)$
32. $z^3 + (-5z^2) + 3z^2 + (-3z^3)$
33. $2.3x + 5y + (-1.7x) + (-4.6y)$
34. $-0.2m + 3.1n + (-2.1n) + (-1.3)m$
35. $2u^2v + (-3uv^2) + 6u^2v + (-2uv^2)$
36. $-xyz + \frac{1}{2}xy + \frac{3}{4}xyz + \frac{3}{2}xy + (-xyz)$
37. $2(u + 2v) + (-3)(2u + v)$
38. $6[r + (-s)] + 5(3r + s)$
39. $-2(3a + b) + 7[a + (-b)]$
40. $-3(6m + n) + (-2)(n + 10m)$
41. $-4(-y + 2z) + (-3)[y + (-5z)]$
42. $-6[r + (-3s)] + (-5)(3s + r)$
43. $2[-3(x + 4y) + (-y)] + 8x$
44. $3[2(-3z + w) + (-z)] + (-6w)$
45. $-2 + (-3)[2(-1 + x) + (-2x)]$
46. $3y + (-2)[3(-y + 1) + y]$

Let the value of a be -1, the value of b be 0.2, the value of c be -2, the value of x be -0.2, and the value of y be 0. Evaluate each expression.

47. $a(b + x)$
48. $c(a + b)$
49. $3a + 4c$
50. $4a + (-2c)$
51. $-a + (5b + c)^2$
52. $-bc + (b + x)^2$
53. $0.3c[-(x^2 + a^2)]$
54. $-2.1a + b(a + c)$
55. $x(b + c)^2$
56. $c^2(a + c)^2$
57. $[a + (-c)][b + (-x)]$
58. $(c^2 + a^2)(b^2 + x^2)$
59. $64x^2[bc^2 + (-1)]$
60. $-36b^2[cx^2 + (-a)]$

3–9 The Reciprocal of a Real Number

Numbers, like $\frac{3}{4}$ and $\frac{4}{3}$, whose product is 1 are called **reciprocals** or **multiplicative inverses** of each other. For example:

1. -2 and $-\frac{1}{2}$ are reciprocals of each other because $-2(-\frac{1}{2}) = 1$.
2. 0.1 is the reciprocal of 10 and 10 is the reciprocal of 0.1 because $0.1 \times 10 = 1$.
3. -1 is its own reciprocal because $(-1)(-1) = 1$.
4. 0 has no reciprocal because the product of 0 and any real number is 0, *not* 1.

The symbol $\frac{1}{a}$ denotes the reciprocal, or multiplicative inverse, of a. For example:

"$\frac{1}{\frac{3}{4}} = \frac{4}{3}$" means "the reciprocal of three-fourths equals four-thirds."

"$\frac{1}{-2} = -\frac{1}{2}$" means "the reciprocal of negative two equals negative one-half."

That *every real number except 0 has a reciprocal* is a basic assumption.

Axiom of Reciprocals

For every real number a except zero, there is a unique real number $\frac{1}{a}$, such that $a \cdot \frac{1}{a} = 1$ and $\frac{1}{a} \cdot a = 1$.

Example 1. Simplify: $(-a)\left(-\frac{1}{a}\right)$ for $a \neq 0$

Solution: By the multiplicative property of -1,

$$-a = (-1)a \quad \text{and} \quad -\frac{1}{a} = (-1)\frac{1}{a}.$$

$$\therefore (-a)\left(-\frac{1}{a}\right) = (-1)a \cdot (-1)\frac{1}{a}$$

$$= \underbrace{(-1)(-1)}_{1} \cdot \underbrace{a \cdot \frac{1}{a}}_{1}$$

$$= 1.$$

$$\therefore (-a)\left(-\frac{1}{a}\right) = 1.$$

The result of Example 1 means that $-a$ and $-\frac{1}{a}$ are reciprocals of each other. Thus,

$$\frac{1}{-a} = -\frac{1}{a}$$

for every nonzero real number a.

Example 2. Simplify each expression:

    a. $(7 \cdot 2)(\frac{1}{7} \cdot \frac{1}{2})$

    b. $(a \cdot b)\left(\frac{1}{a} \cdot \frac{1}{b}\right)$, $a \neq 0$, $b \neq 0$

Solution: **a.** $(7 \cdot 2)(\frac{1}{7} \cdot \frac{1}{2}) = (7 \cdot \frac{1}{7})(2 \cdot \frac{1}{2}) = 1 \cdot 1 = 1$

    b. $(ab)\left(\frac{1}{a} \cdot \frac{1}{b}\right) = \left(a \cdot \frac{1}{a}\right)\left(b \cdot \frac{1}{b}\right) = 1 \cdot 1 = 1$

Example 2 suggests the following fact about the reciprocal of a product that corresponds to the property of the opposite of a sum.

Property of the Reciprocal of a Product

The reciprocal of a product of real numbers, each different from zero, is the product of the reciprocals of the numbers; that is, for all real numbers a and b such that $a \neq 0$ and $b \neq 0$,

$$\frac{1}{ab} = \frac{1}{a} \cdot \frac{1}{b}.$$

EXERCISES

Simplify each expression.

1. $\frac{1}{9}(63)$
2. $\frac{1}{12}(72)$
3. $51(-\frac{1}{3})$
4. $(-\frac{1}{2})(-22)$
5. $72(-\frac{1}{4})(-\frac{1}{6})$
6. $\frac{1}{3}(-90)(-\frac{1}{5})$
7. $\frac{1}{-7} \cdot 42 \cdot \frac{1}{2}$
8. $(-36)(\frac{1}{12})\left(\frac{1}{-3}\right)$
9. $5ab(-\frac{1}{5})$
10. $8xy(-\frac{1}{8})$
11. $(-12b^2)(-\frac{1}{4})$
12. $(-50c^3)(\frac{1}{10})$
13. $-\frac{1}{9}(27k^3)$
14. $-\frac{1}{15}(-150z^8)$
15. $\frac{1}{x}(5xy)$, $x \neq 0$
16. $(4ab)\frac{1}{a}$, $a \neq 0$

17. $\frac{1}{r^3}(10r^3)(-\frac{1}{2})$, $r \neq 0$
18. $-\frac{1}{8}(48s^2)\frac{1}{s^2}$, $s \neq 0$
19. $\frac{1}{2}(10x + 8)$
20. $\frac{1}{3}(12y + 21)$
21. $-\frac{1}{5}(-25c + 70d)$
22. $[8r + (-12s)](-\frac{1}{4})$
23. $[-54mk + (-6k)](-\frac{1}{6})$
24. $-\frac{1}{9}(36uv + 81u)$
25. $\frac{1}{7}(-7a^2 + 7b^2)$
26. $\frac{1}{8}[8x^3 + (-8)y^3]$
27. $\frac{1}{2}(4a + 6b) + \frac{1}{3}(-9a + 3b)$
28. $\frac{1}{5}(-5g + 10h) + \frac{1}{2}[2g + (-8h)]$
29. $6[\frac{1}{2}s + (-\frac{1}{3})t] + (-\frac{1}{7})[7s + (-14t)]$
30. $-3[\frac{1}{3}c + (-\frac{1}{3}d)] + \frac{1}{4}[8c + (-12d)]$
31. $-\frac{1}{6}[24a^2 + (-6)] + (-\frac{1}{4})(16a^2 + 4)$
32. $\frac{1}{5}(-20s^2 + 5) + (-4)(\frac{1}{2}s^2 + \frac{1}{4})$
33. $-3[\frac{1}{2}(4a + 1) + (-\frac{1}{2})] + 7a$
34. $5d + (-\frac{1}{2})[8 + 40(-\frac{1}{5} + \frac{1}{4}d)]$

3–10 More about Flow Charts (Optional)

Consider the following program to compute the squares of the integers from 1 to 99, inclusive.

1. Let n be a variable whose *initial* (first) value is 1.
2. Compute the value of n^2, and let the result be the value of s.
3. Print the values of n and s on one line of the output sheet.
4. If the value of n is 99, stop (your work is done). Otherwise, go to step 5.
5. Add 1 to the value of n, and then go back to step 2.

Notice that the first time we go through steps 2, 3, and 4 in the program, the value of n is 1.

But when we execute step 5, the value of n is increased to 2. Next, the program tells us to go through steps 2, 3, and 4 again, but this time with 2 as the value of n.

After step 5 is executed a second time, the value of n is 3.

Continue repeating steps 2, 3, 4, and 5 until finally we reach step 4 with 99 as the value of n, and then stop.

The result of the program is to compute and print the squares of the integers 1 through 99. Figure 3–9 shows a flow chart of this program.

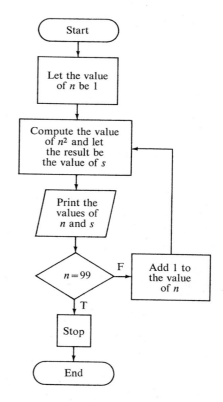

FIGURE 3–9

Figure 3–10 shows the same information as Figure 3–9 on page 69. However, in the assignment boxes of Figure 3–10, left-pointing arrows are used to show assigned values. Thus, in the new flow chart,

"$n \leftarrow 1$" stands for "Let the value of n be 1."

"$s \leftarrow n^2$" stands for "Compute the value of n^2 and let the result be the value of s."

"$n \leftarrow n + 1$" stands for "Add 1 to the value of n."

Note also that in the output box in Figure 3–10, the instruction "Print n and s" means "Print *the values of* n and s."

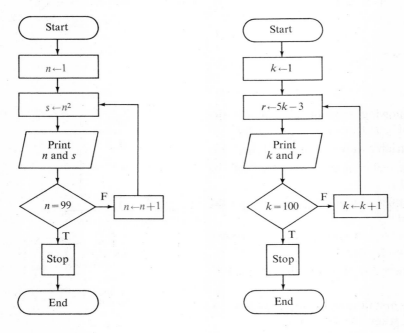

FIGURE 3–10   FIGURE 3–11

Example. Answer the following questions for the program whose flow chart is given in Figure 3–11.

    a. What are the first two values printed?
    b. How many values of k are printed?
    c. What is the last value of $5k - 3$ that is printed?

Solution: **a.** The first two values printed are the initial value of k and the initial value of $5k - 3$; that is, 1 and $5 \times 1 - 3$, or 1 and 2.
    b. 100   **c.** $5 \times 100 - 3$, or 497.

EXERCISES

In Exercises 1–4, the program pictured at the left below is to be executed. In each exercise, display the output of the program for the given values of R and T.

1. $R \leftarrow 3, T \leftarrow 6$
2. $R \leftarrow 15, T \leftarrow 2$
3. $R \leftarrow 650, T \leftarrow \frac{1}{2}$
4. $R \leftarrow 1200, T \leftarrow \frac{1}{3}$

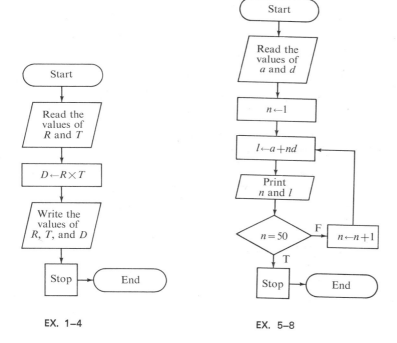

EX. 1–4   EX. 5–8

Exercises 5–8 refer to the flow chart at the right above.

5. What is the value of n when the end of the program is reached?

6. How many values of $a + nd$ are computed by the program?

7. What are the first two values of $a + nd$ if the values of a and d are 5 and 3?

8. What are the first two values of $a + nd$ if the values of a and d are 0 and 7?

Exercises 9–12 refer to the flow chart at the left. In each exercise, tell what is written by the program for the given values of *a* and *b*.

9. $a \leftarrow 3$
   $b \leftarrow 7$

10. $a \leftarrow 5$
    $b \leftarrow 16$

11. $a \leftarrow 4$
    $b \leftarrow 6$

12. $a \leftarrow 2$
    $b \leftarrow -1$

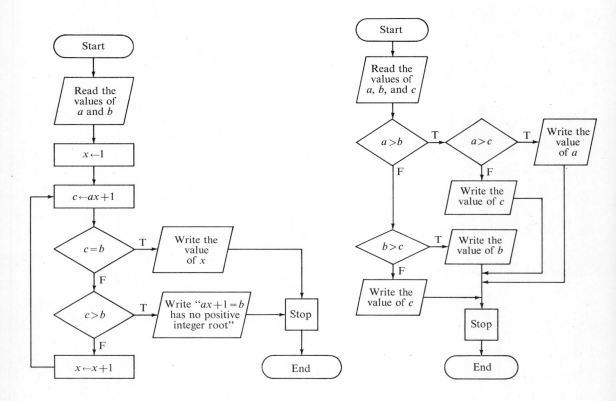

EX. 9–12

EX. 13–18

In Exercises 13–18, name the variable whose value will be written by the program shown in the flow chart at the right for the given values of *a*, *b*, and *c*.

13. $a \leftarrow -2$
    $b \leftarrow 3$
    $c \leftarrow 5$

14. $a \leftarrow 6$
    $b \leftarrow -5$
    $c \leftarrow 6$

15. $a \leftarrow 3$
    $b \leftarrow 7$
    $c \leftarrow -1$

16. $a \leftarrow 0$
    $b \leftarrow -3$
    $c \leftarrow -2$

17. $b < a < c$
18. $a > b; c < b$

Draw a flow chart of the program described in each of the following exercises. Assume that the program is to be executed by a device that can at least read and remember data, add, subtract, multiply, divide, compute powers, compare numbers, and write messages.

19. Read the values of a and b; then compute and write the value of $(a + b)^5$, and stop.
20. Read the values of r and h; then compute and write the value of $\frac{1}{3}\pi r^2 h$, and stop. In the program take the value of π to be 3.1416.
21. Compute and print the values of $n^4 - 1$ for each of the integer values of n from 0 through 15.
22. Compute and write the values of $x^2 + 3$ for values of x from 0 to 1 inclusive, at intervals of 0.05.
23. Suppose that A and B are given as the degree measures of two angles of a triangle. Read the values of A and B, and then write the value of the degree measure of the third angle of the triangle. In case the input data cannot be the measure of two angles of a triangle, write the message "No such triangle."
    Hint: The sum of the degree measures of the angles of a triangle is 180. Therefore, the input data cannot fit a triangle unless $A + B < 180$.
24. Read the values of R, S, and T, and then write the least of these values.

chapter summary

The following statements are true for all real values of each variable except as noted in items 15, 16, and 17.

1. **Axioms of Equality**
   Reflexive property $a = a$
   Symmetric property   If $a = b$, then $b = a$.
   Transitive property   If $a = b$ and $b = c$, then $a = c$.

2. **Axioms of Closure**
   The sum $a + b$ is a unique real number.
   The product ab is a unique real number.

3. **Commutative Axioms** $a + b = b + a$ $ab = ba$

4. **Associative Axioms** $(a + b) + c$ $(ab)c = a(bc)$
                                  $= a + (b + c)$

5. $a + b + c = (a + b) + c$, $a + b + c + d = (a + b + c) + d$, and so on. $abc = (ab)c$, $abcd = (abc)d$, and so on.

6. **Additive Axiom of 0**   There is a unique real number 0 such that

$$a + 0 = a \quad \text{and} \quad 0 + a = a.$$

7. **Axiom of Opposites**   For every a, there is a unique real number $-a$ such that

$$a + (-a) = 0 \quad \text{and} \quad (-a) + a = 0.$$

8. $-(-a) = a$

9. **Property of the Opposite of a Sum** $-(a + b) = (-a) + (-b)$

10. **Distributive Axiom** $a(b + c) = ab + ac$

Also: $(b + c)a = ba + ca$, $ab + ac = a(b + c)$, $ba + ca = (b + c)a$.

11. **Multiplicative Axiom of 1**   There is a unique real number 1 such that $a \cdot 1 = a$ and $1 \cdot a = a$.

12. **Multiplicative Property of 0** $a \cdot 0 = 0$ and $0 \cdot a = 0$

13. **Multiplicative Property of -1** $a(-1) = -a$ and $(-1)a = -a$.

14. **Property of Opposites in Products** $(-a)b = -ab$, $a(-b) = -ab$, $(-a)(-b) = ab$

15. **Axiom of Reciprocals**   For every a different from 0, there is a unique real number $\frac{1}{a}$ such that $a \cdot \frac{1}{a} = 1$ and $\frac{1}{a} \cdot a = 1$.

16. $\frac{1}{\frac{1}{a}} = a$ and $\frac{1}{-a} = -\frac{1}{a}$ ($a \neq 0$)

17. **Property of the Reciprocal of a Product** $\frac{1}{ab} = \frac{1}{a} \cdot \frac{1}{b}$ ($a \neq 0$ and $b \neq 0$)

CHAPTER REVIEW

3–1   Name the closure axiom illustrated by each of these examples.

     1. $2 \cdot 11$ is a real number.   2. $9 + (^-1)$ is a real number.

Which property of equality justifies each of these statements?

3. If $10 \div 2 = 5$, then $5 = 10 \div 2$.
4. If $x = y$ and $y = 15$, then $x = 15$.

3–2 Name the axioms which allow us to make each of the following statements.

5. $1.2 + 2.1 = 2.1 + 1.2$
6. $(2 \times 3) \times 6 = 2 \times (3 \times 6)$

3–3 7. Draw a number line to picture the sum $4 + {}^-3$.

Simplify each expression.

8. $6 + 0$
9. $0 + {}^-8\frac{3}{4}$
10. ${}^-7 + {}^-16$

3–4 Simplify.

11. $-(17 + 3)$
12. $-[14 - (-2)]$
13. $\frac{1}{4} - (-\frac{3}{4})$
14. Given x a member of the set $\{0, 1\}$. Is $-x > 1$?

3–5 State the absolute value of each number.

15. 27
16. -27
17. The graph of $|x| \geq 2$ consists of all points which are at least __?__ units from the __?__.

3–6 Simplify.

18. $7.1 + 2.9 + (10.0)$
19. $3\frac{1}{2} + (-5\frac{1}{2})$

3–7 20. Multiplication is __?__ with respect to addition.

Simplify.

21. $4y + 15y$
22. $2x + 3y + (-x) + 10y$

3–8 23. The __?__ element for multiplication is 1.
24. Whenever we multiply a given real number and 1, the __?__ is the given real number.

Complete.

25. $(-1)(-1) = $ __?__
26. $(-4y)(-9y) = $ __?__
27. $(-3)(5) = $ __?__
28. $3r + (-11r) = $ __?__

3–9 29. The __?__ of 2 is $\frac{1}{2}$.
30. __?__ is the reciprocal of -5.

4 solving equations and problems

4-1 Transforming Equations by Addition

Everyday experience leads one to believe that *if the same number is added to equal numbers, the sums are equal.* A more formal way to state this property is called the **addition property of equality**:

Addition Property of Equality

If a, b, and c are any real numbers such that $a = b$, then

$$a + c = b + c \quad \text{and} \quad c + a = c + b.$$

The addition property of equality follows from facts we already know. The reasoning leading from the given assumption (the *hypothesis*) "a, b, and c are real numbers and $a = b$" to the *conclusion* "$a + c = b + c$ and $c + a = c + b$" is shown in the following sequence of true statements. Notice that the truth of each statement except the first (the given assumption) is guaranteed by one of the properties that we have already learned.

1. a, b, and c are real numbers; $a = b$.   Given
2. $a + c$ is a real number and $c + a$ is a real number.   Axiom of closure for addition
3. $\therefore a + c = b + c$ and $c + a = c + b$.   Substitution principle (substituting b for a in $a + c$ and in $c + a$)

This kind of logical reasoning from known facts and given assumptions to conclusions is called a **direct proof**. Assertions that are proved are called **theorems**. Thus, the addition property just proved is a theorem.

Theorems about real numbers are often discovered in numerical calculations. For instance, look at the following calculation:

$$(8 + 9) + (-9) = 8 + [9 + (-9)]$$
$$= 8 + 0$$
$$= 8$$
$$\therefore (8 + 9) + (-9) = 8.$$

The calculation suggests a useful idea: *to "undo" the addition of a given number to another number, add the opposite of the given number to the sum.* This idea is stated in the following theorem.

For all real numbers *a* and *b*, $(a + b) + (-b) = a$.

This theorem can be used along with the addition property of equality to solve equations. To explain the method, it helps to know that in an equation or an inequality the expressions joined by the equals sign or inequality symbol are called the **members** of the equation or inequality. For example, in the equation

$$\underbrace{x + 9}_{\text{left member}} = \underbrace{-1}_{\text{right member}}$$

the left member is "$x + 9$" and the right member is "-1."

Now study the following sequence of equations:

$$x + 9 = -1 \quad (1)$$
$$x + 9 + (-9) = -1 + (-9) \quad (2)$$
$$\therefore x = -10 \quad (3)$$

To obtain the third equation, add -9 to each member of the first equation (Step 2) and simplify the expressions that result (Step 3). On the other hand, if we add 9 to each member of the third equation, we obtain the first equation:

$$x = -10$$
$$x + 9 = -10 + 9$$
$$\therefore x + 9 = -1$$

Thus, the addition property of equality implies that any root of one of these equations must also satisfy the other equation. Therefore, the equations have the same solution set, namely $\{-10\}$.

Equations having the same solution set over a given set are called **equivalent equations** over that set. To solve an equation, we usually try to change, or **transform**, it into an equivalent equation whose solution set can be found by inspection. The properties of real numbers ensure that each of the following transformations always produces an equivalent equation:

Transformation by Substitution

Substituting for either member of a given equation an expression equivalent to that member.

Transformation by Addition

Adding the same real number to each member of a given equation.

In the following examples and throughout the rest of this book, we use the script letter $\mathcal{R}$ to denote the set of real numbers. Thus,

$$\mathcal{R} = \{\text{the real numbers}\}.$$

Example 1. Solve $y + (-3) = -19$ over $\mathcal{R}$.

Solution:
$$y + (-3) = -19$$
$$y + (-3) + 3 = -19 + 3$$
$$y = -16$$

$\left\{\begin{array}{l}\text{Since } y + (-3) \text{ shows } -3 \\ \text{added to } y, \text{ we add 3 (the} \\ \text{opposite of } -3) \text{ to the sum} \\ \text{to obtain } y.\end{array}\right.$

Check:
$$y + (-3) = -19$$
$$-16 + (-3) \stackrel{?}{=} -19$$
$$-19 = -19$$

∴ the solution set is $\{-16\}$.

Because a numerical mistake may be made in transforming an equation, all work should be checked by substituting each root of the transformed equation in the original equation, as shown in Example 1 above.

Example 2. Solve $17 = 11 + 2a$ over $\mathcal{R}$.

Solution:
$$-11 + 17 = -11 + 11 + 2a$$
$$6 = 2a$$

By inspection, the one and only root of $6 = 2a$ is 3.

Check:
$$17 = 11 + 2a$$
$$17 \stackrel{?}{=} 11 + 2 \cdot 3$$
$$17 \stackrel{?}{=} 11 + 6$$
$$17 = 17$$

∴ the solution set is $\{3\}$.

EXERCISES

Use transformation by addition to solve each equation over $\mathcal{R}$.

1. $x + (-32) = 5$
2. $w + (-17) = 7$
3. $-14 + A = 58$
4. $-32 + B = 76$
5. $75 + k = 75$
6. $42 = m + 42$
7. $r + 30 = 80$
8. $s + 29 = 35$
9. $c + 17 = 0$
10. $d + 31 = 0$
11. $200 = 150 + n$
12. $172 = 2 + d$
13. $f + (-2.1) = 4.3$
14. $g + (0.6) = 1.6$
15. $-\frac{3}{5} + M = \frac{13}{5}$
16. $-\frac{2}{7} + N = \frac{9}{7}$
17. $\frac{3}{4} = L + \frac{1}{4}$
18. $\frac{7}{10} = H + \frac{7}{10}$
19. $-d + 7 = 28$
20. $-k + 5 = 21$
21. $-36 = 12t + (-36)$
22. $\frac{3}{4} = 0.75 + 6s$
23. $-w + (-1) = 4$
24. $-t + 3 = 10$
25. $-3 + 2P = -5$
26. $-1 + (-3)Q = 5$
27. $(x + 1) + (-6) = 4$
28. $(y + 2) + (-8) = 3$
29. $7 + (3 + m) = -12$
30. $14 + (n + 6) = -2$
31. $-1 + [x + (-2)] = 7$
32. $-5 + [y + (-1)] = -4$
33. $(2a + 3) + (-2) = 0$
34. $(3a + 1) + 4 = 5$
★35. $|x| + (-1) = 4$
★36. $|y| + (-2) = -1$
★37. $-3 + |t| = 0$
★38. $-6 + |w| = -4$

4-2 Subtracting Real Numbers

The **difference** $a - b$ between any two real numbers a and b is defined to be the number whose sum with b is a; that is, $a - b$ is the real number satisfying the equation $b + x = a$.

Using only this definition, we compute a difference such as $a - b$ by asking, "What number added to b gives a?" The following example shows how to use the number line to do subtraction problems this way.

Example 1. Simplify: $3 - (-2)$

Solution: On the diagram below, to move from the graph of -2 to the graph of 3 requires a displacement of 5 units to the right. Thus, $3 - (-2) = 5$.

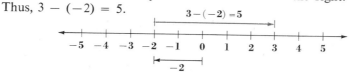

A simple expression for $a - b$ can be found by transforming the equation "$b + x = a$" by addition:

$$
\begin{aligned}
b + x &= a \\
x + b &= a \\
x + b + (-b) &= a + (-b) \\
x + 0 &= a + (-b) \\
\therefore x &= a + (-b)
\end{aligned}
$$

The last equation evidently has just one root, $a + (-b)$. Checking this root in the original equation, we have:

$$
\begin{aligned}
b + x &= a \\
b + a + (-b) &\stackrel{?}{=} a \\
b + (-b) + a &\stackrel{?}{=} a \\
0 + a &\stackrel{?}{=} a \\
a &= a
\end{aligned}
$$

Since the one and only root of "$b + x = a$" is $a + (-b)$, the following theorem has been proved.

Rule for Subtraction

For all real numbers *a* and *b*,

$$a - b = a + (-b).$$

To perform a subtraction, replace the number being subtracted by its opposite, and add.

Since every real number has a unique opposite, if we know b, then we know $-b$. Also, since $a + (-b)$ denotes a sum of real numbers, it represents a real number. Therefore, the rule for subtraction shows that $\mathcal{R}$ is closed under subtraction.

Using this rule, we can always replace a difference by a sum:

Difference	Sum	Value	Check
$3 - 2 =$	$3 + (-2) =$	1	$2 + 1 = 3$
$3 - (-2) =$	$3 + 2 =$	5	$-2 + 5 = 3$
$-3 - 2 =$	$-3 + (-2) =$	-5	$2 + (-5) = -3$
$-3 - (-2) =$	$-3 + 2 =$	-1	$-2 + (-1) = -3$

SOLVING EQUATIONS AND PROBLEMS 81

Example 2. Simplify: $5 - 12 - 4 + 3$

Solution 1:

Step 1	Step 2	Step 3
$5 - 12 = -7$	$-7 - 4 = -11$	$-11 + 3 = -8$

Solution 2:

Step 1	Step 2	Step 3
$5 + 3 = 8$	$-12 + (-4) = -16$	$8 + (-16) = -8$

Expressions for sums such as "$5 + (-x)$" and "$3 + (-2x)$" are usually written as differences: "$5 - x$" and "$3 - 2x$."

Example 3. Simplify: $5 - x + 3 - 2x + 7x$

Solution:
$$5 - x + 3 - 2x + 7x = 5 + 3 + (-x - 2x + 7x)$$
$$= 8 + 4x$$

Example 4. Solve over $\Re$: **a.** $x - 3 = -18$ **b.** $y + 7 = 31$

Solution:

a.
$$x - 3 = -18$$
$$x - 3 + 3 = -18 + 3$$
$$x = -15$$

Check:
$$-15 - 3 \stackrel{?}{=} -18$$
$$-18 = -18$$

∴ the solution set is $\{-15\}$.

b.
$$y + 7 = 31$$
$$y + 7 - 7 = 31 - 7$$
$$y = 24$$

$$24 + 7 \stackrel{?}{=} 31$$
$$31 = 31$$

∴ the solution set is $\{24\}$.

Because the method used in Example 4(b) can be described as *subtracting* 7 from each member of the equation (rather than adding -7 to each member), the method may be called **transformation by subtraction**. Of course, transformation by subtraction is just a special kind of transformation by addition.

The following proof, valid for all real numbers a, b, and c, shows that multiplication of real numbers is distributive with respect to subtraction.

Example 5. To prove: $a(b - c) = ab - ac$

Solution:

Steps	Reasons
a, b, and c are real numbers.	Given
$a(b - c) = a[b + (-c)]$	Rule for subtraction
$ = a(b) + a(-c)$	Distributive axiom
$ = ab + [-(ac)]$	Property of opposites in a product
$ = ab - ac$	Rule for subtraction
∴ $a(b - c) = ab - ac$.	Transitive property of equality

EXERCISES

Simplify each of the following:

1. $4 + (-3) - 7$
2. $0 - (-\frac{8}{7}) + (-\frac{8}{7})$
3. $(217 + 185) - (182 - 27)$
4. $-(413 - 212) + (-182 + 450)$
5. $(x + 7) - (-3 + x)$
6. $(7 - t) - (-3 - t)$

Solve each equation over $\Re$.

7. $t + 34 = 123$
8. $s + 59 = 211$
9. $-13 = x + 28$
10. $-34 = y + 57$
11. $2t + 5 = 5$
12. $7 + 3g = 7$
13. $Q + \frac{1}{2} = -2\frac{1}{2}$
14. $S + \frac{2}{3} = 4\frac{2}{3}$
15. $22 - m = 15$
16. $17 - d = -8$
17. $(n + 7) + 4 = 10$
18. $(3 + c) + 6 = -1$

Write each phrase using algebraic symbols, and simplify.

19. -3 decreased by 8
20. -7 decreased by -3
21. x less $(x - 5)$
22. y less $(7 + y)$
23. Subtract $3\pi + 2$ from $3\pi - 2$.
24. From $5 - 6\pi$ subtract $8 - 6\pi$.
25. Decrease the sum of -8 and 17 by -6.
26. Decrease -18 by the sum of -34 and 12.

Which of the following sets are closed under subtraction? Explain.

27. {natural numbers}
28. {integers}
29. {even integers}
30. {odd integers}

★31. Prove that if $y + 2 = 11$, then $y = 9$.
★32. Prove that if $x + 8 = 12$, then $x = 4$.
★33. Prove that if $x - 3 = 2$, then $x = 5$.
★34. Prove that if $z - 6 = 0$, then $z = 6$.

PROBLEMS

Using real numbers, solve each problem.

1. How much greater is the altitude of Death Valley, California, 282 feet below sea level, than the altitude of the Dead Sea, 1,296 feet below sea level?

2. Find the difference in the altitudes of the Qattara Depression in Egypt, 436 feet below sea level, and the top of Mount McKinley in Alaska, 20,320 feet above sea level.

3. An astronaut entered his space capsule 1 hr. 40 min. before launching time. How long had he been in the capsule 2 hr. 32 min. after launching?

4. Assuming that the Greek mathematician Pythagoras was born in 532 B.C. and died before his birthday in 497 B.C., how old was he when he died?

5. Assuming that the Roman statesman Seneca was born in the year 4 B.C., and died after his birthday in 65 A.D., how old was he when he died? (Assume no year 0.)

6. Pete owed $3.45 to Gerry. After paying Gerry the money he owed, Pete had $2.17 left. How much money had Pete before he paid Gerry?

7. What is the difference between the melting and boiling points of chlorine, if it boils at 34.6° below 0° centigrade and melts at 101.6° below 0° centigrade?

8. Mercury melts at 38.87° below 0° centigrade and boils at 356.9° above 0° centigrade. What is the difference of these temperatures?

9. Mr. Hooker drove a golf ball from a point 180 yards to the east of a hole to a point 47 yards to the west of the hole. How far did the ball travel?

10. Mrs. Adams rode the subway from a point 53 blocks south of Main Street to a point 41 blocks north of Main Street. How many blocks did she travel?

★11. Draw a flow chart of a program to read the value of x from a card and to determine whether x is greater than or equal to zero or whether it is negative. In case $x \geq 0$, the value of $1 + x$ is to be computed and printed. Otherwise, the value of $1 - x$ is to be computed and printed. Assume that more than one card is to be read. What is printed by the program if the value of x is 3? −2? 0?

4–3 Transforming Equations by Multiplication

The **multiplication property of equality** means that when equal numbers are multiplied by the same number, the products are equal. For example, since $7 = 5 + 2$,

$$7 \cdot 3 = (5 + 2)3 \quad \text{and} \quad 3 \cdot 7 = 3(5 + 2).$$

Multiplication Property of Equality

If *a*, *b*, and *c* are any real numbers such that *a* = *b*, then

$$ac = bc \quad \text{and} \quad ca = cb.$$

This property guarantees that the following transformation always produces an equivalent equation.

Transformation by Multiplication

Multiplying each member of a given equation by the same *nonzero* real number.

Example 1. Solve $8m = 72$ over $\Re$.

Solution:
$$8m = 72$$
$$\tfrac{1}{8} \cdot (8m) = \tfrac{1}{8} \cdot 72$$
$$m = 9$$

{Since $8m$ shows m multiplied by 8, multiply the product by $\tfrac{1}{8}$ (the reciprocal of 8) to obtain m.

Check:
$$8(9) \stackrel{?}{=} 72$$
$$72 = 72$$

$\therefore$ the solution set is $\{9\}$.

When "m" is written in place of "$\tfrac{1}{8}(8m)$" and "9" in place of "$\tfrac{1}{8} \cdot 72$" or "$\tfrac{1}{8}(8 \cdot 9)$," we use a very helpful idea: *to "undo" the multiplication of a number by a given number, multiply the product by the reciprocal of the given number.* In general:

For all real numbers a and all *nonzero* real numbers b, $\dfrac{1}{b}(ba) = a$.

In the following examples, and in fact, from here on in this book, assume, unless otherwise stated, that *open sentences are to be solved over $\Re$, the set of real numbers.*

Example 2. $\tfrac{1}{3}t = -5$

Solution:
$$\tfrac{1}{3}t = -5$$
$$3 \cdot \tfrac{1}{3}t = 3 \cdot -5$$
$$t = -15$$

Check:
$$\tfrac{1}{3}(-15) \stackrel{?}{=} -5$$
$$-5 = -5$$

$\therefore$ the solution set is $\{-15\}$.

Example 3. $-5d = 120$

Solution:
$$-5d = 120$$
$$-1(-5d) = -1(120)$$
$$5d = -120$$
$$\tfrac{1}{5} \cdot 5d = \tfrac{1}{5}(-120)$$
$$d = -24$$

Check:
$$-5(-24) \stackrel{?}{=} 120$$
$$120 = 120$$

$\therefore$ the solution set is $\{-24\}$.

Zero cannot be used as a multiplier in transforming an equation. Consider the equation "$\tfrac{1}{3}t = -5$" whose solution set is $\{-15\}$. If we multiplied by 0, the equation "$0 \cdot \frac{1}{3}t = 0(-5)$" would be satisfied by any value of t; its solution set would be $\mathcal{R}$. Thus, the original equation "$\frac{1}{3}t = -5$" and the transformed equation "$0 \cdot \frac{1}{3}t = 0(-5)$" would have different solution sets; they would not be equivalent equations (see page 77). Therefore, *multiplication by zero* cannot be used because it does not produce an equivalent equation.

EXERCISES

Solve each equation over $\mathcal{R}$.

1. $\frac{1}{8}q = 17$
2. $\frac{1}{11}t = 12$
3. $\frac{1}{9}y = -15$
4. $\frac{1}{6}x = -32$
5. $3x = 96$
6. $4x = -128$
7. $13t = -52$
8. $-14k = 84$
9. $5 = -\frac{1}{13}p$
10. $-8 = -\frac{1}{5}b$
11. $\frac{1}{11}m = -3.7$
12. $-\frac{1}{7}h = -2.5$
13. $\frac{1}{2} - d = \frac{7}{2}$
14. $\frac{1}{3}w = \frac{5}{3}$
15. $-\frac{12}{5} = \frac{1}{5}M$
16. $\frac{3}{7} = -\frac{1}{7}K$
17. $0.8 = 0.8h$
18. $-1.7 = 1.7q$
19. $\frac{1}{3}p = 4\frac{1}{3}$
20. $\frac{1}{2}u = 5\frac{1}{2}$
21. $\frac{1}{7}k = -3\frac{2}{7}$
22. $-\frac{1}{8}x = 1\frac{5}{8}$
23. $0 = -4.3c$
24. $-1 = 7.75a$
25. $\frac{1}{2}x + 5 = 20$
26. $\frac{1}{5}g - 6 = 4$
27. $15 = 4z + 11$
28. $2 = -7x + 2$
29. $4 - \frac{1}{3}y = -7$
30. $-1 - \frac{1}{5}t = 2$
★31. $\frac{1}{5}|a| + 1 = -1$
★32. $-2 + \frac{1}{3}|b| = -5$

4–4 Dividing Real Numbers

Why is the statement "$24 \div 6 = 4$" true? We know that $24 \div 6 = 4$ because $6 \times 4 = 24$, and we call $24 \div 6$, or 4, the *quotient* of 24 by 6. In general, the **quotient** $a \div b$ of any real number a by any *nonzero* real number b is the number whose product with b is a; that is, $a \div b$ is the real number satisfying the equation $bx = a$.

The quotient $a \div b$ can also be represented by a fraction:

$$a \div b = \frac{a}{b}$$

We know that

$$\tfrac{7}{2} = 7 \times \tfrac{1}{2}; \qquad \tfrac{3}{5} = 3 \times \tfrac{1}{5}; \qquad \tfrac{8}{4} = 8 \times \tfrac{1}{4}.$$

These statements suggest how to use reciprocals to express any quotient as a product.

Rule for Division

For all real numbers a and all nonzero real numbers b,

$$a \div b = a \times \frac{1}{b}.$$

To perform a division, replace the divisor by its reciprocal, and multiply.

We can show that this rule is correct by transforming the equation "$bx = a$" by multiplication:

$$bx = a$$

$$\frac{1}{b} \cdot bx = \frac{1}{b} \cdot a \qquad (b \neq 0)$$

$$1 \cdot x = a \cdot \frac{1}{b}$$

$$\therefore x = a \cdot \frac{1}{b}$$

The last equation has just one root, $a \cdot \dfrac{1}{b}$. Checking this root in the original equation, we have:

$$bx = a$$

$$b\left(a \cdot \frac{1}{b}\right) \stackrel{?}{=} a$$

$$a\left(b \cdot \frac{1}{b}\right) \stackrel{?}{=} a$$

$$a \cdot 1 \stackrel{?}{=} a$$

$$a = a$$

Since the one and only root of "$bx = a$" is $a \cdot \dfrac{1}{b}$, it follows that

$$a \div b = a \cdot \frac{1}{b}, \qquad \text{or} \qquad \frac{a}{b} = a \cdot \frac{1}{b}.$$

Since every nonzero real number has a unique reciprocal, if we know b

($b \neq 0$), then we know $\frac{1}{b}$. Since $a \cdot \frac{1}{b}$ denotes a product of real numbers, it represents a real number. Therefore, this theorem implies that *the set of real numbers is closed under division*, **excluding division by zero**.

Using this theorem, any quotient can be replaced by a product.

Quotient	Product	Value	Check
$24 \div 6$ or $\frac{24}{6}$	$= 24 \times \frac{1}{6}$	$= 4$	$6 \times 4 = 24$
$24 \div (-6)$ or $\frac{24}{-6}$	$= 24 \times -\frac{1}{6}$	$= -4$	$-6 \times (-4) = 24$
$-24 \div 6$ or $\frac{-24}{6}$	$= -24 \times \frac{1}{6}$	$= -4$	$6 \times (-4) = -24$
$-24 \div (-6)$ or $\frac{-24}{-6}$	$= -24 \times -\frac{1}{6}$	$= 4$	$-6 \times (4) = -24$

Why must division by zero be excluded in the set of real numbers? For any number a, $\frac{a}{0} = c$ would mean that $a = 0 \times c$. If $a \neq 0$, *no* value of c can make the last equation a true statement, since $0 \times c = 0$ for each value of c. But if $a = 0$, *every* value of c makes the equation a true statement. Therefore, a "quotient" $\frac{a}{0}$ either would have *no* value or would have an *infinite set* of values. Therefore, *division by zero has no meaning in the set of real numbers*.

In particular, we cannot divide zero by zero. But we can divide zero by any other number. Consider these examples:

$$\tfrac{0}{9} = 0 \cdot \tfrac{1}{9} = 0$$
$$0 \div (-8) = 0 \cdot -\tfrac{1}{8} = 0$$

For $a \neq 0$, $0 \div a = 0 \cdot \frac{1}{a} = 0$. Thus, *the quotient of zero divided by any nonzero number is zero*.

Example. Solve each equation: **a.** $4x = -56$ **b.** $-15 = -\frac{x}{3}$

Solution: **a.** $4x = -56$

$$\frac{4x}{4} = \frac{-56}{4}$$

$$x = -14$$

Check: $4(-14) \stackrel{?}{=} -56$

$-56 = -56$

∴ the solution set is $\{-14\}$.

b. $-15 = -\frac{x}{3}$

$-3 \cdot (-15) = -3 \cdot \left(\frac{x}{-3}\right)$

$45 = x$

$-15 \stackrel{?}{=} -\frac{45}{3}$

$-15 = -15$

∴ the solution set is $\{45\}$.

Because the method used in part (a) of the preceding example can be described as dividing each member of the equation by 4 (rather than multiplying each member by $\frac{1}{4}$), the method may be called **transformation by division**. Of course, transformation by division is just a special case of transformation by multiplication.

EXERCISES

State the value of each quotient.

1. $125 \div (-25)$
2. $-96 \div 16$
3. $0 \div (-15)$
4. $0 \div 49$
5. $-17 \div (-\frac{1}{3})$
6. $-1.44 \div (1.2)$
7. $-0.09 \div 0.3$
8. $-16 \div (-\frac{1}{8})$
9. $\dfrac{5}{-\frac{1}{4}}$
10. $\dfrac{8}{-\frac{1}{2}}$
11. $-6.96 \div (24)$
12. $16.8 \div (-12)$
13. $0 \div (-\frac{3}{8})$
14. $\dfrac{0}{1.89}$
15. $-7x \div \frac{1}{3}$
16. $8z \div (-\frac{1}{11})$

Solve each equation.

17. $18y = 360$
18. $33x = 165$
19. $-270 = -20t$
20. $234 = -78s$
21. $-63a = 378$
22. $21x = -252$
23. $1.5q = -2.25$
24. $-1.3p = -1.69$

Find the average of the numbers in each set given in Exercises 25–28. (The **average** is the quotient of the sum of the numbers by the number of the numbers.)

25. $\{-9, 14, -33, 22, -44\}$
26. $\{-91, -39, 45, 80\}$
27. $\{-67, -25, -8, 8, 25, 67\}$
28. $\{-42, 0, -19, 42, -56\}$

If the value of x is -2, y is -1, z is 3, and w is 6, evaluate each expression.

29. $\dfrac{3xy}{w}$
30. $\dfrac{w}{xz}$
31. $\dfrac{xz^2}{w}$
32. $\dfrac{4w}{x^2z}$
33. $\dfrac{wz}{xy^2}$
34. $\dfrac{wy}{xz}$
35. $\dfrac{w+z}{y^2}$
36. $\dfrac{w^2}{z^2 - 3y}$
37. $\dfrac{z-x}{z+x}$
38. $\dfrac{w+2y}{x^2}$
39. $\dfrac{y^2 - 4x}{w+z}$
40. $\dfrac{w-z}{y^2 - x}$
41. $\dfrac{(z+1)^2 - w}{x - 4y}$
42. $\left(\dfrac{w}{z}\right)^2 - \left(\dfrac{x}{y}\right)^2$
43. $\dfrac{x^2 + 2xy + y^2}{(w-z)^2}$
44. $\dfrac{x^2 + 2(w+y)}{z^2 - w + y}$

Determine which of the following sets are closed under division, excluding division by zero.

45. $\{1\}$
46. $\{-1, 0, 1\}$
47. {the positive numbers}
48. $\{-1\}$
49. $\{\frac{1}{2}, 1, 2\}$
50. {the negative numbers}

Give the reason justifying each step in the following proofs.

★51. *To prove:* If a, b, and c are any real numbers such that $a = b$ and $c \neq 0$, then $\dfrac{a}{c} = \dfrac{b}{c}$.

1. a, b, and c are real numbers; $a = b$; $c \neq 0$.
2. $\dfrac{1}{c}$ is a real number.
3. $a \cdot \dfrac{1}{c} = b \cdot \dfrac{1}{c}$
4. $\dfrac{a}{c} = \dfrac{b}{c}$

★52. *To prove:* If a, b, and c are any real numbers such that $c \neq 0$ and $\dfrac{a}{c} = \dfrac{b}{c}$, then $a = b$.

1. a, b, and c are real numbers; $c \neq 0$; $\dfrac{a}{c} = \dfrac{b}{c}$.
2. $\dfrac{a}{c} \cdot c = \dfrac{b}{c} \cdot c$
3. $\left(a \cdot \dfrac{1}{c}\right) c = \left(b \cdot \dfrac{1}{c}\right) c$
4. $a \left(\dfrac{1}{c} \cdot c\right) = b \left(\dfrac{1}{c} \cdot c\right)$
5. $a \cdot 1 = b \cdot 1$
6. $a = b$

4–5 Using Several Transformations

We know that $9 + 4 - 4 = 9$ and $9 - 4 + 4 = 9$. In general,

$$a + b - b = a \quad \text{and} \quad a - b + b = a.$$

Because the operations of adding and subtracting the same number are "opposite" in effect, we can "undo" the result of one of these operations by "doing" the opposite operation. We call addition of a given number and subtraction of the same number *inverse operations*. Similarly, multiplication and division are called inverse operations.

The following example shows that in solving an equation, we *undo* the *operations* used in building the equation, but in *reverse order*.

Building an Equation		*Solving the Equation*	
$t = 4$	Given	$4t - t + 5 = 17$	Given
$3 \cdot t = 3 \cdot 4$	Multiply by 3	$3t + 5 = 17$	Substitute $3t$ for $4t - t$
$3t = 12$			
$3t + 5 = 12 + 5$	Add 5	$3t + 5 - 5 = 17 - 5$	Subtract 5
$3t + 5 = 17$		$3t = 12$	
		$\dfrac{3t}{3} = \dfrac{12}{3}$	Divide by 3
$4t - t + 5 = 17$	Substitute $4t - t$ for $3t$	$t = 4$	

The following steps are usually helpful in transforming an equation into an equivalent equation that can be solved by inspection.

1. **Simplify each member of the equation.**
2. **If there are indicated additions or subtractions, use the inverse operations to undo them.**
3. **If there are indicated multiplications or divisions involving the variable, use the inverse operations to undo them.**
4. **If the transformed equation can be solved by inspection, check its root in the given equation.**

Example 1. Solve: $12x - 3 - 2x = 23 + 19$

Solution:

1. Copy the equation; simplify each member.

$$12x - 3 - 2x = 23 + 19$$
$$10x - 3 = 42$$

2. Add 3 to each member.

$$10x - 3 + 3 = 42 + 3$$
$$10x = 45$$

3. Divide each member by 10.

$$\frac{10x}{10} = \frac{45}{10}$$
$$x = 4.5$$

Check:
$$12(4.5) - 3 - 2(4.5) \stackrel{?}{=} 23 + 19$$
$$54.0 - 3 - 9.0 \stackrel{?}{=} 42$$
$$42 = 42$$

∴ the solution set is $\{4.5\}$.

Example 2. Solve: $38 = 2 - 3(y - 5)$

Solution:

1. Copy the equation; use the distributive axiom and simplify the right member.

$$38 = 2 - 3(y - 5)$$
$$38 = 2 - 3y + 15$$
$$38 = 17 - 3y$$

2. Subtract 17 from each member.

$$38 - 17 = 17 - 3y - 17$$
$$21 = -3y$$

3. Divide each member by -3.

$$\frac{21}{-3} = \frac{-3y}{-3}$$
$$-7 = y$$

4. Check is left to you.

EXERCISES

Solve each equation.

1. $3x + 2 = 17$
2. $5z - 3 = 17$
3. $\dfrac{y}{2} - 3 = 6$
4. $\dfrac{z}{3} + 5 = -1$
5. $7x - 5x = -18$
6. $11t - 8t = -3$
7. $-2 = 3z - z$
8. $-8 = 5a - 3a$
9. $3w + 1 - 2w = 3$
10. $4t - 3 - 3t = -5$
11. $2 = 6x - 3 - x$
12. $-7 = 2y - 3 - 3y$
13. $b - \tfrac{1}{2}b - 4 = 0$
14. $\tfrac{7}{6}c - c - 6 = 0$
15. $5.5r - r = 90$
16. $8.4s - s = 14.8$
17. $99 - 4p - 6p = -1$
18. $-36 - b - 4b = 64$
19. $2x - 1 + x + 1 = 27$
20. $-2 + p + 2p + 2 = -12$
21. $7 = 6g - 5g - 1$
22. $0 = 7m + 4 - 3m$
23. $0 = -y - 3y - 16$
24. $0 = 35 - 5x - 2x$
25. $18k - 6k - 14k = 6$
26. $11h - 5h + 16 = 4$
27. $2h - 3h + 4 - h = 10$
28. $5y + 10 - 3y - y = -7$
29. $x + (x + 1) + (x + 2) = 9$
30. $(y - 2) + (y - 1) + y = -3$
31. $19n - 11 - 12n - 5 = 5$
32. $-7h - 17 - 6 + 4h = 17$
33. $-2t + 3(2 + t) = -10$
34. $2(a + 5) - a = -7$
35. $2(x + 3) - 6x - 4 = 8$
36. $4z + 3(z + 7) = -14$
37. $-5(1 + w) + 4w = 8$
38. $-3(2 - y) + 2y = 7$
39. $5a - (1 - a) + 4 = 9$
40. $-7 - (-3 - 2g) - 10 = 0$
41. $1.5k - (k + 2.4) = 17.4$
42. $3s - (0.5s + 7.2) = 32.8$
43. $5(m + 2) - 4(m + 1) - 3 = 0$
44. $74 - 2[(35 + 2x) - 3] = -6$
45. $3(8n - 2) - 3(1 - n) + 8 = 8$
46. $2a - [(3a + 4) - 5] - a = 5$
47. $-2[z + 3(5 - z)] - 5(z - 7) = 0$
48. $6[x - 2(2x + 3) + 1] + 3(5 + 6x) + x = 0$
★49. $7(|x| - 2) - 3|x| - 16 = 2$
★50. $2|t| - (|t| - 1) = 7$

4–6 Using Equations to Solve Problems

In a "word problem" we are told how certain numbers are related to one another. If we can translate these numerical relationships into an equation, then we can solve the problem by solving the equation.

Example 1. The length of the service module containing the engines of the Apollo 11 spacecraft was 4 feet shorter than twice the length of the command module housing the crew. Also, the lunar module was 1 foot longer than the command module. If the overall length of Apollo 11 was 45 feet, how long was the command module?

Solution:

1. The first step in solving this problem is to select a symbol to represent the number we wish to find: the number of feet in the length of the command module.

    Let x represent this number. We know that the length of the service module was 4 feet less than 2 times the length of the command module, so $2x - 4$ represents the length of the service module. Also, the lunar module (LM) was 1 foot longer than the command module, so that the LM was $x + 1$ feet long. The replacement set for x must be the set of positive real numbers because x represents a nonzero number of feet.

2. The second step is to write an open sentence. We do this by setting one expression for the length of Apollo 11 equal to another.

length of command module	added to	length of service module	added to	length of lunar module	equals	length of Apollo 11
x	$+$	$2x - 4$	$+$	$x + 1$	$=$	45

3. The third step is to solve the open sentence.

    $$x + 2x - 4 + x + 1 = 45$$
    $$4x - 3 = 45$$
    $$4x - 3 + 3 = 45 + 3$$
    $$4x = 48$$
    $$\frac{4x}{4} = \frac{48}{4}$$
    $$x = 12$$

    ∴ the one and only root of the equation is 12. Hence, the only possible length of the command module is 12 feet.

4. The fourth step is to check the results with the words of the problem.

Length of command module	12
Length of service module (4 feet less than 2 times 12 feet)	20
Length of lunar module (1 foot more than 12 feet)	13
Length of Apollo	45

The steps taken to solve the preceding problem suggest the following plan that usually helps in solving any problem.

Plan for Solving a Word Problem

After carefully reading the problem, decide what numbers are asked for. Then take these steps:

1. **Choose a variable with an appropriate replacement set, and use the variable in representing each described number.**
2. **Write an open sentence by using facts given in the problem.**
3. **Find the solution set of the open sentence.**
4. **Check the results with the words of the problem.**

PROBLEMS

Whenever a sketch will help solve a problem, draw one.

1. The area of a rectangle 6 yards wide is 78 square yards. Find the length of the rectangle.
2. The perimeter of a square is 124 meters. How long is each side of the square?
3. An airplane's altimeter reads 2,546 feet. What is the airplane's altitude if this reading is 3.5 feet less than the true reading?
4. In checking a patient's pulse rate at 6:00 A.M., Nurse Perlman found that the rate was 7 beats less per minute than the rate recorded at 8:00 P.M. the previous night. If Miss Perlman found the pulse rate to be 72 beats per minute, what was the rate recorded the previous night?
5. On Sunday morning, a copy of the *Times* costs 30 cents more than a copy of the *News*. Mr. Donnelly, who buys both papers, spends 70 cents for the Sunday editions. How much does a copy of each paper cost on Sunday?
6. Together, a house and lot cost $40,000. The house cost seven times as much as the lot. How much did the lot cost? the house?
7. In an election for town clerk, 584 people voted for one or the other of the two candidates. The winner received 122 votes more than his opponent. How many people voted for the winner?
8. Professor Landers took 55 minutes to drive from his home to the University and back. The return drive took 7 minutes less than the trip to the University. How long did it take him each way?
9. Water is a compound made up of 8 parts by weight of oxygen and 1 part by weight of hydrogen. How many grams of hydrogen are there in 225 grams of water?

10. On a 600-mile trip, Mr. Seda traveled by automobile and airplane. If he traveled seven times farther by airplane than by automobile, how far did he travel by automobile?

11. Mr. Brown will need 186 ft. of fence to enclose his rectangular yard. If the length of his yard is 9 feet more than the width, what are the dimensions of his yard?

12. The width of a rectangle is 7 centimeters less than the length. If the perimeter is 410 centimeters, what are the dimensions of the rectangle?

13. The sum of twice a number and 27 is 51. Find the number.

14. The sum of three times a number and 11 is -13. Find the number.

15. Four times a number diminished by 27 is 45. Find the number.

16. Find a number such that 93 less than twice the number is 59.

17. One number is 22 less than a second number. If the greater number is subtracted from twice the lesser number, the difference is 16. Find the lesser number.

18. The difference of two numbers is 17. If twice the greater number is subtracted from 5 times the lesser number, the difference is 2. Find the greater number.

19. A 98-foot length of TV cable is cut so that one piece is 10 feet shorter than twice the length of the other piece. Find the length of the shorter piece.

20. The entertainment portion of a 30-minute TV program lasted 4 minutes longer than 4 times the portion devoted to advertising. How many minutes of the program were devoted to advertising?

21. Mr. Richards bought 4 times as many six-cent stamps as ten-cent stamps, and 3 times as many five-cent stamps as ten-cent ones. If he paid $7.35 for all the stamps that he bought, how many stamps of each denomination did he buy?

22. In a day, Machine X produces twice as many plastic bowls as Machine Y. Machine Z produces 50 more bowls than Machine Y. If the total production is 6,170 bowls, how many bowls does each produce?

23. Information in numerical units called "bytes" is stored in the memory of an electronic computer. The storage capacity of Computer II is twice that of Computer I, while Computer III has a capacity four times as great as that of Computer II. If the total storage capacity of the three machines is 88,000 bytes, find the capacity of each computer.

24. The record low temperature at the United States installation at Antarctica is 5.5° warmer than the record low at the Russian station. The average of the two record low temperatures is minus 118.65° F. What is the official United States record low?

★25. The coefficients a, b, and c in the equation "$ax + b = c$" are given on a card. Draw a flow chart of a program to determine whether $a = 0$ or not. In case $a = 0$, the program should print the message, "Not a linear equation." Otherwise, the program should compute and print the root of the equation. Assume that cards for several equations are to be processed.

4–7 Equations Having the Variable in Both Members

In the equation
$$2p = 63 - 5p,$$

the variable appears in both members. Is it permissible to add "$5p$" to each member?

For every real number p, "$5p$" denotes a product of real numbers, and, therefore, it represents a real number. Because the same real number may be added to each member of an equation, it is permissible to add "$5p$" to each member and the solution set of the equation is not changed by doing so. Thus, we may solve the equation as follows:

$$2p = 63 - 5p$$
$$2p + 5p = 63 - 5p + 5p$$
$$7p = 63$$
$$\frac{7p}{7} = \frac{63}{7}$$
$$p = 9$$

Check:
$$2p = 63 - 5p$$
$$2 \cdot 9 \stackrel{?}{=} 63 - 5 \cdot 9$$
$$18 \stackrel{?}{=} 63 - 45$$
$$18 = 18$$

∴ the solution set is {9}.

The next example shows that not every equation has a real root.

Example 1. Solve: $1 - (2x + 8) = (x + 9) - 3x$

Solution:
$$1 - (2x + 8) = (x + 9) - 3x$$
$$1 - 2x - 8 = -2x + 9$$
$$-2x - 7 = -2x + 9$$
$$-2x - 7 + 2x = -2x + 9 + 2x$$
$$-7 = 9$$

Since the given equation is equivalent to the false statement "$-7 = 9$," the *equation has no root*. The solution set is ∅.

Example 2 involves an equation that is satisfied by every real number.

Example 2. Solve: $\frac{3}{5}(5 - 10n) - 2 = 1 - 6n$

Solution:
$$\frac{3}{5}(5 - 10n) - 2 = 1 - 6n$$
$$\frac{3}{5} \cdot 5 - \frac{3}{5} \cdot 10n - 2 = 1 - 6n$$
$$3 - 6n - 2 = 1 - 6n$$
$$1 - 6n = 1 - 6n$$
$$1 - 6n + 6n = 1 - 6n + 6n$$
$$1 = 1$$

Since the given equation is equivalent to the true statement "$1 = 1$," the equation is satisfied by every real number. The solution set is $\mathfrak{R}$.

Any equation which is a true statement for every numerical replacement of the variable(s) in the equation is called an **identity**. Thus, "$\frac{3}{5}(5 - 10n) - 2 = 1 - 6n$" is an identity.

In many fields of applied mathematics, equations called **formulas** represent the numerical relationships between physical measurements. Recall, for example, such formulas as:

area of a rectangle = length of rectangle × width of rectangle
$$A = lw$$

distance traveled = constant rate × time traveled
$$d = rt$$

and

simple interest = principal × rate of interest × time
$$I = Prt$$

In applying formulas, we often have to find equivalent equations in which a particular variable is expressed in terms of other variables.

Example 3. Solve: $A = P + Prt$ for r

Solution:
$$A = P + Prt$$
$$A - P = P + Prt - P$$
$$A - P = Prt$$
$$\frac{A - P}{Pt} = \frac{Prt}{Pt}$$
$$\frac{A - P}{Pt} = r$$

or $\quad r = \dfrac{A - P}{Pt}, \; (P \neq 0, t \neq 0)$

Notice that the result in Example 3 is given in the form of an equation rather than as a statement of a solution set. This is customary in working with formulas. Also, because Pt was used as a divisor, neither P not t can equal 0.

Example 4. Jim is exactly 9 years older than his sister, Myra. Last year, he was twice as old as Myra was. How old is Jim?

Solution: 1. Let y = Myra's age in years now. (The symbol = here is used to mean "represent.")

The facts in the problem are arranged in the chart below.

	Myra	Jim
Age now	y	$y + 9$
Age last year (1 year previous)	$y - 1$	$y + 9 - 1 = y + 8$

2. Last year:

Jim's age   was   twice Myra's age
$y + 8$ = $2(y - 1)$

3. $$y + 8 = 2y - 2$$
$$10 = y$$

4. If Myra is 10 years old, Jim is $10 + 9$, or 19, years old. Was Jim twice as old as Myra last year? That is:

$$19 - 1 \stackrel{?}{=} 2(10 - 1)$$
$$18 \stackrel{?}{=} 2(9)$$
$$18 = 18$$

∴ Jim is 19 years old.

EXERCISES

Solve each equation in one variable. If the equation is an identity, state this fact. If the equation contains more than one variable, solve for the variable in color.

1. $8q = 35 + 3q$
2. $11a = 8a + 42$
3. $7b = 54 - 2b$
4. $9x = 84 - 5x$
5. $21y = 16y - 30$
6. $31t = 25t - 48$

7. $28 - 2m = 2m$
9. $168 - 12p = 0$
11. $195 + 18z = 15z$
13. $4a = 2a - 12$
15. $2x - 45 = -x$
17. $4t + 6 = 2t$
19. $22 - x = 7x - 2$
21. $8 + 2z = z + 15$
23. $3z - 5 = z + 11$
25. $4b + 33 = 9b + 5$
27. $3x + 7 - x = 5x - 1$
29. $P = 1 + 2w$
31. $12w - 8t = 4u$
33. $2x + 3y = 1$
35. $M = \dfrac{x + y + z}{3}$
37. $-3(2t - 5) = 3(t - 1)$
39. $-3r + 2(r + 1) = 7$
41. $T = 2\pi rh + 2\pi r^2, r \neq 0$
43. $a = S(1 - r), S \neq 0$

8. $9n - 70 = 2n$
10. $0 = 196 - 7s$
12. $182 + 53c = 40c$
14. $5w = w - 64$
16. $-r = 93 - 4r$
18. $-5z - 8 = -3z$
20. $12m - 5 = 2 - 5m$
22. $27 + 5r = 9 + 3r$
24. $-2w + 7 = 5w - 21$
26. $17 - 4h = 14 - h$
28. $2y - 3 + 5y = y + 15$
30. $A = \pi + 2h$
32. $24a - 3v = 5v$
34. $5x - 2y = 1$
36. $P = l + 2w + 2h$
38. $4(6 - z) = -2(3z + 1)$
40. $-2n + 3(2n - 1) = 29$
42. $T = mg - mf, m \neq 0$
44. $S = \dfrac{n}{2}(a + l), n \neq 0$

45. $2(5x + 1) + 16 = 4 + 3(x - 7)$
46. $6(y - 3) + 5 = 23 + 2(y - 2)$
47. $5(2z - 4) + 6 = -2(7 - 5z)$
48. $5x + (x + 15) - 3(2x + 4) = 3$
49. $10t + (5t - 4) = 9t + (6t + 2)$
50. $4(4y + 6) + 8 = -12 - 16(1 - y)$
51. $2.5z + 0.3(480 - z) = 3.2$
52. $0.5x - 0.25(7 - 4x) = 0.1x - 0.7(4 + x)$
53. $4(x - a) + 4a = 7(b + 3) - 21$
54. $-3(c + y) + 3c = 6(c - 2) + 12$
55. $-2[x - 3(2 - x)] - 4 = 3(x + 1) - 6$
56. $8[3 + 2(2y + 3)] = -2[16 - (y + 7)]$
57. $5[z - 2(3z - 6)] = 3(24 - 9z)$
58. $2[4t + 7 - 3(2t + 3)] = 8(3t - 4) - 7(2t - 3)$
59. $3[2(2r + 3) - (r + 1) - (r + 2)] = 0$
60. $2[3(y - 4) - 2(5y + 1)] = 3(y - 2) - 4(2y + 1)$
61. $7[2x + 2(3 - 2x) + 1] = 4 + 5(2 - 3x) - x$
62. $-3[2(y - 5) - y] = 2[3y - 5(-3 + y)] + y$

PROBLEMS

Solve each of the following problems. Draw a sketch, if needed.

1. Find a number which equals its opposite increased by 8.
2. Find a number which equals its opposite decreased by 12.
3. Multiplying a certain number by 2 yields the same result as adding 15 to 3 times the number. Find the number.
4. Five times a certain number is twelve more than twice the number. What is the number?
5. The sum of two numbers is 17. Six times the smaller number is two less than twice the greater. What are the numbers?
6. The difference between two numbers is 8. Five times the smaller number is 4 more than three times the greater. What are the numbers?
7. The sum of two numbers is 15. Four times the smaller number is 60 less than twice the greater number. Find the numbers.
8. Express 11 as a sum of two addends, such that the larger addend is four times the sum of 9 and the smaller addend.
9. One house is four times as old as another. Eighteen years from now it will be twice as old. How old are the houses?
10. Mrs. Lee is three times as old as her daughter, Sue. Ten years from now she will be twice as old as Sue. How old are Mrs. Lee and Sue?
11. A molecule of sucrose contains 8 more oxygen atoms than a molecule of silver nitrate, or 1 fewer than 4 times as many as a molecule of silver nitrate. How many oxygen atoms does a molecule of each of these compounds contain?
12. The length of one rectangle is 3 feet more than the length of another rectangle. The width of the first rectangle is 5 feet. The width of the second is 2 feet. If the area of the first rectangle is 120 square feet greater than the area of the second, find the length of each rectangle.
13. The perimeter of a rectangle is 24 meters. Three times the length is 9 meters less than 6 times the width. Find the dimensions of the rectangle.
14. The perimeter of an isosceles triangle is 28 centimeters. Five times the length of the base is 4 centimeters more than 7 times the length of each of the congruent sides. How long is each side of the triangle?
15. The length of one rectangle is 4 inches greater than the width of a second rectangle. The width of the first rectangle is 6 inches and the length of the second rectangle is 9 inches. The area of the second rectangle is 3 square inches less than the area of the first rectangle. Find the dimensions of the first rectangle.
16. An apple has 29 more calories than a peach and 13 fewer calories than a banana. If 3 apples have 43 fewer calories than 2 bananas and 2 peaches, how many calories does an apple have?

17. A cup of coffee contains 20 more milligrams of caffeine than a cup of tea and 85 more milligrams of caffeine than the average cola drink. If one cup of tea and four cola drinks contain the same amount of caffeine as one cup of coffee, how much caffeine is there in one cup of coffee?

18. One year the folk-rock group The Gypsy Traders made two 45 r.p.m. records and one $33\frac{1}{3}$ r.p.m. record. During that year one of the 45's sold 50,000 more records than the other 45 and 30,000 more than 14 times as many records as the $33\frac{1}{3}$. If the best-selling 45 sold 745,000 more records than the $33\frac{1}{3}$, how many of each recording were sold?

19. If the Washington Monument were 85 feet taller, it would be just as tall as the Gateway Arch in St. Louis. If the height of the Washington Monument is 240 feet more than one-half the height of the Gateway Arch, how tall is each of these structures?

20. The Suez Canal, which is about 104 miles long, was opened 45 years before the Panama Canal (length, about 51 miles). As of 1959, the Suez Canal had been operating twice as long as the Panama Canal. In what year was the Suez Canal opened? the Panama Canal?

21. In tennis, the length of a doubles court and a singles court are the same, but the width of a doubles court is 9 feet greater than that of a singles court. The length of a singles court is 6 feet greater than twice the width of a doubles court, and its width is 1 foot greater than one-third the length of a doubles court. What are the dimensions of a singles court?

22. George Washington was born 11 years before Thomas Jefferson. In 1770 Washington's age was 3 years more than 7 times the age of Jefferson in 1748. How old was each man in 1750?

23. On a rendezvous mission, a spacecraft traveling in a circular orbit makes a forward burn in order to transfer into the orbit in which a second spacecraft is traveling. The apogee (highest point) of the first orbit is 5 miles higher than the perigee (lowest point). The apogee of the second orbit is 18 miles higher than the apogee of the first orbit, and the perigee of the second orbit is 19 miles higher than the perigee of the first orbit. Six times the number of miles in the apogee of the first orbit is 100 miles more than five times the number of miles in the perigee of the second orbit. Find the apogee of the first orbit.

24. At one time, 4 pounds of onions cost the same as 2 pounds of string beans. At the same time, 1 pound of string beans cost 3 times as much as a pound of potatoes, while 1 pound of onions cost 4 cents less than 2 pounds of potatoes. What was the cost of 1 pound of each vegetable?

★25. The coefficients a, b, c, and d in the equation "$ax + b = cx + d$" are given on a card. Draw a flow chart of a program to compute and print the unique root (if any) of the equation. (*Hint:* The equation has a unique root if $a - c \neq 0$; it has no root if $a - c = 0$ and $d - b \neq 0$; it has all real numbers as roots if $a - c = 0$ and $d - b = 0$.) Assume that cards for several equations are to be processed.

4-8 Equations and Functions

An equation such as "$y = 2x - 1$" can be used to assign to each number in the domain of the variable x another number, the value of y. For example, if $x \in \{3, 4, 5\}$, then "$y = 2x - 1$" makes the assignments, or pairings, shown below:

$$
\begin{array}{cccc}
x : & 3 & 4 & 5 \\
& \downarrow & \downarrow & \downarrow \\
y = 2x - 1 : & 2 \cdot 3 - 1 = 5 & 2 \cdot 4 - 1 = 7 & 2 \cdot 5 - 1 = 9
\end{array}
$$

The given equation pairs each member of $\{3, 4, 5\}$ with a single member of $\{5, 7, 9\}$. This example illustrates the mathematical idea of a *function*.

A **function** consists of two sets D and R together with a rule that assigns to each element of D exactly one element of R. The set D is called the **domain** of the function, and R is called the **range** of the function. Thus, in the preceding example the domain of the function is $\{3, 4, 5\}$ and the range is $\{5, 7, 9\}$. Figure 4-1 pictures this function.

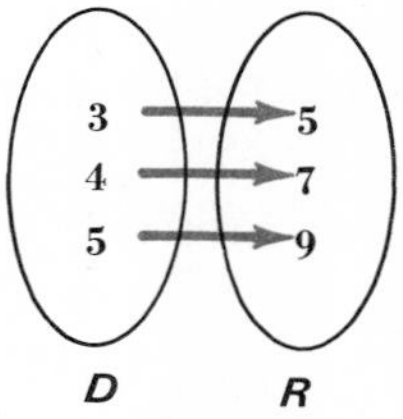

Functions are often named by single letters such as f, g, H, s, and so on. The arrow notation

$$f: x \rightarrow 2x - 1$$

is read "the function f that pairs x with $2x - 1$." Of course, to specify the function completely, we must also identify the domain D of the function. The numbers assigned by the rule then form the range.

FIGURE 4-1

Example 1. Specify in roster form the range of

$$g: x \rightarrow x^2 + 1 \quad \text{if} \quad D = \{-1, 0, 1, 2\}.$$

Solution: In "$x^2 + 1$," replace x with the name of each member of D to find the members of the range, R.

x	$x^2 + 1$
-1	$(-1)^2 + 1 = 2$
0	$0^2 + 1 = 1$
1	$1^2 + 1 = 2$
2	$2^2 + 1 = 5$

$\therefore R = \{1, 2, 5\}$.

Notice that the function g assigns to both -1 and 1 the number 2. However, in specifying the range of g, 2 is named only once.

Members of the range are called the **values** of the function. For instance, in Example 1, the values of g are 1, 2, and 5.

To state that
$$g: x \rightarrow x^2 + 1$$
assigns to the number 2 the number 5, we use the equation
$$g(2) = 5,$$
which may be read: "g at two equals five," or "g of two equals five," or "the value of g at two equals five." Notice that $g(2)$ does *not* name the product of g and 2! It names the *number* that g assigns to 2.

Example 2. Given $H: t \rightarrow t^3$ with $D = \{0, 1, 2\}$, find $H(0)$, $H(1)$, and $H(2)$.

Solution: First write the equation
$$H(t) = t^3.$$
Then:
$$H(0) = 0^3 = 0$$
$$H(1) = 1^3 = 1$$
$$H(2) = 2^3 = 8$$

$\therefore H(0) = 0, H(1) = 1, H(2) = 8.$

The variable used in defining a function does not matter. For instance,
$$m: z \rightarrow z^3 \quad \text{with } D = \{0, 1, 2\}$$
is the same function as H in Example 2. Both H and m pair each member of $\{0, 1, 2\}$ with its cube.

It is not necessary that a function be defined by a single equation or formula, like the functions in Examples 1 and 2. Functions that cannot be defined so simply are also important. For instance, the function that assigns the number 1 to each positive number, 0 to 0, and -1 to each negative number is very useful. We can define it in symbols this way:

$$s: s(x) = \begin{cases} 1 \text{ if } x > 0 \\ 0 \text{ if } x = 0 \\ -1 \text{ if } x < 0, x \in \mathcal{R} \end{cases}$$

To read these symbols, say: "The function s with domain $\mathcal{R}$ such that s at x equals one if x is greater than zero, zero if x equals zero, and negative one if x is less than zero."

EXERCISES

Use arrow notation to represent the function whose values are given by the expression shown. Assume that the domain is $\mathcal{R}$.

1. $f; 4x - 3$
2. $g; 3m^2 + 2m + 1$

State the range of the specified function.

3. $g: y \to y + 2, D = \{0, 1, 2\}$
4. $h: z \to z - 1, D = \{3, 4, 5\}$
5. $f: n \to 2n + 1, D = \{-1, 2, 3\}$
6. $f: x \to 2x^2, D = \{0, 1, 2\}$

Specify in roster form the range of the given function.

7. $f: x \to 2x + 5, D = \{0, -1, 2\}$
8. $g: z \to 4z - 3, D = \{-1, 2, 3\}$
9. $h: y \to y^2 - 5, D = \{3, -4, -5\}$
10. $f: t \to 2t^2 + 1, D = \{0, -1, -2\}$
11. $l: n \to \dfrac{n + 3}{2}, D = \{1, 3, 5\}$
12. $s: x \to \dfrac{3}{x + 1}, D = \{2, 5, 8\}$
13. $g: x \to x(x + 2), D = \{0, 5, -10\}$
14. $h: z \to (z + 1)(z + 2), D = \{0, -5, 10\}$
15. $f: x \to x^2 - 2x + 3, D = \{8, -9, -10\}$
16. $s: t \to 2t^2 - t + 1, D = \{5, -6, -7\}$

Given $f: f(x) = (x + 1)^2$, find:

17. $f(1)$
18. $f(3)$
19. $f(-5)$
20. $f(-4)$

Given $h: z \to (z^2 + 1)^2$, find:

21. $h(1)$
22. $h(3)$
23. $h(3) - h(1)$
24. $h(3) + h(1)$

Given $f: x \to (x + 1)^2 + 1$ and $g: z \to z^2 + 1$, find:

25. $f(3)$
26. $g(3)$
27. $f(3) - g(3)$
28. $f(3) + g(3)$

State the values at -1, 0, and $\tfrac{1}{2}$ of the function whose values are given by the indicated formulas. Assume that $D = \mathcal{R}$ for each function.

29. $F(x) = \begin{cases} 3 \text{ if } x \geq 0 \\ -3 \text{ if } x < 0 \end{cases}$

30. $T(c) = \begin{cases} 1 \text{ if } c \text{ is an integer} \\ 0 \text{ if } c \text{ is not an integer} \end{cases}$

31. $f(a) = \begin{cases} a \text{ if } a \leq 0 \\ 2a \text{ if } a > 0 \end{cases}$

32. $f(x) = \begin{cases} x \text{ if } x \geq 0 \\ -x \text{ if } x < 0 \end{cases}$

★Given $f: x \rightarrow 3x + 2$ and $h: y \rightarrow \dfrac{y - 2}{3}$, find:

33. $f[h(3)]$ *Hint:* First find $h(3)$.
34. $h[f(3)]$ 35. $f[h(5)]$ 36. $h[f(5)]$

chapter summary

1. A **theorem** is a statement which has been proved through logical reasoning from known facts and given assumptions.

2. In the following two theorems, a, b, and c represent any three real numbers.
   Addition property of equality: If $a = b$, then $a + c = b + c$.
   Multiplication property of equality: If $a = b$, then $ac = bc$.

3. **Rule for subtraction:** $a - b = a + (-b)$.

4. **Rule for division:** $a \div b = a \cdot \dfrac{1}{b}$.

5. To solve an equation **transform** it into an **equivalent equation** by undoing each indicated operation with the appropriate **inverse operation (addition — subtraction; multiplication by a nonzero number — division by a nonzero number)** until an equation is obtained which shows the solution set. To **check** the tentative value, **substitute** the value for the variable in the **original equation** and see whether or not it satisfies the equation by evaluating each member separately.

6. Suggested steps in solving an equation as outlined in item 5:
   (a) Simplify each member.
   (b) Undo any indicated additions or subtractions.
   (c) Undo any indicated multiplications or divisions involving the variable.
   (d) Check the tentative answer in the original equation.

7. In solving word problems, first read the problem carefully and decide what numbers are asked for. Then:
   1. Choose a variable with an appropriate replacement set and use the variable in representing each described number.
   2. Form an open sentence by using facts given in the problem.
   3. Find the solution set of the open sentence.
   4. Check the answer with the words of the problem.

8. A **function** consists of two sets D and R together with a rule that assigns to each element of D exactly one element of R. The set D is called the **domain** of the function, and R is called the **range** of the function. Members of R are the **values** of the function.

CHAPTER REVIEW

4–1 Solve:

1. $t - 17 = 22$
2. $x - 0.03 = 0.3$
3. $-\frac{4}{7} + y = \frac{10}{7}$
4. $|p| - 2 = 4$

4–2 Solve:

5. $-27 = q + 21$
6. $n + \frac{3}{5} = 4$
7. $(c + 8) + 3 = 14$
8. $Z + 0.07 = 0.008$

4–3 Solve:

9. $-13r = 91$
10. $\frac{1}{3}h = \frac{7}{3}$
11. $\frac{1}{6} = 3k$
12. $\frac{3}{2}q = 72$

4–4 Solve:

13. $-12 = -4x$
14. $\frac{3}{4}x = \frac{9}{8}$
15. $17t = -408$
16. $14x = -7$

4–5 Solve:

17. $4t - 2 + 3t + 8 = -15$
18. $-3r + 2(4r + 11) = -3$
19. $-1.8r + 5.7 + 3.3r = 1.2$
20. $\frac{5}{4}k - 3(\frac{1}{4}k + 6) = 12$

4–6

21. A Ritewell pen and pencil set costs $3.39, the pen costing $0.69 more than the pencil. Find the cost of each.

22. During the summer Pam's older brother, Phil, earned $46 less than twice the amount Pam earned. If their total earnings amounted to $818, how much did each earn?

4–7

23. $2(3h - 5) + 12 = 5(2h + 3) + 9h$

24. The width of a rectangle is 4 inches less than twice the width of a smaller rectangle. The larger rectangle is 15 inches long; the smaller is 16 inches long. If the area of the larger rectangle is 108 square inches more than the area of the smaller, find the width of the larger rectangle.

4–8

25. Specify in roster form the range of the given function.
    $f: x \to (x - 1)^2, D = \{-1, 0, 1, 2\}$.

26. Given $g: x \to \dfrac{1}{1 + x}$, find:

    a. $g(-2)$
    b. $g(0)$
    c. $g(2)$

5 solving inequalities; more problems

5–1 Axioms of Order

In comparing real numbers, we make the following basic assumption.

Axiom of Comparison

For all real numbers *a* and *b*, one and only one of the following statements is true:

$$a < b, \quad a = b, \quad b < a.$$

For example, only the first of the following statements is true.

$$-5 < 2; \quad -5 = 2; \quad 2 < -5.$$

Now suppose that we have a horizontal number line with positive direction toward the right. Suppose also that we know two facts about the graphs of three numbers *a*, *b*, and *c* (Figure 5–1):

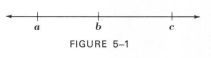

FIGURE 5–1

1. The graph of *a* is left of the graph of *b* (that is, $a < b$).
2. The graph of *b* is left of the graph of *c* (that is, $b < c$).

Figure 5–1 suggests that the graph of *a* is also left of the graph of *c* (that is, $a < c$). This sort of thinking makes the following assumption reasonable.

Transitive Axiom of Order

For all real numbers *a*, *b*, and *c*:
1. **If $a < b$ and $b < c$, then $a < c$; similarly,**
2. **If $a > b$ and $b > c$, then $a > c$.**

On the number lines shown in Figures 5-2 and 5-3, the graph of −5 is left of the graph of 2. In each figure, if we move 4 units from the graph of −5 and also 4 units *in the same direction* from the graph of 2, we arrive at points *in the same order* on the line as the graphs of −5 and 2. Thus, Figures 5-2 and 5-3 picture the following statements:

$$-5 < 2 \qquad\qquad -5 < 2$$
$$-5 + 4 < 2 + 4 \qquad -5 + (-4) < 2 + (-4)$$
$$\text{or} \quad -1 < 6 \qquad \text{or} \qquad -9 < -2$$

FIGURE 5-2   FIGURE 5-3

These statements suggest the next assumption to make.

Addition Axiom of Order

For all real numbers *a*, *b*, and *c*:
1. **If $a < b$, then $a + c < b + c$; similarly:**
2. **If $a > b$, then $a + c > b + c$.**

To see what happens when each member of the inequality "$-5 < 2$" is multiplied by a nonzero real number, consider the following two cases:

Case 1. Multiply by 2.   *Case 2.* Multiply by −2.

$$-5 < 2 \qquad\qquad\qquad\qquad -5 < 2$$
$$-5 \cdot 2 \underline{\ ?\ } 2 \cdot 2 \qquad\qquad -5(-2) \underline{\ ?\ } 2(-2)$$
$$-10 \underline{\ ?\ } 4 \qquad\qquad\qquad 10 \underline{\ ?\ } -4$$
$$-10 < 4 \qquad\qquad\qquad 10 > -4$$
$$\therefore -5 \cdot 2 < 2 \cdot 2. \qquad\qquad \therefore -5(-2) > 2(-2).$$

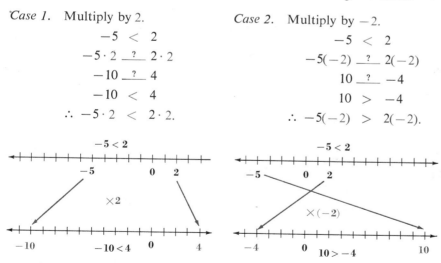

The two cases suggest that *multiplying each member of an inequality* by
(1) a positive number *preserves* the *direction, order,* or *sense* of the inequality;
(2) a negative number *reverses* the *direction* of the inequality.
These results illustrate our next axiom.

Multiplication Axiom of Order

For all real numbers *a*, *b*, and *c*:
1. If $a < b$ and $c > 0$, then $ac < bc$; similarly, if $a > b$ and $c > 0$, then $ac > bc$.
2. If $a < b$ and $c < 0$, then $ac > bc$; similarly, if $a > b$ and $c < 0$, then $ac < bc$.

When we multiply any inequality by a nonzero real number, we *must take into account the direction associated with the multiplier*. The following shows what happens when the multiplier is zero.

$$-5 < 2$$
$$-5 \cdot 0 \underline{?} 2 \cdot 0$$
$$0 \underline{?} 0$$
$$0 = 0 \qquad \therefore -5 \cdot 0 = 2 \cdot 0.$$

Because of the multiplicative property of zero, *multiplying each member of any inequality by 0 produces the identity "0 = 0."*

The axioms that have been stated guarantee that the following transformations of a given inequality always produce an **equivalent inequality**, that is, one with the same solution set. As was noted in the development of transformations that produce equivalent equations, transformation by subtraction is a special kind of transformation by addition (page 81) and transformation by division is a special case of transformation by multiplication (page 88).

Transformations That Produce an Equivalent Inequality

1. **Substituting for either member of the inequality an expression equivalent to that member.**
2. **Adding to (or subtracting from) each member the same real number.**
3. **Multiplying (or dividing) each member by the same positive number.**
4. **Multiplying (or dividing) each member by the same negative number and reversing the direction of the inequality.**

Example. Solve $7z - 2 \leq 3(z + 3) + 1$ over $\mathcal{R}$, and graph its solution set.

Solution:
1. Copy the inequality. $\qquad 7z - 2 \leq 3(z + 3) + 1$
2. Use the distributive axiom $\qquad 7z - 2 \leq 3z + 9 + 1$
   to simplify the right member. $\qquad 7z - 2 \leq 3z + 10$
3. Add 2 to each member. $\qquad 7z - 2 + 2 \leq 3z + 10 + 2$
$$7z \leq 3z + 12$$

4. Subtract $3z$ from each member.

$$7z - 3z \leq 3z + 12 - 3z$$
$$4z \leq 12$$

5. Divide each member by 4.

$$\frac{4z}{4} \leq \frac{12}{4}$$
$$z \leq 3$$

∴ the solution set is {all real numbers less than or equal to 3}.

6. Draw the graph of the solution set.

EXERCISES

Solve each inequality. In each of Exercises 1–22, show also the graph of the solution set.

1. $x - 40 < -100$
2. $y + 25 > 30$
3. $14z \geq -98$
4. $15t > 225$
5. $2z - 3 > 9$
6. $5t + 2 \leq 7$
7. $1 + \frac{x}{5} \leq -1$
8. $2 + \frac{z}{3} < -2$
9. $-2n + 6 > -4$
10. $-13t + 1 \geq -38$
11. $1 - \frac{y}{3} < 1$
12. $2 - \frac{n}{4} \geq 2$
13. $-3y + 7 + 5y \leq 1$
14. $4x + 3 - 2x < -1$
15. $2(a - 1) < 8$
16. $3(b + 5) > 9$
17. $\frac{3}{4}x - \frac{1}{2} \geq \frac{1}{4}$
18. $\frac{3}{2}u + \frac{1}{6} \leq \frac{7}{6}$
19. $2z - 1 \leq 3 - 2z$
20. $-3s + 6 > s - 30$
21. $16 - 8n \leq n - 20$
22. $21 - 15x < -8x - 7$
23. $4(t - 2) > 5(t - 3)$
24. $3(y + 12) \geq 6(y - 4)$
25. $5(4x + 3) - 7(3x - 4) \leq 10$
26. $2(6n - 8) - 9(n + 4) > 2n$
27. $6(z - 5) \leq 15 + 5(7 - 2z)$
28. $3(2r + 3) - (3r + 2) > 12$
29. $-2(t - 5) - 1 \leq 5t + 7(1 - t)$
30. $3(y - 4) - (6y + 2) \geq 5 - 3y$
31. $4(s + 1) - 2(s + 3) \geq 2(s - 1)$
32. $5(g - 4) - (g + 2) \leq 4(g - 4) - 6$

Transform each inequality into an equivalent inequality in which one member is the variable in color.

33. $4(3x - a) < 6x + 7a$
34. $3(y - 6b) > 2(6b - y)$
35. $9(t + 4c) \leq 8(t - c)$
36. $d - 3(2z - 3d) \geq 2(3z - d)$

5-2 Intersection and Union of Sets

Most of the work in this book deals with sets whose members are real numbers. Therefore, we say that the set of all real numbers, ℜ, is the *universe* of the discussion. In general, the **universe**, or **universal set**, of a discussion is the overall set that includes all the elements in all of the sets under consideration.

Figure 5–4 uses a diagram* to picture a universe U and a set A whose elements all belong to U. In the figure, the region inside and on the rectangle represents U, while the region inside and on the circle represents A. Because every member of A is also a member of U, A is called a **subset** of U. In symbols, we write

$$A \subset U,$$

and say "A is a subset of U."

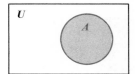

FIGURE 5–4

Example 1. Let $U = \{0, 2, 5\}$. Specify in roster form all subsets of U that have exactly two members.

Solution: $\{0, 2\}$, $\{0, 5\}$, and $\{2, 5\}$.

Is $\{0, 2, 5\}$ a subset of $\{0, 2, 5\}$? Since every member of $\{0, 2, 5\}$ is, indeed, a member of $\{0, 2, 5\}$, the statement "$\{0, 2, 5\} \subset \{0, 2, 5\}$" is certainly true. In fact, *every set is a subset of itself.*

The empty set ∅ is also a subset of $\{0, 2, 5\}$. In working with sets, it is useful to agree that *∅ is a subset of every set.*

Figure 5–5 pictures relationships among the sets:

$U = \{1, 2, 3, 4, 5, 6, 7, 8\}$
$S = \{1, 2, 3, 4\}$
$T = \{1, 3, 6\}$
$M = \{7, 8\}$

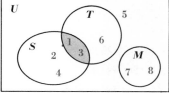

FIGURE 5–5

The set represented by shading in Figure 5–5 is $\{1, 3\}$ and consists of the elements that $\{1, 2, 3, 4\}$ and $\{1, 3, 6\}$ have in common. We call $\{1, 3\}$ the *intersection* of $\{1, 2, 3, 4\}$ and $\{1, 3, 6\}$, and write

$$\{1, 2, 3, 4\} \cap \{1, 3, 6\} = \{1, 3\}.$$

---

* Diagrams that picture set relationships are called Venn diagrams in honor of the English mathematician John Venn (1834–1923).

In general, for any two sets S and T, the set consisting of the elements belonging to *both* S and T is called the **intersection** of S and T, denoted by $S \cap T$. In a Venn diagram, $S \cap T$ is pictured by the overlap of the regions representing S and T.

The regions T and M do not overlap because $T = \{1, 3, 6\}$ and $M = \{7, 8\}$ do not have any common members. Sets, such as T and M, which have no elements in common are called **disjoint sets**. $T \cap M = \emptyset$.

The shaded region in Figure 5-6 represents the set consisting of all the elements which belong to *at least one* of the following sets: $S = \{1, 2, 3, 4, 5\}$ and $T = \{1, 3, 5, 7, 9\}$. This set contains all the elements of S together with all the elements of T. We call it the *union* of the two sets and write

$\{1, 2, 3, 4, 5\} \cup \{1, 3, 5, 7, 9\} = \{1, 2, 3, 4, 5, 7, 9\}$.

In general, for any two sets S and T, the set consisting of all the elements belonging to *at least one* of the sets S and T is called the **union** of S and T, denoted by $S \cup T$.

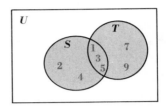

$S \cup T = \{1, 2, 3, 4, 5, 7, 9\}$

FIGURE 5-6

Example 2. If $A = \{1, 2, 3, 4, 5\}$, $A \cap B = \{1, 3\}$, and $A \cup B = \{1, 2, 3, 4, 5, 6, 7\}$, draw a Venn diagram showing A and B. Identify B by roster.

Solution:

1. Draw a diagram. Mark it to show the elements of $A \cap B$.

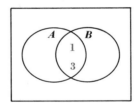

2. Mark it to show the remaining elements of A.

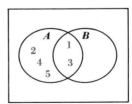

3. Indicate, as members of B (but not A), the elements of $A \cup B$ not shown in Steps 1 and 2.
∴ $B = \{1, 3, 6, 7\}$.

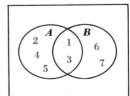

EXERCISES

Let $U = \{-5, 0, 10\}$. Specify by roster all of the subsets of U that have:

1. one element
2. three elements
3. no element
4. at least 2 elements

Specify the intersection and union of the given sets. If the given sets are disjoint, state that fact.

5. $\{-1, 3, 5, 7\}$; $\{3, 4, 5\}$
6. $\{-2, -4, -6, -8\}$; $\{-6, -7, -8\}$
7. $\{-3, 0, 3, 4\}$; $\{-3, 4\}$
8. $\{0, 6\}$; $\{-6, -3, 0, 3, 6\}$
9. $\{0, 2, 4\}$; $\{1, 3, 5\}$
10. $\{-10, -5, 0\}$; $\{5, 10\}$
11. {the even natural numbers}; $\{1, 2, 3\}$
12. {the even natural numbers}; {the odd natural numbers}
13. {the positive numbers}; {the negative numbers}
14. $\mathcal{R}$; $\emptyset$

In Exercises 15–26, let A be the first set given and B, the second. Graph: **a.** A; **b.** B; **c.** $A \cap B$; **d.** $A \cup B$. In case a required set is the empty set, state that fact and omit the graph.

Example. $A = \{\text{natural numbers less than } 7\}$;
$B = \{\text{natural numbers between } 3\frac{1}{2} \text{ and } 9\frac{1}{2}\}$.

Solution:

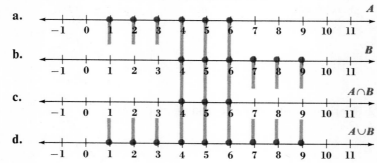

15. {natural numbers less than or equal to 4};
    {2, 6, and the natural numbers between 2 and 6}
16. {the whole numbers less than or equal to 3};
    {7, and the natural numbers less than 7}
17. {natural numbers less than 6};
    {even whole numbers less than 11}
18. {odd whole numbers less than 10};
    {natural numbers less than 5}

19. {negative integers greater than -5};
    {-2, 2, and the integers between -2 and 2}
20. {-3, and the integers between -3 and 1};
    {nonnegative integers less than 4}
21. {the real numbers between -1 and 1};
    {the positive real numbers}
22. {the real numbers between -2 and 2};
    {the negative real numbers}
23. {4, and the real numbers greater than 4};
    {the real numbers less than or equal to 6}
24. {the real numbers greater than or equal to -1};
    {2 and the real numbers less than 2}
25. {the real numbers less than -2};
    {the real numbers less than -3}
26. {the real numbers greater than 0};
    {the real numbers greater than 2}

For each of the following exercises, copy the Venn diagram at the right. Then shade the region representing the set named.

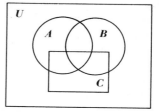

27. $A \cup C$
28. $A \cap C$
29. $(A \cap B) \cap C$
30. $(A \cup B) \cup C$
31. $A \cup (B \cap C)$
32. $A \cap (B \cup C)$
33. $(A \cup B) \cap (A \cup C)$
34. $(A \cap B) \cup (A \cap C)$

Let R, S, and T be subsets of $U = \{1, 2, 3, 4, 5, 6\}$. Specify S by roster in each exercise.

35. $R = \{1, 3, 5\}$, $R \cap S = \{3\}$, and $R \cup S = \{1, 3, 5, 6\}$
36. $R = \{2, 4, 6\}$, $R \cap S = \{2, 6\}$, and $R \cup S = \{1, 2, 3, 4, 6\}$
37. $R = \{3, 5\}$, $S \cap T = \{3\}$, $S \cup T = \{1, 2, 3, 4, 5, 6\}$,
    $R \cup T = \{1, 2, 3, 4, 5\}$
38. $R \cap R = \emptyset$, $T \cup R = \{6\}$, $S \cup T = U$, $S \cap T = \emptyset$

5–3 Combining Inequalities

Often we are interested in variables whose values must satisfy *two* inequalities *at the same time*. For example, the solution set of "$-2 < x < 5$" consists of those values of x for which *both* of the inequalities in the conjunction "$-2 < x$ and $x < 5$" are true.

Similarly, to solve the open sentence "$-3 < y + 2 \le 6$" we must find the values of y for which the conjunction of inequalities

$$-3 < y + 2 \quad \text{and} \quad y + 2 \le 6$$

is true. Transforming each of these inequalities by subtracting 2 from each member, we find:

$$
\begin{array}{lcl}
-3 < y + 2 & \text{and} & y + 2 \le 6 \\
-3 - 2 < y + 2 - 2 & \mid & y + 2 - 2 \le 6 - 2 \\
-5 < y & \text{and} & y \le 4
\end{array}
$$

Thus, "$-3 < y + 2 \le 6$" is equivalent to "$-5 < y \le 4$." Its solution set is $\{4$, and all the real numbers between -5 and $4\}$. The graph is shown in Figure 5-7.

FIGURE 5-7

In Figure 5-8, the graph of "$-3 < y + 2 \le 6$" consists of the points that belong to *both* the graph of "$-5 < y$" *and* the graph of "$y \le 4$." In fact, the graph of "$-3 < y + 2 \le 6$" is the intersection of the other two graphs! Similarly, the solution set of "$-3 < y + 2 \le 6$" is the intersection of the solution sets of "$-5 < y$" and "$y \le 4$."

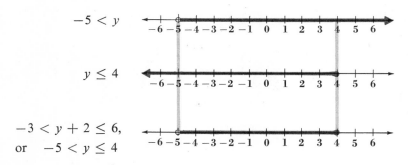

FIGURE 5-8

Identify the *union* of the solution sets of "$-5 < y$" and "$y \le 4$." Since every real number satisfies *at least one* of these inequalities, the union of their solution sets is $\mathcal{R}$, the set of all real numbers. Thus, for every real number y,

$$-5 < y \quad \text{or} \quad y \le 4$$

is true. A sentence formed by joining two sentences by the word or is called a **disjunction**. For a disjunction to be true, *at least one* of the joined sentences must be a true statement.

Example. Solve the open sentence and draw its graph:

$$2t + 3 > 7 \quad \text{or} \quad 2t + 3 < -7$$

Solution: Copy the joined inequalities and transform them as follows:

$$2t + 3 > 7 \qquad \qquad 2t + 3 < -7$$
$$2t + 3 - 3 > 7 - 3 \qquad 2t + 3 - 3 < -7 - 3$$
$$2t > 4 \qquad \qquad 2t < -10$$
$$\frac{2t}{2} > \frac{4}{2} \qquad \qquad \frac{2t}{2} < \frac{-10}{2}$$
$$t > 2 \quad \text{or} \quad t < -5$$

∴ the solution set is {all real numbers that are greater than 2 or less than -5}.

In dealing with two inequalities, it is essential to decide whether we want the set of numbers satisfying *both* of them or the set satisfying *at least one* of them. For example, the set satisfying *both* of the inequalities "$x \leq 3$" and "$x > 3$" is the *empty set*, whereas the set of numbers satisfying *at least one* of them is the set of *real numbers*!

EXERCISES

Solve each open sentence over $\mathcal{R}$ and graph each solution set that is not ∅.

1. $-2 \leq x + 3 < 4$
2. $1 \leq y + 7 < 6$
3. $-5 \leq -2 + a \leq 0$
4. $-4 < -1 + b < 3$
5. $2 < 3t + 2 < 14$
6. $-15 \leq 4b - 5 \leq -9$
7. $y + 2 > 5$ or $y + 2 < -5$
8. $3 + x \leq -4$ or $3 + x \geq 4$
9. $2r - 1 \leq -1$ or $2r - 1 \geq 1$
10. $6m - 3 > 9$ or $6m - 3 < -9$
11. $2 \geq -3 - 5p \geq -18$
12. $5 > 4 - 3u > -13$
13. $3 - 6c < -15$ or $3 - 6c \geq 15$
14. $7 - 2w > 3$ or $7 - 2w \leq -3$
15. $z - 2 > -1$ and $z + 5 \leq 9$
16. $g + 1 \leq -3$ and $g - 2 \geq -8$
17. $v + 2 \geq -6$ or $v - 3 < 2$
18. $5 - y < 2$ or $4 + y > 7$

19. $7 + 3q < 1$ or $-12 < q - 1$
20. $4 < \frac{1}{2}d + 4$ or $3d - 8 \leq d$
21. $9 - 2m > 11$ and $5m < 2m + 9$
22. $13x \leq 7x - 12$ and $1 - 4x > 13$
23. $3z + 4 \geq z + 10$ or $3z - 3 \geq 2z - 9$
24. $8 - 2k > 2k$ or $5k - 1 < 2k + 14$
25. $2r - 1 \leq 2r + 8 \leq 2r + 4$
26. $3 + 4k < 4k + 7 < 1 + 4k$
27. $x - 1 < 2x + 3 < x + 4$
28. $7 - 3y \geq 6 - 4y \geq 4 - 3y$

5–4 Absolute Value in Open Sentences

Equations and inequalities involving absolute value occur in many parts of mathematics.

Example 1. Solve: $|6m - 3| = 9$

Solution: $|6m - 3| = 9$ is equivalent to the *disjunction:*

$$
\begin{array}{lcl}
6m - 3 = -9 & & 6m - 3 = 9 \\
6m - 3 + 3 = -9 + 3 & & 6m - 3 + 3 = 9 + 3 \\
6m = -6 & & 6m = 12 \\
\dfrac{6m}{6} = \dfrac{-6}{6} & & \dfrac{6m}{6} = \dfrac{12}{6} \\
m = -1 & \text{or} & m = 2
\end{array}
$$

∴ the solution set is $\{-1, 2\}$.

Example 2. Solve: $|6m - 3| < 9$

Solution: $|6m - 3| < 9$ is equivalent to the *conjunction:*

$$-9 < 6m - 3 < 9$$
$$-9 + 3 < 6m - 3 + 3 < 9 + 3$$
$$-6 < 6m < 12$$
$$\frac{-6}{6} < \frac{6m}{6} < \frac{12}{6}$$
$$-1 < m < 2$$

∴ the solution set is {all real numbers between -1 and 2}.

Example 3. Solve: $|6m - 3| > 9$

Solution: $|6m - 3| > 9$ is equivalent to the *disjunction:*

$$6m - 3 < -9 \quad \text{or} \quad 6m - 3 > 9$$

Completing the solution is left to you. You should find that the solution set is {all real numbers that are less than -1 or greater than 2}.

EXERCISES

Solve each open sentence and draw its graph.

1. $|x - 2| = 4$
2. $|y + 3| = 7$
3. $|a| \geq 3$
4. $|m| < 2$
5. $|t - 6| < 5$
6. $|8 + b| > 0$
7. $|2 - y| \leq 1$
8. $|3 - z| \geq 4$
9. $|1 - 3t| = 2$
10. $|1 - 2m| = 9$
11. $|2v - 9| \leq 1$
12. $|3l - 7| < 9$
13. $|3 - 3x| \geq 0$
14. $|5 - 2y| < 0$
15. $|7 - 2p| < 7$
16. $|12 - 3d| \geq 12$
17. $3|r| + 2 = 8$
18. $3 + 2|a| > -1$

Specify and graph the solution set of each open sentence.

19. $|1 - (2 - a)| = 5$
20. $|8 - (r - 1)| = 6$
21. $|1 + 2(y - 1)| < 3$
22. $|5 - 3(2 - z)| \geq 0$
23. $4 - 3(2 - |s|) \geq 7$
24. $-6 + 4(1 - 3|y|) \leq 20$
25. $5(1 + |c - 3|) - 9 \leq -4$
26. $4(2 + |1 - m|) + 10 \geq 18$
★27. $|2d + 1| = d - 4$
★28. $|3r + 2| = r + 5$
★29. $|3 - 5t| \geq 1 + 3t$
★30. $|2w + 6| < 4w - 1$

5-5 Problems about Integers

On page 16, we saw that the set of integers is the infinite set

$$J = \{\ldots, -3, -2, -1, 0, 1, 2, 3, \ldots\}.$$

If we count by ones from any given integer, then we obtain **consecutive integers.** In general, if n is any integer, then $n + 1$, $n + 2$, and $n + 3$ are the next three consecutive integers in increasing order. Similarly, $n - 1$, $n - 2$, $n - 3$, are the preceding three consecutive integers in decreasing order.

Example 1. Find three consecutive integers whose sum is 54.

Solution: 1. Let $n =$ the first integer.
Then $n + 1 =$ the second integer and $n + 2 =$ the third integer.

2. The sum of the integers is 54.
$$n + (n + 1) + (n + 2) = 54.$$

3. $\quad 3n + 3 = 54$
$3n + 3 - 3 = 54 - 3$
$3n = 51$
$$\frac{3n}{3} = \frac{51}{3}$$
$n = 17$

$\therefore n + 1 = 18 \quad$ and $\quad n + 2 = 19.$

Check: Is the sum of 17, 18, and 19 equal to 54?
$$17 + 18 + 19 \stackrel{?}{=} 54$$
$$54 = 54$$

$\therefore$ the three consecutive integers are 17, 18, and 19.

We call 21 a *multiple* of 3 because $21 = 3 \times 7$. In general, the product of any real number and an *integer* is called a **multiple** of the real number.
The multiples of 2 are the *even integers:*

$$\ldots, -6, -4, -2, 0, 2, 4, 6 \ldots$$

If we count by *twos* from an even integer, we obtain **consecutive even integers**. The integers that are not even are the odd integers:

$$\ldots, -5, -3, -1, 1, 3, 5, \ldots$$

Counting by two from an odd integer yields **consecutive odd integers**. Thus, if a is even, then $a + 2$ is the next greater even integer and $a - 2$ is the preceding smaller even integer. In case a is odd, then $a - 2$, a, and $a + 2$ are three consecutive odd integers in increasing order.

In the next example, an inequality is used to solve a problem about consecutive even integers.

Example 2. Find all sets of four consecutive positive even integers such that the greatest integer in the set is greater than twice the least integer in the set.

Solution: 1. Let x = the least of the positive even integers in the set; then the next three consecutive even integers are:

$$x + 2,\ x + 4,\ \text{and}\ x + 6.$$

2. The greatest integer is greater than twice the least integer

	$x + 6$	>	$2x$
3.	$x + 6 - x$	>	$2x - x$
	6	>	x

Since the least even integer must be less than 6 and must be positive, the only choices for the least integer are 2 and 4.

∴ the only possible sets are $A = \{2, 4, 6, 8\}$ and $B = \{4, 6, 8, 10\}$.

Check: In each set, is the greatest integer greater than twice the least?
In A: $8 \underline{\ ?\ } 2 \cdot 2;\ 8 > 4$.
In B: $10 \underline{\ ?\ } 2 \cdot 4;\ 10 > 8$.
∴ the required sets are $\{2, 4, 6, 8\}$ and $\{4, 6, 8, 10\}$.

EXERCISES

1. The sum of two consecutive integers is 45. Find the numbers.
2. The sum of two consecutive odd integers is 36. Find the numbers.
3. Find three consecutive integers whose sum is -33.
4. Find three consecutive odd integers whose sum is 87.
5. Find four consecutive integers whose sum is -102.
6. Find four consecutive even integers whose sum is 76.
7. Find three consecutive integers, if the sum of the first and third is 146.
8. Find four consecutive integers if the sum of the second and fourth is 58.

9. Find the two least consecutive odd integers whose sum is greater than 16.
10. Find the two greatest consecutive even integers whose sum is less than 60.
11. The larger of two consecutive even integers is six less than twice the smaller. Find the numbers.
12. The smaller of two consecutive integers is one more than twice the larger. Find the numbers.
13. Find four consecutive integers such that four times the second diminished by twice the fourth is 10.
14. Find four consecutive odd integers such that the third is the sum of twice the second and the fourth.
15. Four times the smaller of two consecutive even integers is less than three times the larger. What are the largest possible values for the integers?
16. Three consecutive integers are such that their sum is more than 20 decreased by twice the second integer. What are the smallest possible values for the integers?
17. The smaller of two consecutive integers is less than 4 more than half the larger. Find the largest possible values for the integers.
18. Three consecutive integers are such that the sum of the second and third is greater than half the first diminished by 6. What are the smallest possible values for the integers?
★ 19. Find all sets of three consecutive multiples of 4 whose sum is between -84 and -36.
★ 20. Find the three greatest consecutive multiples of 7 such that the middle one is greater than their sum decreased by 50.
★ 21. The coefficients a, b, and c in the inequality "$ax + b > c$" are given on a card. Draw a flow chart of a program to solve the inequality. Assume that $a \neq 0$ and that cards for several inequalities are to be processed.

5–6 Uniform-Motion Problems

An object that moves in a straight line without changing its speed is said to be in **uniform motion**. In solving problems about uniform motion, diagrams and charts often help organize the given facts. The basic formula in such problems is:

$$\text{distance} = \text{rate} \times \text{time}$$
$$d = r \cdot t$$

SOLVING INEQUALITIES; MORE PROBLEMS

Example 1. (Motion in Opposite Directions) A submarine left a surface ship and cruised due south at a constant rate of 28 knots (nautical miles per hour). If the surface ship started off at the same time on a course due north at a constant rate of 22 knots, in how many hours will the ships be 125 nautical miles apart?

Solution:

1. Let t = the number of hours traveled.
2. Make a sketch illustrating the given facts. Arrange the facts in a chart.

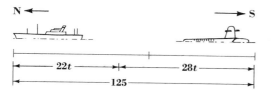

	r	t	d
Submarine	28	t	$28t$
Surface ship	22	t	$22t$

The submarine's rate is 28 knots. The surface ship's rate is 22 knots. Distance apart is 125 nautical miles. Each travels same number of hours.

Submarine's distance + Surface ship's distance = Total distance

$$28t + 22t = 125$$

3.
$$28t + 22t = 125$$
$$50t = 125$$
$$t = 2\tfrac{1}{2}$$

4. To check the $2\tfrac{1}{2}$ hour time, answer the question, "How far did each ship travel?"

Submarine traveled $\quad 28 \cdot 2\tfrac{1}{2} = 70$ miles
Surface ship traveled $\quad 22 \cdot 2\tfrac{1}{2} = 55$ miles
The sum of their distances $\quad \overset{?}{=} 125$ miles
$\quad\quad\quad\quad\quad\quad\quad\quad\quad\quad 125 = 125$

∴ the ships will be 125 nautical miles apart in $2\tfrac{1}{2}$ hours.

Example 2. (Motion in the Same Direction) Mr. Jones left Elmsville at 8:00 A.M. one morning and drove to Bond City at a constant rate of 50 miles per hour (mph). At 10:00 A.M. the same day, Sgt. Holliday of the Highway Patrol left Elmsville. Following the same route, he arrived in Bond City at the same time as Mr. Jones. If both men arrived in Bond City at 3:00 P.M., at what constant rate had Sgt. Holliday traveled?

(Solution on page 122.)

122 CHAPTER FIVE

Solution: 1. Let r = the rate of Sgt. Holliday in mph.
2. Make a sketch showing the given facts.
Make a chart of the facts in the given problem.

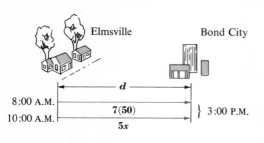

	r	t	d
Mr. Jones	50	7	350
Sgt. Holliday	x	5	$5x$

$$\underbrace{\text{Distance of}}_{350\text{ Mr. Jones}} = \underbrace{\text{Distance of}}_{5x\text{ Sgt. Holliday}}$$

Steps 3 and 4 are left to you.

Example 3. (Round trip) To pick up a rare serum needed to treat a very sick child, Major Evans flew from Tinden Air Force Base to Capital City and back. He maintained a constant speed of 480 miles per hour to the city, and 600 miles per hour back to the base. If the actual flying time for the round trip was $1\frac{1}{4}$ hours, how long did the flight to Capital City take? How far is Tinden Air Force Base from Capital City?

Solution: 1. Let t = the number of hours in flying time from TAFB to Capital City. Then $\frac{5}{4} - t$ = the number of hours in flying time from Capital City to TAFB.
2. The given facts are these: (a) The total time is $1\frac{1}{4}$ or $\frac{5}{4}$ hours. (b) The rate in one direction is 480 mph. (c) The rate in the opposite direction is 600 mph. (d) The number of miles covered in each direction is the same.

	r	t	d
To Capital City	480	t	$480t$
To TAFB	600	$\frac{5}{4} - t$	$600(\frac{5}{4} - t)$

$$\underbrace{\text{Distance to}}_{480t\text{ Capital City}} = \underbrace{\text{Distance to}}_{600(\frac{5}{4} - t)\text{ TAFB}}$$

Steps 3 and 4 are left to you. Remember that when we have found a value for t, we must use this value to answer the problem's second question, which is about a *distance*.

PROBLEMS

Make a drawing and a chart for each problem. Then form an open sentence, solve it, check your result in the words of the problem, and, finally, answer the question asked in the problem.

1. The steamship Empress Anne sailing due west at 32 knots passed the freighter Oregon which was sailing due east at 24 knots. In how many hours after the meeting will the ships be 448 nautical miles apart?

2. Majestic Airlines flight 324, flying due north at a groundspeed of 460 miles per hour, passed Omega Airlines flight 117 flying due south at a groundspeed of 530 miles per hour. In how many hours after passing each other will these two planes be 1980 miles apart?

3. Tim and a friend left a campsite on a trip down the river in a canoe, maintaining a constant speed of 4 miles per hour. 4 hours later, Tim's brother set out after them in a motorboat carrying the camping supplies. The motorboat traveled at a rate of 12 miles per hour. How long after he started did Tim's brother overtake him?

4. A motorist traveling 55 miles per hour is being pursued by a highway patrol car traveling 65 miles per hour. If the patrol car is 4 miles behind the motorist, how long will it take the patrol car to overtake the motorist?

5. It takes a passenger train 2 hours less time than it takes a freight train to make the trip from Central City to Clear Creek. If the passenger train averages 60 miles per hour on the trip while the freight train averages 40 miles per hour, how far is it from Central City to Clear Creek?

6. A freight train leaves Eastburg at 6:00 A.M. traveling toward Bechton at 42 miles per hour. At 8:00 A.M., a passenger train leaves Eastburg on a parallel track traveling toward Bechton at 70 miles per hour. If the trains arrive at Bechton at the same time, how far is it from Eastburg to Bechton?

7. Marie rode her bicycle from her home to the bicycle shop in town and then walked back home. If she averaged 6 miles per hour riding and 3 miles per hour walking, how far is it from her home to the bicycle shop if her traveling time totaled 1 hour?

8. A ski lift carries a skier up a slope at the rate of 120 feet per minute and he returns from the top to the bottom on a path parallel to the lift at an average rate of 280 feet per minute. How long is the lift if the round trip traveling time is 20 minutes?

9. An airplane on a search mission flies due east from an airport, turns and flies due west back to the airport. The plane cruises at 200 miles per hour when flying east, and 250 miles per hour when flying west. What is the farthest point from the airport the plane can reach if it can remain in the air for 9 hours?

10. On the West River, Larry paddles a canoe at the rate of 6 miles per hour downstream and 2 miles per hour upstream. If he makes a trip down the river and then back to his departure point in $1\frac{1}{2}$ hours, and if the river is flowing at the rate of 2 mph, how far down the river does he go?

11. A steamship radios the Coast Guard that an ill passenger must be flown to a hospital as soon as possible. If the steamship is 180 nautical miles from the Coast Guard station and sailing directly towards it at the time a helicopter is dispatched from the station, and if the rate of the helicopter is 90 knots while the rate of the ship is 30 knots, how long will it take the helicopter to reach the ship?

12. Two ships are sailing toward each other and are 120 nautical miles apart. If the rate of one ship is 4 knots greater than the rate of the other, and if they meet in 3 hours, find the rate of each ship.

13. Two cars bound for Buffalo on the New York State Thruway leave a service area at the same time. Their average speeds differ by 10 miles an hour. Six hours later the slower car reached an exchange that the faster car had reached an hour earlier. Find the average speed of each car.

14. At midnight, two river steamers are 100 miles apart. They pass Natchez at 5 A.M. headed in opposite directions. If the northbound boat steams at an average speed of 8 miles per hour, find the speed of the southbound boat.

15. After an airplane had been flying for 2 hours, a change in wind increased the plane's groundspeed by 30 miles per hour. If the entire trip of 570 miles took $3\frac{1}{2}$ hours, how far did the plane go the first two hours?

16. Mr. Ferris can drive from Mountain View to Tuxford at 45 miles per hour in time to meet an appointment. If he takes 30 minutes out of the trip to eat, however, he must drive at 60 miles per hour to make the appointment. How far is it from Mountain View to Tuxford?

17. A private plane had been flying for 2 hours when it encountered head winds which reduced its average speed by 20 miles an hour. If it took the plane 5 hours to travel 640 miles, find its average speed before encountering head winds.

18. Some members of the Rocky Mountain Outing Club hiked to an overnight campsite at the rate of 3 miles per hour. The following morning they returned on horseback over the same route at 10 miles per hour. The total time spent in going and returning was $6\frac{1}{2}$ hours. Find the distance to the campsite.

★ 19. It ordinarily takes a bus 20 minutes to travel the 12 miles from the City Hall to Central City to the airport. However, during the rush hours from 4:00 P.M. to 6:00 P.M., the bus will travel only 10 miles in the same length of time. If busses leave City Hall on the hour and every 15 minutes thereafter, and if it takes a person 20 minutes from the time he leaves a bus to get on an airplane, what is the last bus Mr. Samuels can take from City Hall to make a plane leaving at 5:50 P.M.?

★20. An aircraft landing at Capital City between the hours of 5:00 P.M. and 7:00 P.M. must expect to "hold" (fly in circles to await a landing clearance) for an average time of 20 minutes. Amalgamated Airlines flight 227 from Los Angeles arrives at Capital City airport 1 hour late due to headwinds which reduced its groundspeed from 500 to 400 miles per hour. If the flight leaves Los Angeles on schedule at 12:50 P.M. Capital City time, when can it expect to land at Capital City?

5-7 Mixture and Other Problems

A merchant often mixes goods of two or more kinds in order to sell a blend at a given price. Similarly, a chemist often mixes solutions of different strengths of a chemical to obtain a solution of desired strength. All such problems are solved in the same way. The sum of the values or weights of the original ingredients must equal the value or weight of the final mixture.

Example 1. The registration fee at a convention was $5.00 for members and $9.00 for nonmembers. If receipts at the convention were $4450, and 850 persons were in attendance, how many members were registered?

Solution: 1. Let $n =$ number of members present.
Then $850 - n =$ number of nonmembers present.

2.

	Number	Fee in dollars	Receipts
Members registered	n	5	$5n$
Nonmembers registered	$850 - n$	9	$9(850 - n)$
Total registered	850	—	4450

Receipts from members + Receipts from nonmembers = Total Receipts

$$5n + 9(850 - n) = 4450$$

Steps 3 and 4 are left for you.

Not every problem has a solution. Consider the following:

Example 2. Find two consecutive integers whose sum is 46.

Solution:
1. Let x = the first integer.
   Then $x + 1$ = the second integer.
2. The sum of the integers is 46.
   $$x + (x + 1) = 46$$
3. $2x + 1 = 46$
   $2x = 45$
   $x = 22\frac{1}{2}$
   $x + 1 = 23\frac{1}{2}$
4. The numbers obtained, $22\frac{1}{2}$ and $23\frac{1}{2}$, are not *integers*. Therefore, they do not satisfy the problem's requirements. The facts of the problem do not fit with one another; they are inconsistent. In any pair of consecutive integers, one is an even integer and the other is odd. Therefore, their sum is odd and cannot be 46!

In reading a problem, be on the lookout for inconsistent facts. Also be alert to recognize problems in which not enough facts are given to obtain a definite answer. Most of the problems in the following set can be solved. But some of them fail to have solutions either because their facts are inconsistent or because they give too few facts.

PROBLEMS

1. Joe has an equal number of dimes and quarters. If he has $2.10 in all, how many coins of each type does he have?
2. Mr. Cirillo bought some six-cent and ten-cent stamps. He bought 25 stamps in all, and paid $1.70 for them. How many stamps of each kind did he buy?
3. Airline fares for a flight from Century City to Glendale are $30 for first class and $25 for tourist class. If a flight carried 52 passengers for a total fare of $1360, how many first class passengers were on the flight?
4. The Leesburg Little Theatre holds 110 persons. Adult admissions are $1.50, and children can attend for 80¢. If the theatre is full, and total receipts for admissions are $158, how many adults are present?
5. How many pounds of candy worth 90¢ per pound must be added to 10 pounds of a candy worth 60¢ per pound to form a mixture worth 80¢ per pound?

6. How many pounds of coffee worth 50¢ per pound must be added to 15 pounds of coffee worth 80¢ a pound to form a mixture worth 60¢ a pound?

7. Mrs. Tully bought 4 more bags of peanuts selling at 35 cents a bag than of cashews selling at 45 cents a bag. She spent 3 dollars for the nuts. How many bags of peanuts did she buy?

8. A man buys 23 more 6-cent stamps than he does 10-cent stamps. How many of each kind of stamp does he buy, if the total cost of the stamps is $4.38?

9. A boy has 12 more quarters than dimes in his savings bank. How many of each kind of coin has he if this bank contains $12.45?

10. Find the least two consecutive positive integers whose difference is 6.

11. The degree measures of the angles of a triangle are consecutive odd integers. What are the degree measures of the angles?

12. Nancy is four times as old as her sister, Fran. Three years ago, the sum of their ages was 4 years. How old was Fran 3 years ago?

13. In going to a medical meeting, Dr. Lloyd made a trip of 500 miles. He traveled by train for $1\frac{1}{2}$ hours and by automobile for the rest of the trip. The average speed of the train was 15 miles per hour more than that of the automobile. Find the average speed of the automobile.

14. Mary bought 3 times as many 10-cent stamps as she did 6-cent stamps and 6 more 6-cent stamps than she did 45-cent special delivery stamps. How many of each kind of stamp did she buy if her total expenditure for stamps was $8.64?

15. A drive-in cashier took $20 in bills to the bank to get change. She asked for twice as many dimes as nickels and three times as many quarters as nickels. Was the bank teller able to grant her request?

16. Jack and Jill, together, have 19 paperback books. If Jack lost 3 of his books, but Jill doubled her supply, the two of them, together, would have 40 books. How many does each have now?

17. A gourmet-foods shop sells almonds for $2.10 a pound, pecans for $1.90 a pound, and peanuts for $1.10 a pound. The manager makes a mixture of these nuts to sell for $1.50 a pound. He uses one-half as many pounds of almonds as pecans. How much of each does he use?

18. A tea merchant prepares a blend of 30 pounds of tea to sell for $1.95 per pound. For the blend, he uses two types of tea, one selling at $2.15 per pound, the other at $2.05 per pound. How much of each type should he use for the blend?

19. For two weeks Earl collected nickels, dimes, and quarters. He found that he had twice as many quarters as dimes and $4.65 in all. How many coins of each type did he collect?

20. Tom works at The China Box packing dishes for shipment. He receives 5 cents for each piece he packs successfully and is fined 12 cents for each piece he breaks. If he handles 187 pieces and is paid $8.16, how many pieces did he break?

21. An enthusiastic alumnus agreed to contribute $2 for each $5 his college raised above $10,000 in its annual fund drive. If $23,000 was raised in the drive, how much did this alumnus then contribute?

22. Find the least two positive integers whose sum is an even integer and whose difference is an odd integer.

23. Find two even integers whose product is a positive integer and whose sum is zero.

24. How much coffee worth $0.84 a pound should be mixed with coffee worth $1.02 a pound to produce a mixture worth $0.96 a pound?

25. Eighteen-carat gold contains 18 parts by weight of gold and 6 parts of other metals. Fourteen-carat gold contains 14 parts of gold and 10 parts of other metals. How much eighteen-carat gold must be mixed with fourteen-carat gold to obtain an alloy containing 17 parts of gold and 7 parts of other metals?

★26. Draw a flow chart of a program to compute and print the different ways of making a dollar's worth of change in nickels and dimes. (*Hint:* If d dimes and n nickels are used, then $n = \frac{1}{5}(100 - 10d)$ where $d \in \{0, 1, 2, \ldots, 10\}$.)

★27. Redraw the flow chart for Problem 26 if quarters as well as nickels and dimes may be used to make the change for one dollar.

chapter summary

1. **Axiom of Comparison:** For all real numbers a and b, one and only one of the following statements is true: $a < b$, $a = b$, $b < a$.

2. **Transitive Axiom of Order:** For all real numbers a, b, and c, (1) if $a < b$ and $b < c$, then $a < c$, and (2) if $a > b$ and $b > c$, then $a > c$.

3. **Addition Axiom of Order:** For all real numbers a, b, and c, (1) if $a < b$, then $a + c < b + c$ and (2) if $a > b$, then $a + c > b + c$.

4. **Multiplication Axiom of Order:** For all real numbers a, b, and c, (1) if $a < b$ and $c > 0$, then $ac < bc$; similarly, if $a > b$ and $c > 0$, then $ac > bc$. Also, (2) if $a < b$ and $c < 0$, then $ac > bc$; similarly, if $a > b$ and $c < 0$, then $ac < bc$.

5. Set relationships can be pictured by means of Venn diagrams. The relationships include **subset**, **union**, and **intersection**. **Disjoint sets** have no elements in common.

CHAPTER REVIEW

5–1 **1.** Solve and then graph the solution set: $7 - 5x > 30$

5–2 **2.** Given $A = \{2, 4, 6, 8, 10\}$ and $B = \{1, 3, 5, 7, 9\}$. State (1) $A \cap B$; (2) $A \cup B$.

5–3 **3.** Solve this open sentence and draw its graph: $y + 1 > 3$ *or* $6y + 5 < 2$

5–4 **4.** Solve: $|2x - 5| = 7$ **5.** Solve: $|2x - 5| > 7$

5–5 **6.** Find 3 consecutive integers whose sum is 96.

5–6 **7.** At 8 A.M. two planes leave St. Louis. One flies west at 350 miles per hour. The other flies east at 400 miles per hour. At what time will they be 1500 miles apart?

5–7 **8.** Admission to the skating rink is $1.25 in the evening and 75¢ during the day. On a day where 430 persons bought admission tickets the total receipts were $387.50. How many people bought tickets in the evening?

6 operations with polynomials

6–1 Adding Polynomials

Each of the terms 8, y, $5x^2$, and $-3ab$ is called a *monomial*. A **monomial** is a term which is either a numeral (8), or a variable (y), or an indicated product of a numeral and one or more variables ($5x^2$ or $-3ab$). (Compare this definition of *monomial* with the definition of *term* on page 21.)

An indicated sum of monomials is called a **polynomial**. (The prefix "poly" means many.) A polynomial such as

$$7x^4 + 3x^3 + (-5x^2) + 2x + (-5)$$

is usually written as

$$7x^4 + 3x^3 - 5x^2 + 2x - 5.$$

$2x^2 + 3$ is a polynomial of two terms, which is called a **binomial** since "bi" means two; $x^2 - 2xy + y^2$ is a polynomial of three terms, which is called a **trinomial** since "tri" means three. A monomial is considered to be a polynomial of one term.

The **degree of a monomial in a variable** is the number of times that the variable occurs as a factor in the monomial. For example,

$$3x^2y^5z$$

is of degree 2 in x, 5 in y, and 1 in z. The **degree of a monomial** is the sum of the degrees in each of its variables. Thus, the degree of $3x^2y^5z$ is 8, because $2 + 5 + 1 = 8$. If a monomial other than 0 contains no variables, **its degree is zero**. The monomial 0 has **no degree**.

A polynomial having no two terms *similar* (page 62) is said to be **simplified** (page 61) or to be in **simple form**. The **degree of a polynomial** in simple form

is the same as the greatest of the degrees of its terms. The degrees of the terms of
$$15x^4 - 7x^3y^2 + 2xy + 4$$
are, in order, 4, 5, 2, and 0. Thus, the degree of this polynomial is 5. Since $3x^4 + 8x^2 + y - 3x^4$ contains the similar terms $3x^4$ and $-3x^4$, it can be simplified to $8x^2 + y$; therefore, its degree is 2, not 4.

To add two polynomials such as
$$6r^2s + 11 \quad \text{and} \quad 3r^2s - 5r + 3,$$
write the sum
$$(6r^2s + 11) + (3r^2s - 5r + 3)$$
and then simplify it by adding the similar terms:
$$(6r^2s + 11) + (3r^2s - 5r + 3) = (6 + 3)r^2s - 5r + (11 + 3)$$
$$= 9r^2s - 5r + 14$$

To add two polynomials, write the sum and add the similar terms.

The addition may also be done vertically:

$$\begin{array}{r} 6r^2s + 11 \\ 3r^2s - 5r + 3 \\ \hline 9r^2s - 5r + 14 \end{array}$$

Addition of polynomials can be checked by assigning a numerical value to each variable and evaluating the expressions. For example, let 2 be the value of r and 3 be the value of s.

$$\begin{array}{l}
6r^2s + 11 \to 6 \cdot 4 \cdot 3 + 11 = 83 \\
3r^2s - 5r + 3 \to 3 \cdot 4 \cdot 3 - (5 \cdot 2) + 3 = 29 \\
\hline
9r^2s - 5r + 14 \to 9 \cdot 4 \cdot 3 - (5 \cdot 2) + 14 \stackrel{?}{=} 112 \\
 112 = 112
\end{array}$$

Though it is convenient to choose small numbers for checking, explain why 1 and 0 are not suitable values to choose for checking.

In working with polynomials, it is convenient to arrange the terms in order of either decreasing degree or increasing degree in a particular variable.

In order of decreasing degree in m: $\quad m^3 + 6m^2 + 12m + 6$

In order of increasing degree in y: $\quad 16 - 8y^2 + y^4$

In order of decreasing degree in r: $\quad 32r^5 + 7r^4s - 2r^2s^2 - 18s^3$

132 CHAPTER SIX

EXERCISES

Simplify each expression for a sum.

1. $4m - 2n$
   $3m + 3n$

2. $-3t + s$
   $2t + 4s$

3. $2x - 5y$
   $-3x - 2y$

4. $3z - 2u$
   $-8z + 7u$

5. $3x^2 - 2x$
   $-x^2 + 3x$

6. $5y^2 - 2xy$
   $10y^2 + 18xy$

7. $21rs - 13s$
   $-15rs - 7s$

8. $83u^2 - 17v^2$
   $-15u^2 + 26v^2$

9. $7.2y - 3.1x$
   $-0.6y + 8.3x$

10. $-4.2w + 0.7v$
    $3.9w - 1.2v$

11. $\frac{3}{7}t - \frac{4}{13}s$
    $-\frac{4}{7}t + \frac{7}{13}s$

12. $z^3 - t^3$
    $\frac{1}{5}z^3 + \frac{2}{3}t^3$

13. $x^2 - 3x + 2$
    $3x^2 \quad - 5$
    $\quad 4x + 8$

14. $x^2 \quad + 8$
    $3z + 15$
    $-6x^2 - 2z - 7$

15. $3x^3 - 4x^2 + 7x$
    $\quad 2x^2 - 6x + 7$

16. $-3y^4 + 3x^3 - 5x^2 + x + 3$
    $\quad - 4x^3 + 2x^2 - x - 3$

Check each of the following sums. When values are given for the variables, use them. When a sum is incorrect, write the correct sum.

17. $7x - 2$
    $2x + 5$
    $9x + 3$
    Let $x = 3$.

18. $2z - 15$
    $-7z + 8$
    $-9z - 7$
    Let $z = 2$.

19. $2t + 5$
    $t - 7$
    $-4t + 8$
    $-t + 6$
    Let $t = 2$.

20. $3x + y - 2$
    $5x - 2y$
    $\quad 2y + 6$
    $8x + y + 8$
    Let $x = 2, y = 3$.

21. $2m^2 + 3n$
    $\quad 6n - 5$
    $-m^2 - n - 6$
    $m^2 + 8n - 11$
    Let $m = 5, n = 3$.

22. $z^2 - 3z + 2$
    $-z^2 + z - 5$
    $2z^2 + 5z + 2$
    $2z^2 + 9z - 1$
    Let $z = 2$.

Simplify each expression for a sum.

23. $(n + 3) + (2n + 5)$
24. $(2m - 7) + (3m + 7)$
25. $(\frac{1}{2}z - \frac{3}{2}) + (\frac{1}{4}z + \frac{1}{2})$
26. $(\frac{2}{3}t + 8) + (-\frac{7}{3}t - \frac{7}{2})$
27. $(3.1x^2 + 0.1) + (1.2x^2 - 2.3)$
28. $(-8.1y^2 - 2.2) + (3.8y^2 + 5.1)$
29. $(3z^3 - z^2) + (z^3 - 4z)$
30. $(5n^4 + 3n^2) + (n^3 - 3n^2)$
31. $(4x^3 - 2x^2 + 3x + 1) + (x^3 + 3x^2 - 5x - 1)$
32. $(-2z^3 + z^2 + 5z - 2) + (3z^3 - z^2 - 5z + 2)$
33. $(a^4 - 3a^2 + 2a - 1) + (2a^4 - a^3 + a^2 - 2a - 2)$
34. $(2b^5 + b^4 - 2b + 3) + (2b^4 + 3b^3 - b^2 + 2b - 2)$
35. $(-4t^6 + 2t^4 - 3t^3 + 2t^2 - 1) + (7t^6 + t^5 - 2t^4 + 3t^3 - 2t^2 + t - 1)$

6-2 Subtracting Polynomials

To subtract one polynomial (the *subtrahend*) from another (the *minuend*), use the same procedure as in subtracting real numbers (page 80). That is, add the *opposite* of each term of the subtrahend. For example,

$$(12z + 3) - (2z - 4) = 12z + 3 + (-2z) + 4 = 10z + 7.$$

Example 1. $(17r^2 - 5r + 2) - (9r^2 + r - 5)$

Solution 1: $(17r^2 - 5r + 2) - (9r^2 + r - 5)$
$= 17r^2 - 5r + 2 - 9r^2 - r + 5$
$= 8r^2 - 6r + 7$

Solution 2:
$$\begin{array}{r} 17r^2 - 5r + 2 \\ 9r^2 + r - 5 \\ \hline 8r^2 - 6r + 7 \end{array}$$

In general:

To subtract one polynomial from another, add the opposite of each term of the polynomial being subtracted (the subtrahend).

Grouping symbols may appear in an equation to indicate addition or subtraction of a polynomial.

Example 2. Solve: $y^2 + 2y + (y - 5 + 3y) = 3y - (-7 + 3y - y^2)$

Solution: $y^2 + 2y + (y - 5 + 3y) = 3y - (-7 + 3y - y^2)$

Rewrite without parentheses → $y^2 + 2y + y - 5 + 3y = 3y + 7 - 3y + y^2$

Add similar terms → $y^2 + 6y - 5 = 0 + 7 + y^2$

Subtract y^2 from each member → $6y - 5 = 7$

Add 5 to each member → $6y = 12$

Divide each member by 6 → $y = 2$

Check: $y^2 + 2y + (y - 5 + 3y) = 3y - (-7 + 3y - y^2)$
$2^2 + 2 \cdot 2 + (2 - 5 + 3 \cdot 2) \stackrel{?}{=} 3 \cdot 2 - (-7 + 3 \cdot 2 - 2^2)$
$4 + 4 + (2 - 5 + 6) \stackrel{?}{=} 6 - (-7 + 6 - 4)$
$8 + (3) \stackrel{?}{=} 6 - (-5)$
$11 = 11$

∴ the solution set is {2}.

Note: In the solution use was made of the fact that adding the same polynomial to each member of an open sentence produces an equivalent sentence.

EXERCISES

Subtract. Check by evaluation, using 2 for x, 3 for y, 4 for a, 5 for b.

1. $\begin{array}{r} 4a + 4 \\ 2a + 1 \\ \hline \end{array}$
2. $\begin{array}{r} 2b + 3 \\ b + 2 \\ \hline \end{array}$
3. $\begin{array}{r} 4y + 3 \\ 2y - 5 \\ \hline \end{array}$
4. $\begin{array}{r} x - 3 \\ 2x + 2 \\ \hline \end{array}$
5. $\begin{array}{r} 4x - 5y \\ -3x - 2y \\ \hline \end{array}$
6. $\begin{array}{r} 7a + b \\ -a - 2b \\ \hline \end{array}$
7. $\begin{array}{r} x^2 - 3x + 1 \\ x^2 - 2x - 2 \\ \hline \end{array}$
8. $\begin{array}{r} 2y^2 - 3y + 1 \\ -y^2 + y + 2 \\ \hline \end{array}$
9. $\begin{array}{r} 2a^2 - 3a + 5 \\ a^2 - 2 \\ \hline \end{array}$
10. $\begin{array}{r} 2b^2 + 5b - 5 \\ 3b^2 - 4b \\ \hline \end{array}$
11. $\begin{array}{r} ax + by \\ -2ax + by \\ \hline \end{array}$
12. $\begin{array}{r} xy - ab \\ xy - 3ab \\ \hline \end{array}$

Simplify.

13. $(3r + 2s) - (r + s)$
14. $(2p + 3q) - (2p - q)$
15. $(-5u + v) - (-4u - v)$
16. $(-2x - 5) - (-x + 7)$
17. $(z^2 - 3z + 2) - (-z^2 - 2z + 2)$
18. $(t^3 - 2t^2 + 3) - (2t^3 + 3t^2 - 2t)$
19. $(2x^4 - 3x^2 + 1) - (x^4 - 2x^2 - x + 2)$
20. $(3y^4 + 2y^3 - 3y) - (2y^4 + 2y^3 - 4y - 2)$

Solve each equation.

21. $2x - (3x + 2) = 5$
22. $-3y - (7 - 5y) = 15$
23. $(4t - 2) - t = -20$
24. $(4 - 3r) - 2r = -21$
25. $(2s - 5) - (3s + 2) = 5 - 3s$
26. $(2y + 3) - (4 - 2y) = -10 + y$
27. $(-3k + 4) - (k - 6) = -3k - 11$
28. $(5z - 3) - (z - 8) = 2z - 7$
29. $(4 - 2x) - (x - 5) = (x + 2) - (3 - x)$
30. $(-2 - 3a) - (4 - a) = (3 - a) - (5 - a)$
31. $(p^2 - 2p + 1) - (2p^2 + 3p + 5) = 2 - 3p - p^2$
32. $(2s^2 - 5s - 2) - (s^2 - s + 8) = s^2 + 2s - 28$
33. $(4 - 6z + 3z^2) - (14 + z - z^2) = 4z^2 + z + 70$
34. $(-3 - 7x - 5x^2) - (8 + x - x^2) = 49 - 4x - 4x^2$
35. $2t - [3 - (t + 2)] = 29$
36. $4n - [n - (6 + 2n)] = 44$
37. $-2x - [3x - (14 - x)] = -42 + x$
38. $-7y - [8 - (y + 24)] = -2y - 22$
39. $-[3x - (5 - 2x) + 3] = -49 - 2x$
40. $-[17 - (3z - 5) + 6z] = 2z + 18$

When $f(x)$ and $g(x)$ are as given, simplify $f(x) - g(x)$.

Example. $f(x) = 3x^2 + 2x - 1; g(x) = 2 + 3x - x^2$

Solution:
$$\begin{aligned} f(x) - g(x) &= 3x^2 + 2x - 1 - (2 + 3x - x^2) \\ &= 3x^2 + 2x - 1 - 2 - 3x + x^2 \\ &= 4x^2 - x - 3. \end{aligned}$$

41. $f(x) = 5x^2 - 3x + 1; g(x) = -3x^2 + x - 5$
42. $f(x) = 7 - 3x - 15x^2; g(x) = 6 + 2x - 11x^2$
43. $f(x) = 3 - [x + (x - 1)]; g(x) = x - (2 - x^2)$
44. $f(x) = x - [3 - (x + 2)]; g(x) = (3 - x) - (x + 1)$
45. $f(x) = x^2 - [2 - (x + 1)]; g(x) = x^2 - [3 - (2 - 3x)]$
46. $f(x) = 3x - [2x^2 - (x + 1)]; g(x) = 7 - [5x^2 - (x + 7)]$

6–3 The Product of Powers

Recall (page 24) that b^4 (read "b to the fourth" or "b-fourth") stands for $b \cdot b \cdot b \cdot b$. Therefore:

$$b^4 \cdot b^2 = \underbrace{\overbrace{(b \cdot b \cdot b \cdot b)}^{\text{4 factors}} \cdot \overbrace{(b \cdot b)}^{\text{2 factors}}}_{\text{6 factors}} = b^6 = b^{4+2}$$

Thus, in multiplying these two *powers of the same base*, the exponent can be found by retaining the base and adding the exponents of the factors. In general, for positive integral exponents m and n:

$$b^m \cdot b^n = \underbrace{\overbrace{(b \cdot b \cdots \cdots b)}^{m \text{ factors}} \overbrace{(b \cdot b \cdots \cdots b)}^{n \text{ factors}}}_{m + n \text{ factors}} = b^{m+n}$$

This result is usually stated as follows:

Rule of Exponents for Multiplication

For all positive integers m and n: $b^m \cdot b^n = b^{m+n}$

To multiply monomials, we may use this rule of exponents together with the commutative and associative axioms of multiplication to determine the numerical factor and the variable factors of the product. Thus:

$$(6x^2y)(-5x^5y^4) = (6 \cdot -5)(x^2 \cdot x^5)(y \cdot y^4) = -30x^{2+5}y^{1+4} = -30x^7y^5$$

Note that this rule of exponents applies only when the bases of the powers are the *same*. We cannot use it, for example, to simplify the product x^8y^9 because the bases of the powers x^8 and y^9 are *different*. Thus, the product x^8y^9 cannot be simplified.

EXERCISES

Simplify each expression.

1. $(2yz)(3y^2)$
2. $(5ab)(4a^2)$
3. $(-2m^2n)(mn^3)$
4. $(pq^3)(-5p^4q^2)$
5. $(-2x)(x^2y)(y)$
6. $(-3z)(4zt)(t^3)$
7. $(-2y)(-2y)(-2y)$
8. $(-4z)(4z)(4z)$
9. $(-3a^2b)(-2ab)(5b^4)$
10. $(-5k^2m)(-3km^2)(k^2m^4)$
11. $(0.2x)(5x^2y)(-xyz^3)$
12. $(0.3z^2)(12z^4r)(5zr^5)$
13. $(\frac{1}{3}m^2n)(\frac{1}{2}mn^5)$
14. $(-\frac{1}{4}s^3t^2)(\frac{1}{2}s^2t)$
15. $-\frac{1}{5}u(u^5v^5)(-5uv^3)$
16. $\frac{2}{3}t^2(rs^2t)(6r^2s)$
17. $(-3a^2b^2c)(-3a^2b^2c)(-3a^2b^2c)$
18. $(\frac{1}{2}xy^2z^3)(\frac{1}{2}xy^2z^3)(\frac{1}{2}xy^2z^3)$
19. $(x^n)(x)$
20. $(y^n)(y^n)$
21. $(2z^{n+1})(z)$
22. $(3t^{n-2})(-2t^{3-n})$
23. $(-2r)(r^2)(-r^3) + (3r^2)(r^4)$
24. $(-2x)(xy^2)(-5xz^2) + (-x^2)(2xy)(4yz^2)$
25. $(5h^2)(-3hk^2)(k^2) - (-3h)(2h^2k)(k^3)$
26. $(a^3bc)(-2b^2c)(3c^2) + (-a^2b^2c^2)(2a)(-3bc^2)$
27. $(5mn^2)(-3m^3n)(p^2) + (8mp)(3np)(m^3n^2)$
28. $(u^2vw)(-uv^2w)(-uvw^2) + (4u^2v^2w^2)(2u^2v^2w^2)$
★29. $y^4(3y + 2) + 2y^2(y^3 - 2y^2)$
★30. $t^3(t - t^2) - 2t^2(t^3 + t^2)$

6–4 The Power of a Product

Consider $3x^3$ and $(3x)^3$; they are not equal unless x has the value 0. We have:

$$3x^3 = 3 \cdot x \cdot x \cdot x$$
$$(3x)^3 = 3x \cdot 3x \cdot 3x = 27x^3$$

In general, for every positive integral exponent m,

$$(ab)^m = \overbrace{(ab)(ab) \cdots (ab)}^{(ab) \text{ is a factor } m \text{ times}}$$
$$\therefore (ab)^m = \underbrace{(a \cdot a \cdots \cdot a)}_{m \text{ factors}}\underbrace{(b \cdot b \cdots \cdot b)}_{m \text{ factors}} = a^m b^m$$

This result may be summarized as follows:

Rule of Exponents for a Power of a Product

For every positive integer m: $(ab)^m = a^m b^m$

For example:

$$(-2z)^4 = (-2)^4 z^4 = 16z^4 \quad \text{and} \quad (5pq)^2 = 5^2 p^2 q^2 = 25p^2 q^2$$

Of course, the base of a power may itself be a power:

$$(b^2)^3 = b^2 \cdot b^2 \cdot b^2 = b^{2+2+2} = b^6 = b^{2 \cdot 3}$$

In general,

$$(b^m)^n = \overbrace{(b^m)(b^m) \cdots (b^m)}^{b^m \text{ is a factor } n \text{ times}}$$
$$\therefore (b^m)^n = \underbrace{b^{m+m+\cdots+m}}_{n \text{ terms}} = b^{mn}$$

This result gives the following:

Rule of Exponents for a Power of a Power

For all positive integers m and n: $(b^m)^n = b^{mn}$

Both of these rules are used in the following illustration:

$$(-6s^6 t^4)^3 = (-6)^3 (s^6)^3 (t^4)^3 = -216 s^{18} t^{12}$$

EXERCISES

Simplify each expression.

1. $(2a)^4$
2. $(3z)^3$
3. $(-4t^2)^2$
4. $(-2n^3)^2$
5. $3z(2z)^2$
6. $4b(3b)^3$
7. $-2s(st)^2$
8. $-3p(p^2 q)^3$

9. $(x^2y)^2(xy^2)^3$
10. $(c^3d)^3(cd^2)^2$
11. $(-\frac{1}{2}a^2b)^2(4ab^3)^2$
12. $(-\frac{1}{3}mn)^3(9mn^2)^2$
13. $(-rs)^2(2r^2s)^3(0.5s)$
14. $(yz^2)^2(-4y^2)^3(0.25z^2)$
15. $(c^4k)^2(-3k)^3(\frac{1}{3}c^2)^2$
16. $(-2x^2y)^3(\frac{1}{4}xy)^2(-2y^2)^2$
17. $(xz)^n(x^2z)$
18. $(a^nb^m)^2(a^2b)$
19. $(2ab^2)^3 + (2ab^2)^2(6ab^2)$
20. $(3u)(u^2v)^3 + (2u)^2(-u^5v^3)$
21. $(-2r^3)^2(rs^2t) - (-r^2)^2(rst)(-r^2s)$
22. $(-6y^2z^3)^2(2yz) - (3yz^2)^3(-12y^2z)$
23. $(\frac{1}{3}m^2n)^4(-9mn^2)^2 - (3mn)^2(\frac{1}{9}m^2n^2)(m^3n^2)^2$
24. $(\frac{2}{3}pq^2)^3(-9p^2q)^2 - (-pq^2)^2(2pq)^3(5p^2q)$
★25. $-3x^2y(y-2) + 2xy^2(x-3)$ ★26. $5ab^2(ab-b^2) - 2a^2(b^3+b)$

6–5 Multiplying a Polynomial by a Monomial

The distributive axiom, together with the rules of exponents for multiplication, enables us to multiply any polynomial by a monomial. For example:

$$3a(4a^2 + 2) = 3a(4a^2) + 3a(2) = 12a^3 + 6a$$

This result is illustrated in the figure:

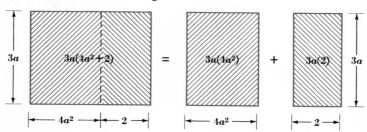

The largest rectangle can be separated into the other two. Its area is equal to the sum of the areas of the other two rectangles.

We may multiply either horizontally or vertically, as shown here.

$$-6x^3(4x^2 - 2x + 1) = -24x^5 + 12x^4 - 6x^3$$

$$\begin{array}{r} 4x^2 - 2x + 1 \\ -6x^3 \\ \hline -24x^5 + 12x^4 - 6x^3 \end{array}$$

In general:

To multiply a polynomial by a monomial, use the distributive axiom: multiply each term of the polynomial by the monomial, and then add the products.

EXERCISES

In each exercise, apply the distributive axiom.

1. $3(a^2 + 2ab + 2b^2)$
2. $5(2z^2 - 2z + 6)$
3. $-4(3 - 2x - 3x^2)$
4. $-2(y^2 - 4y + 7)$
5. $2x^2(x + xy - 3y^2)$
6. $-4t^3(3r + 2rt - 4t^2)$
7. $3x^2y(4 + 2y - xy^2)$
8. $6n^2t(4nt - 3nt^2 + 4t^3)$
9. $(6p^2 - 4pq + 3q^2)(-2pq)$
10. $(4k^2 + 3kn + 2n^2)(-5k)$
11. $3xz^2(7 + 5x - 3z + 2x^3z)$
12. $5r^2s(3 - 2r + 7s - 5rs^2)$
13. $-8c^3d^2(c^2 + d^2 - 4c - 5d)$
14. $-6p^3r(3p - 4pr + 7r^2 - 2p^2r^2)$

Solve each equation over $\mathcal{R}$. (Recall pages 77–78 and 84.)

15. $3n + 2(n - 5) = 35$
16. $2z + 5(6 - z) = 3$
17. $-7u + 5(2u - 3) = -39$
18. $-9t + 3(t - 4) = 42$
19. $2(x + 5) - 3(x - 4) = 7$
20. $5(2p - 3) - 2(5 - p) = -7$
21. $-7(y - 3) + 5(3y - 5) = 44$
22. $-10(3 - 4n) - 7(5n + 3) = -51$
23. $5p - 7(3 - p) = 6 + 2(p + 5)$
24. $6k - 5(3k + 2) = 5(k - 1) - 8$
25. $\frac{1}{4}(3z + 8) + \frac{3}{4}(z - 4) = \frac{1}{2}(z + 4) - 2$
26. $\frac{1}{2}(6z + 4) - \frac{2}{3}(9 - 3z) = 2(z + 1) + 3$
27. $2 - 5[3t + 2(6 - t)] = -7(t - 4)$
28. $7 + 3[-s - 3(s + 5)] = 5(7 - 2s) + 5$
★29. $2x + x[3 - (x + 1)] = x(2 - x) + 12$
★30. $7z - 4[3 + z(5 - z)] = 2z(6 + 2z) + 8$

PROBLEMS

Write a polynomial that represents the number described.

Example. The area of a rectangle that is 20 feet longer than it is wide.

Solution: Let w = number of feet in the width.
Then $w + 20$ = number of feet in the length.
Area = $w(w + 20) = w^2 + 20w$ (square feet).

1. The area of a rectangle that is 5 inches longer than it is wide.
2. The area of a rectangle whose width is 7 inches shorter than its length.
3. The area of a triangle in which the length of the altitude is 3 feet less than the length of the base.
4. The area of a triangle in which the length of the base is 5 feet greater than the length of its altitude.

5. The total distance traveled by an automobile that travels at an average speed of 40 miles per hour for 2 hours and then increases its speed so that its average speed during the next hour is x miles per hour more.

6. The total distance traveled by an automobile that travels at an average speed of s miles per hour for 2 hours and then increases its speed so that its average speed during the next hour is 10 miles per hour more.

7. The total distance traveled if I travel for 3 hours at $(40 + x)$ miles per hour and then for 4 hours at $(100 + x)$ miles per hour.

8. The total distance you travel in flying 2 hours at $(300 + x)$ miles per hour and then 4 hours at $1\frac{1}{2}$ times that rate.

9. A plane traveled for 2 hours at an average airspeed (speed relative to the air) of v miles per hour with an average tail wind of 20 miles per hour. Then the plane traveled against a wind averaging 5 miles an hour for 2 more hours. Write a polynomial in simple form for the total land distance covered.

10. A man can row x miles per hour in still water. He rowed down a river with a current of 4 miles per hour for half an hour and then up a creek against a current of 2 miles an hour for half an hour. Write a polynomial in simple form for the total distance he rowed.

11. The dimensions of a rectangular playground are 40 feet by 24 feet. The walk around it is of uniform width. The playground and walk have a combined area of 2240 square feet. Write a polynomial in simple form for the total area.

12. A picture is 15 inches wide and 20 inches long. It is enclosed in a frame of uniform width. The area of the frame is $\frac{2}{3}$ that of the picture. Write a polynomial in simple form to express the total area.

6-6 Multiplying Two Polynomials

To express the product $(3y + 2)(6y + 1)$ as a polynomial, first treat $(6y + 1)$ as a number to be multiplied by the binomial $3y + 2$ and apply the distributive axiom. Then apply the distributive axiom to each of the products. Finally, simplify the resulting polynomial.

$(3y + 2)(6y + 1)$
$= (3y)(6y + 1) \quad + 2(6y + 1)$
$= (3y)(6y) + (3y)(1) + 2(6y) + 2(1)$
$= 18y^2 \quad + 3y \quad + 12y \quad + 2$
$= 18y^2 \quad + \quad 15y \quad + 2$

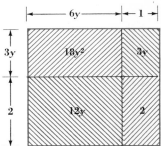

The figure illustrates this product.

To multiply one polynomial by another, use the distributive axiom: multiply each term of one polynomial by each term of the other, and then add the products.

Usually it is convenient to set up the multiplication of polynomials in vertical form, and to work from left to right, thus:

$$\begin{array}{r} 6y + 1 \\ 3y + 2 \\ \hline \end{array}$$

This is $3y(6y + 1) \longrightarrow 18y^2 + 3y$
This is $2(6y + 1) \longrightarrow 12y + 2$
This is $(3y + 2)(6y + 1) \longrightarrow 18y^2 + 15y + 2$

To check the accuracy of your multiplication, evaluate the factors and their product, using any numbers except 0 and 1.

EXERCISES

Express each product as a polynomial in simple form.

1. $(a + 1)(a + 2)$
2. $(c + 3)(c + 4)$
3. $(x - 2)(x + 2)$
4. $(y + 1)(y - 1)$
5. $(z - 3)(z + 6)$
6. $(p - 2)(p - 2)$
7. $(2a + 1)(a + 2)$
8. $(b + 4)(3b + 1)$
9. $(2n + 3)(n - 5)$
10. $(3n - 1)(n + 2)$
11. $(3x - 3)(x - 5)$
12. $(7y - 2)(y - 3)$
13. $(2r + 3)(3r + 2)$
14. $(5t - 2)(3t + 1)$
15. $(6d + 5)(3d - 2)$
16. $(7a - 6)(2a - 5)$
17. $(2x - y)(x + y)$
18. $(3z - u)(2z + 3u)$
19. $(2x + 7z)(2x - 7z)$
20. $(3y + 2v)(3y + 2v)$
21. $(2x - 3t)(2x - 3t)$
22. $(4a + b)(4a + b)$
23. $(0.2z + 1)(1.4z - 2)$
24. $(2.1n + 6)(10n - 0.3)$
25. $(x^2 - y)(x^2 + y)$
26. $(2x^2 + 4)(x^2 - 3)$
27. $(a^2 - b^2)(a^2 - b^2)$
28. $(p^2 + q^2)(p^2 + q^2)$
29. $(x + 1)(x^2 + 3x - 2)$
30. $(z + 2)(z^2 - 3z + 5)$
31. $(2x - 1)(x^2 + 3x + 5)$
32. $(a - 2)(3a^2 + 5a - 1)$
33. $(x - y)(x^2 + xy + y^2)$
34. $(r + s)(r^2 + ry + s^2)$
35. $(x + 1)(2x - 3) + (x + 1)(3x - 1)$
36. $(y + 2)(y - 2) + (2y + 1)(2y + 1)$
37. $(2a + b)(3a - b) - (a - b)(2a - 3b)$

38. $(3c - d)(2c + 3d) - (4c + d)(2c - d)$
39. $(x^2 + 2x - 1)(x^2 - x + 3)$
40. $(2y^2 + y - 2)(y^2 + 3y + 5)$
41. $(2r - s + t)(r + 2s - t)$
42. $(a + 2b + 3c)(2a - b + 2c)$
43. $(t - 1)(t^4 + t^3 + t^2 + t + 1)$
44. $(k + 1)(k^4 - k^3 + k^2 - k + 1)$

Use the properties of real numbers to prove that each of the following is true for all real numbers *a, b, c,* and *d*.

45. $(a + b)(c + d) = ac + bc + ad + bd$
46. $(a - b)(c - d) = ac - bc - ad + bd$
47. $(a + b)(a - b) = a^2 - b^2$
48. $(a + b)(a + b) = a^2 + 2ab + b^2$

6–7 Problems about Areas

The solution of problems concerned with areas often requires the ability to multiply polynomials. In analyzing such problems, sketches are quite helpful.

Example. The length of a page in a book is 2 inches greater than the width of the page. A book designer finds that if the length is increased by 2 inches and the width by 1 inch, the area of the page is increased by 19 square inches. What are the dimensions of the original page?

Solution: 1.   Let w = number of inches in width of the page.

Then $w + 2$ = number of inches in length of the page.

2. area of enlarged page − area of original page = amount of increase in area

$(w + 1)(w + 4) - w(w + 2) = 19$

3. Solve the equation.

$(w + 1)(w + 4) - w(w + 2) = 19$
$w^2 + 5w + 4 - w^2 - 2w = 19$
$3w + 4 = 19$
$3w = 15$
$w = 5$
$w + 2 = 7$

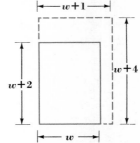

Check: 4. (1) Is the length 2 inches greater than the width?

$$5 + 2 \stackrel{?}{=} 7$$
$$7 = 7$$

(2) Does the area increase by 19 square inches if the length is increased by 2 inches and the width by 1 inch?

$$\underbrace{\text{Area of original page}}_{\downarrow} + 19 = \underbrace{\text{area of enlarged page}}_{\downarrow}$$
$$5 \times 7 + 19 \stackrel{?}{=} 6 \times 9$$
$$35 + 19 \stackrel{?}{=} 54$$
$$54 = 54$$

Width of original page: 5 inches
Length of original page: 7 inches

PROBLEMS

1. The rectangular base of a radio receiver is twice as long as it is wide. A design engineer finds that in order to fit the receiver into the space available on a space vehicle, he must reduce the length by 3 inches and the width by 1 inch. If this change reduces the area of the base by 27 square inches, what were the original dimensions of the receiver?

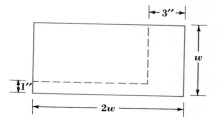

2. The McKay gangplow will plow a strip 3 times as long as it is wide in a given time. The Durley gangplow is 2 feet wider than the McKay plow and will plow 68 square feet more in the same length of time. If the Durley plow can be pulled 4 feet farther than the McKay plow in the given time period, how wide is each plow?

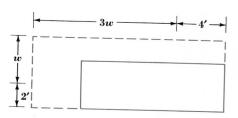

For each problem, make a sketch and solve.

3. A rectangular television picture tube is 3 inches longer than it is wide. If the length and width of the tube are each increased by 2 inches, the picture area is increased by 54 square inches. What are the dimensions of the original tube?

4. A rectangular swimming pool is 5 feet longer than it is wide. If a concrete walk 2 feet wide is placed around the pool, the area covered by the pool and the walk is 156 square feet greater than the area covered by the pool alone. What are the dimensions of the pool?

5. A rectangular picture is 2 inches longer than it is wide. When a frame 3 inches wide is placed around the picture, the area covered by the picture and frame is 168 square inches greater than the area covered by the picture alone. What are the dimensions of the picture?

6. The area occupied by an unframed rectangular picture is 64 square inches less than the area occupied by the picture mounted in a frame 2 inches wide. What are the dimensions of the picture if it is 4 inches longer than it is wide?

7. Two flat pans, one square and the other rectangular in shape, both have the same surface area. If the length of the rectangular pan is 4 inches greater and the width 3 inches shorter than the length of one side of the square pan, what are the dimensions of each pan?

8. Mrs. Carter had a rectangular piece of material with which to make a cover for a couch. The piece was 1 yard longer than 3 times its width. She cut 1 yard off its length and $\frac{1}{2}$ yard off its width to form a new rectangular piece. If the pieces she cut off had a total area of 5 square yards, what were the dimensions of the original piece of material?

9. Mr. Balter wanted to make a plywood rectangular table top that would be 3 feet longer than it was wide. He found that by making the top 1 foot narrower and 2 feet shorter than planned, he could save $2.60 on the plywood. If the plywood he was using cost 20 cents per square foot, what were Mr. Balter's original dimensions for his table top?

10. Steve's first aquarium had a rectangular base that was 8 inches longer than it was wide. He replaced this with a larger aquarium that was 2 inches wider and 4 inches longer than the first one. He found that if he filled each to a depth of 10 inches, 840 cubic inches more water was required to fill the new aquarium than the old. What were the dimensions of the base of the original aquarium?

11. A cross section of a circular concrete pipe has a wall 2 inches thick, and a cross-sectional area of $138\frac{2}{7}$ square inches. Find the inner diameter of the pipe. Use $\frac{22}{7}$ for π.

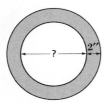 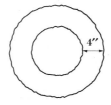

EX. 11 EX. 12

12. When the radius of a circular ripple in a pond increases by 4 inches, the area it encloses increases by 352 square inches. What is the radius of the ripple after the increase? Use $\frac{22}{7}$ for π.

6–8 Powers of Polynomials

The area of a square is found by using the formula $A = s^2$, the area of a circle by using $A = \pi r^2$, and the volume of a cube by using $V = s^3$.

The figure shows a cube in which each edge s is given by $(3y - 2)$. The base of this cube then has an area of

$(3y - 2)^2$, read "the square of $3y - 2$."

The volume is

$(3y - 2)^3$, read "the cube of $3y - 2$."

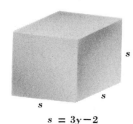

$s = 3y - 2$

In each of these expressions, the exponent shows how many times the polynomial is to be used as a factor. To find the product of these factors expressed as a sum of monomials, it is necessary to perform the indicated multiplication. This computation is called **expanding the expression**. The expansion of $(3y - 2)^2$ and $(3y - 2)^3$ is performed as follows:

$(3y - 2)^2$:
$$\begin{array}{r} 3y - 2 \\ 3y - 2 \\ \hline 9y^2 - 6y \\ -6y + 4 \\ \hline 9y^2 - 12y + 4 = (3y - 2)^2 \end{array}$$

$(3y - 2)^3$:
$$\begin{array}{r} 9y^2 - 12y + 4 \\ 3y - 2 \\ \hline 27y^3 - 36y^2 + 12y \\ -18y^2 + 24y - 8 \\ \hline 27y^3 - 54y^2 + 36y - 8 = (3y - 2)^3 \end{array}$$

EXERCISES

Expand each power. Check your results by numerical evaluation.

1. $(x + 1)^2$
2. $(y + 2)^2$
3. $(a - 3)^2$
4. $(b - 4)^2$
5. $(a + b)^2$
6. $(a - b)^2$
7. $(2x - z)^2$
8. $(2 - 3y)^2$
9. $(a + b)^3$
10. $(a - b)^3$
11. $(2x + z)^3$
12. $(3x - 2y)^3$
13. $(x + \frac{1}{2})^2$
14. $(y - \frac{1}{3})^2$
15. $(a - 0.3)^2$
16. $(b + 1.2)^2$
17. $2x(x + 5)^2$
18. $-3y(y - 2)^2$
19. $(x - y)(x + y)^2$
20. $(a + 2b)(a - 3b)^2$
21. $4x^2 + (x + 2)^2$
22. $3z^2 - (z - 2)^2$
23. $(x + y + z)^2$
24. $(x - y + z)^2$
25. $(x + y)^4$
26. $(x - y)^4$
27. $[(a + b)^2 - (a - b)^2]^2$
28. $[(a + b)^2 + 3(a + b) - 8ab]^2$
29. $(c + d)^3 - (c - d)^3$
30. $(p - q)^3 + (p + q)^3$
31. $(x^n + 3)^3$
32. $(2 - y^n)^3$

PROBLEMS

For each problem, make a sketch where helpful, and solve.

1. The length of a side of one square is 3 inches greater than the length of a side of a second square. If the area of the first square exceeds the area of the second by 51 square inches, find the lengths of the sides of each square.

2. After a square picture is framed with a two-inch border, the picture and frame occupy 96 square inches more of wall space than did the picture alone. Find the dimensions of the picture.

3. The squares of two consecutive integers differ by 27. Find the integers.

4. The squares of two consecutive even integers differ by 52. Find the integers.

5. The difference of the squares of two consecutive odd integers is 40. Find the integers.

6. The squares of two consecutive integers differ by 95. Find the integers.

7. The product of two consecutive integers exceeds the square of the lesser integer by 15. Find the integers.

8. The product of two consecutive odd integers is 30 less than the square of the greater integer. Find the integers.

9. A square window is framed by two rectangular wooden shutters. Each shutter is as long as one side of the window and its width is 18 inches less than its length. If the area covered by one of the shutters is 594 square inches less than the area covered by the window, find the dimensions of one shutter.

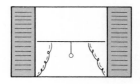

10. A body falling under the influence of gravity falls approximately $16t^2$ feet in t seconds. At the end of what second after starting would a body have fallen 256 feet?

11. A one-inch thick circular metal casting has a circular hole in the center with radius 2 inches less than the outer radius of the casting. If the casting is made of a metal weighing $\frac{7}{18}$ pounds per cubic inch, and the casting weighs $14\frac{2}{3}$ pounds, find the interior and exterior radius of the casting. (Use $\pi \doteq \frac{22}{7}$.)

12. A concrete walk around a circular fishpond is 3 feet wide. At $2 per square foot, the total cost of the walk came to $433. Find the area of the surface of the pond. (Use $\pi = 3.14$.)

6–9 The Quotient of Powers

Recall (page 86) that dividing by a number is the same as multiplying by the reciprocal of the number. Thus,

$$\frac{xy}{cd} = xy \cdot \frac{1}{cd}.$$

We also know (page 68) that

$$\frac{1}{cd} = \frac{1}{c} \cdot \frac{1}{d}.$$

Putting these facts together, we can reason as follows:

$$\begin{aligned}
\frac{xy}{cd} &= xy\left(\frac{1}{c} \cdot \frac{1}{d}\right) && \text{Substitution principle} \\
&= \left(x \cdot \frac{1}{c}\right)\left(y \cdot \frac{1}{d}\right) && \text{Commutative and associative axioms of multiplication} \\
&= \frac{x}{c} \cdot \frac{y}{d} && \text{Rule for division}
\end{aligned}$$

This gives the following theorem:

Property of Quotients

For all real numbers x and y, and nonzero real numbers c and d:

$$\frac{xy}{cd} = \frac{x}{c} \cdot \frac{y}{d}$$

In particular, if $c = 1$, we have

$$\frac{xy}{d} = x \cdot \frac{y}{d};$$

and if $x = 1$, we have

$$\frac{y}{cd} = \frac{1}{c} \cdot \frac{y}{d}.$$

For example:

$$\frac{36 \cdot 35}{6 \cdot 7} = \frac{36}{6} \cdot \frac{35}{7} = 6 \cdot 5 = 30$$

$$\frac{11 \cdot 21}{7} = 11 \cdot \frac{21}{7} = 11 \cdot 3 = 33$$

$$\frac{27}{2 \cdot 9} = \frac{1}{2} \cdot \frac{27}{9} = \frac{1}{2} \cdot 3 = \frac{3}{2}$$

This property of quotients is helpful in simplifying quotients of powers. For example:

$$\frac{a^7}{a^3} = \frac{a^4 \cdot a^3}{a^3} = a^4 \cdot \frac{a^3}{a^3} = a^4 \cdot 1$$

$$\therefore \frac{a^7}{a^3} = a^4 = a^{7-3}$$

Notice that we could have found the exponent in the quotient by retaining the base and subtracting the exponent in the denominator from the exponent in the numerator. Thus, whenever $m > n$, we may write $\left(\text{since } \frac{b^n}{b^n} = 1 \right)$:

$$\frac{b^m}{b^n} = \frac{b^{m-n} \cdot b^n}{b^n} = b^{m-n} \cdot \frac{b^n}{b^n} = b^{m-n}$$

On the other hand:

$$\frac{a^5}{a^9} = \frac{a^5}{a^4 \cdot a^5} = \frac{1}{a^4} \cdot \frac{a^5}{a^5} = \frac{1}{a^4} \cdot 1$$

$$\therefore \frac{a^5}{a^9} = \frac{1}{a^4} = \frac{1}{a^{9-5}}$$

Thus, whenever $m < n$, we may write:

$$\frac{b^m}{b^n} = \frac{b^m}{b^{n-m} \cdot b^m} = \frac{1}{b^{n-m}} \cdot \frac{b^m}{b^m} = \frac{1}{b^{n-m}}$$

These are two important rules:

Rules of Exponents for Division

For all positive integers m and n and nonzero real numbers b:

$$\text{If } m > n, \text{ then } \frac{b^m}{b^n} = b^{m-n}.$$

$$\text{If } m < n, \text{ then } \frac{b^m}{b^n} = \frac{1}{b^{n-m}}.$$

When dividing monomials, use these rules together with the property of quotients on page 147. Thus:

$$\frac{12u^5v^3}{-2uv^2} = \frac{12}{-2} \cdot \frac{u^5}{u} \cdot \frac{v^3}{v^2} = -6 \cdot u^{5-1} \cdot v^{3-2} = -6u^4v$$

$$\frac{-5x^7y^5}{-30x^2y^8} = \frac{-5}{-30} \cdot \frac{x^7}{x^2} \cdot \frac{y^5}{y^8} = \frac{1}{6} \cdot x^{7-2} \cdot \frac{1}{y^{8-5}} = \frac{x^5}{6y^3}$$

Example. $\dfrac{x^2y^3}{x^2y} + \dfrac{6xy^5}{-3xy^3}$

Solution: $\dfrac{x^2y^3}{x^2y} + \dfrac{6xy^5}{-3xy^3} = \dfrac{x^2}{x^2} \cdot \dfrac{y}{y} \cdot y^2 + \dfrac{6}{-3} \cdot \dfrac{x}{x} \cdot \dfrac{y^3}{y^3} \cdot y^2$

$= (1) \cdot (1) \cdot y^2 + (-2) \cdot (1) \cdot (1) \cdot y^2$

$= y^2 + (-2)y^2 = -y^2$

EXERCISES

Simplify each expression, assuming that no divisor is equal to 0.

1. $\dfrac{t^{10}}{t^3}$
2. $\dfrac{s^{12}}{s^3}$
3. $\dfrac{3n^8}{n^8}$
4. $\dfrac{12m^7}{4m^6}$

5. $\dfrac{24x^5y^4}{-6x^3y^3}$
6. $\dfrac{-32t^7s^8}{4t^7s^3}$
7. $\dfrac{-14m^3n^2}{-28m^3n}$
8. $\dfrac{-10a^{10}b^5}{-2a^{12}b^5}$

9. $\dfrac{64p^5q^7}{8p^{10}q^{10}}$
10. $\dfrac{-7x^8y^6}{56x^6y^8}$
11. $\dfrac{48r^{12}s^{10}}{16r^{24}s^{10}}$
12. $\dfrac{14t^{11}s^5}{-42t^{12}s^4}$

13. $\dfrac{(3xy)^2}{6xy^3}$
14. $\dfrac{16a^5b^2}{(2ab)^3}$
15. $\dfrac{(4m^5n^2)^2}{-(2m^2n^2)^3}$
16. $\dfrac{-(3c^3d^2)^3}{6(c^2d^3)^3}$

17. $\dfrac{(-3)^3(x^2)^3(y^2)^2}{(-9)^2(x^3)^3(y^3)^3}$
18. $\dfrac{(-4)^2(a^3)^2(b^4)^2}{(-2)^3(a^3)^3(b^2)^4}$

19. $\dfrac{3t^2}{t} + \dfrac{8t^4}{2t^3}$
20. $\dfrac{70k^5}{10k^2} - \dfrac{36k^4}{6k}$

21. $\dfrac{21c^2d}{3d} - \dfrac{18c^3d^2}{6cd^2}$
22. $\dfrac{45a^3b^2}{9ab} - \dfrac{52a^2b^5}{4b^4}$

23. $\dfrac{24x^5y^3}{-3xy} + \dfrac{8x^4y^6}{2y^4}$
24. $\dfrac{38p^3q^5}{19pq^2} - \dfrac{15p^4q^4}{-3p^2q}$

25. $\dfrac{3c^2d + 9c^2d}{4c} - 2cd$
26. $8x^2y + \dfrac{8x^4y^2 - 12x^4y^2}{x^2y}$

27. $\dfrac{3a^3b^2 + 15a^3b^2}{2ab^2} + \dfrac{4a^6b^3 - 10a^6b^3}{2a^4b^3}$

28. $\dfrac{4rs^3 + 16rs^3}{4s^2} - \dfrac{3r^3s^5 + 11r^3s^5}{2r^2s^4}$

29. $\dfrac{14m^5n^4 - 8m^5n^4}{6mn^2} - \dfrac{8m^6n^2 + m^6n^2}{3m^2}$

30. $\dfrac{-11c^5d^4 + 5c^5d^4}{2c^2d^2} - \dfrac{3c^6d^3 + 5c^6d^3}{4c^3d}$

6–10 Zero and Negative Exponents

In the rules of exponents for division on page 148, $m \neq n$. However, if we were to apply those rules to evaluate $\dfrac{b^m}{b^m}$, we would obtain two expressions:

$$\frac{b^m}{b^m} = b^{m-m} \cdot 1 = b^0 \qquad \text{and} \qquad \frac{b^m}{b^m} = \frac{1}{b^{m-m}} = \frac{1}{b^0}$$

Of course, $\dfrac{b^m}{b^m} = 1$, and so it appears that b^0 ought to be 1.

From these considerations, we make this definition of b^0 (read "b exponent zero"):

b^0 **is 1 for every nonzero real number** b.

We shall not use the expression 0^0.

If we are to apply the rule $\dfrac{b^m}{b^n} = b^{m-n}$ when $m < n$, that is, when $m - n$ represents a negative number, we must give meaning to powers with negative exponents. Suppose we apply the rule as follows:

$$\frac{b^5}{b^2} = b^{5-2} = b^3 \qquad \frac{b^2}{b^5} = b^{2-5} = b^{-3}$$

Since $\dfrac{b^5}{b^2}$ and $\dfrac{b^2}{b^5}$ are reciprocals, b^3 and b^{-3} should be reciprocals; that is, it should be true that $b^3 = \dfrac{1}{b^{-3}}$ and $b^{-3} = \dfrac{1}{b^3}$.

In general, we define powers with negative exponents as follows:

$$b^n = \frac{1}{b^{-n}} \qquad \text{and} \qquad b^{-n} = \frac{1}{b^n} \quad (b \neq 0)$$

The rules of exponents may now be extended to include negative exponents. For example, $a^3 \cdot a^{-2} = a^{3+(-2)} = a^1 = a$.

Examples:

1. $a^3 \cdot b^{-2} = a^3 \cdot \dfrac{1}{b^2} = \dfrac{a^3}{b^2}$

2. $\dfrac{x^5}{y^{-3}} = x^5 \cdot \dfrac{1}{y^{-3}} = x^5 y^3$

3. $\dfrac{x^0 y^{-2}}{z^2} = x^0 \cdot y^{-2} \cdot \dfrac{1}{z^2} = 1 \cdot \dfrac{1}{y^2} \cdot \dfrac{1}{z^2} = \dfrac{1}{y^2 z^2}$

4. $\dfrac{12 \times 10^8}{3 \times 10^{11}} = 4 \times 10^{-3} = 0.004$

OPERATIONS WITH POLYNOMIALS

EXERCISES

Write an equivalent expression using only positive exponents. Assume that 0 is not a member of the replacement set of any of the variables.

1. xy^{-3}
2. $a^{-2}b$
3. $\dfrac{r^{-2}}{s}$
4. $\dfrac{t}{s^{-2}}$
5. $\dfrac{x^0 z^{-2}}{u^{-2}}$
6. $\dfrac{(-2)^0 y^{-4}}{z^{-3}}$
7. $(2z)^0$
8. $(x^2 z)^0$
9. $(r^{-2}s)^{-3}$
10. $(x^2 y^{-3})^{-2}$
11. $\left(\dfrac{2}{z}\right)^{-1}$
12. $\left(\dfrac{x^2 y}{2}\right)^{-1}$

Express each fraction as a product of powers.

Example 1. $\dfrac{x^{-3}}{y^2} = x^{-3} y^{-2}$

13. $\dfrac{a^2}{b^2}$
14. $\dfrac{t^3}{s^4}$
15. $\dfrac{x^{-4}}{y^{-3}}$
16. $\dfrac{m^2}{n^{-4}}$
17. $\dfrac{a^{-3}b^2}{cd^{-4}}$
18. $\dfrac{pq^{-3}}{m^2 n}$

State a polynomial in simple form equivalent to each expression.

Example 2. $\dfrac{2x}{x^{-3}} + 3^0 x^4 = 2x^4 + x^4 = 3x^4$

19. $3y^2 + \dfrac{1}{y^{-2}}$
20. $4t^3 + \dfrac{2t^2}{t^{-1}}$
21. $n^0 m^3 - \dfrac{m^5}{m^2}$
22. $\dfrac{x^{-3}}{x^{-7}} + \dfrac{4x^2}{x^{-2}}$
23. $\dfrac{1}{(xy)^{-3}} + \dfrac{4x^3 y}{y^{-2}}$
24. $\dfrac{-2}{(rs^2)^{-1}} + \dfrac{5rs}{s^{-1}}$
25. $\dfrac{3xz^{-1}}{z^{-1}} - \dfrac{5x^{-2}z}{x^{-3}z}$
26. $\dfrac{-4a^3}{-2a^{-2}b^{-3}} - \dfrac{7ab}{a^{-4}b^{-2}}$
27. $\dfrac{8 \times 10^7}{4 \times 10^{-2}}$
28. $\dfrac{1.6 \times 10^{-3}}{0.4 \times 10^{-6}}$
29. $\left(\tfrac{3}{5}\right)^{-2} + \left(\tfrac{4}{3}\right)^2$
30. $\left(\tfrac{4}{7}\right)^2 - \left(\tfrac{7}{4}\right)^{-2}$
31. $(x + 2y)^0 + \dfrac{1}{(x-y)^{-1}}$
32. $(a+2)^2 - \dfrac{1}{(a-3^{-1})}$

6-11 Dividing a Polynomial by a Monomial

One way to simplify the expression $(84 + 28) \div 7$ is to use the distributive axiom:

$$\frac{84 + 28}{7} = \frac{1}{7}(84 + 28) = \frac{1}{7}(84) + \frac{1}{7}(28) = 12 + 4 = 16$$

Similarly, we may simplify the algebraic expression $(ax + ay) \div a$:

$$\begin{aligned}
\frac{ax + ay}{a} &= \frac{1}{a}(ax + ay) \\
&= \frac{1}{a}(ax) + \frac{1}{a}(ay) \\
&= \left(\frac{1}{a} \cdot a\right)x + \left(\frac{1}{a} \cdot a\right)y \\
&= 1x + 1y \\
&= x + y
\end{aligned}$$

The effect of this procedure is that of *dividing* each term of the polynomial $ax + ay$ by the monomial a.

Example 1. $\dfrac{16x^4 + 12x^3 - 8x^2}{4x^2}$

Solution: $\dfrac{16x^4 + 12x^3 - 8x^2}{4x^2} = \dfrac{16x^4}{4x^2} + \dfrac{12x^3}{4x^2} - \dfrac{8x^2}{4x^2}$

$= 4x^2 + 3x - 2$

Example 2. $\dfrac{bt^2 + t - b}{bt}$

Solution: $\dfrac{bt^2 + t - b}{bt} = \dfrac{bt^2}{bt} + \dfrac{t}{bt} - \dfrac{b}{bt} = t + \dfrac{1}{b} - \dfrac{1}{t}$

Note that $16x^4 + 12x^3 - 8x^2$ is evenly divisible by $4x^2$, but $bt^2 + t - b$ is not evenly divisible by bt. We say that one polynomial is **evenly divisible** or simply **divisible** by another polynomial if the quotient is also a polynomial.

To divide a polynomial by a monomial, divide each term of the polynomial by the monomial, and then add the quotients.

EXERCISES

Simplify each expression by performing the indicated division.

1. $\dfrac{6a + 12}{3}$
2. $\dfrac{8t + 16}{4}$
3. $\dfrac{18 - 9z}{9}$
4. $\dfrac{24 - 15k}{5}$
5. $\dfrac{11z^2 + 22z}{11z}$
6. $\dfrac{14k^2 - 12k}{2k}$
7. $\dfrac{8p^3 - 2p^2}{2p^2}$
8. $\dfrac{11g^2 + 33g^3}{11g^2}$
9. $\dfrac{3t + 5}{t}$
10. $\dfrac{8x + 6}{x}$
11. $\dfrac{12k + 3}{6k}$
12. $\dfrac{14 - 3t}{3t}$
13. $\dfrac{3m - n}{2n}$
14. $\dfrac{6k - 3r}{2k}$
15. $\dfrac{8x + 4y}{2xy}$
16. $\dfrac{3a - 6b}{3ab}$
17. $\dfrac{16x^2 - 12x + 8}{-4}$
18. $\dfrac{27z^2 + 18z - 36}{-9}$
19. $\dfrac{3x^2y + 6xy^2 - 9x^2y^2}{3xy}$
20. $\dfrac{40pq^2 + 30p^2q - 20p^2q^2}{10pq}$
21. $\dfrac{8z^5 - 32z^4 + 16z^3}{-8z^4}$
22. $\dfrac{28u^7 - 16u^5 + 20u^3}{-4u^5}$
23. $\dfrac{14x^2 - 18x}{2x} + \dfrac{15x^2 - 25x}{5x}$
24. $\dfrac{25y^3 - 15y^2 + 30y}{-5y} + \dfrac{8y^5 - 3y^3}{y^3}$
25. $\dfrac{35k^2t - 28kt + 7kt^2}{7kt} - \dfrac{15k^2t^2 - 21k^2t}{3k^2t}$
26. $\dfrac{40cd^2 - 32c^2d + 24c^2d^2}{-8cd} + \dfrac{4c^3d - 12c^2d^2}{3c^2d}$
27. $\dfrac{21x^4y^2 - 14x^3y}{-7x^2y} + \dfrac{16x^5y^3 + 8x^4y^2}{4x^3y^2} + \dfrac{x^4y^4 - x^5y^4}{-x^3y^3}$
28. $\dfrac{-4a^3b + 8ab^2}{-2ab} - \dfrac{12a^4b^3 + 9a^2b^4}{3a^2b^3} + \dfrac{2a^3b^3 - 4ab^4}{2ab^3}$
29. If $\dfrac{x + y}{y} = 2.6$, find $\dfrac{x}{y}$.
30. If $\dfrac{2a + 4b}{2b} = 3.9$, find $\dfrac{a}{b}$.

Solve each equation over $\mathcal{R}$.

31. $\dfrac{6x^2 - 4x}{2x} = 3x - 5$
32. $\dfrac{8y^2 + 6y}{2y} = y + 3$
33. $\dfrac{18n - 9n^2}{3n} = \dfrac{n^3 + 2n^2}{n^2}$
34. $\dfrac{14z - 21z^2}{7z} = \dfrac{2z^3 + 4z^2}{2z^2}$

6–12 Dividing a Polynomial by a Polynomial

The adjoining example shows the use of the division algorithm (that is, procedure for dividing) in arithmetic.

Divide 334 by 15.

1. The method uses repeated subtraction. First 20(15), then 2(15) is subtracted from 334.

2. The distributive axiom helps shorten the number of steps. Without it you would have to subtract 15 from 334 twenty-two times. The division ends when the remainder is either zero or a positive number less than the divisor.

3. The check is a transformation of the division equation,

$$334 - (22)(15) = 4,$$

to

$$334 = (22)(15) + 4.$$

$$\begin{array}{r} \text{Quotient} \\ 2 \Big\} \;\; 22 \\ 20 \Big\} \\ \text{Divisor} \to 15\overline{)334} \leftarrow \text{Dividend} \\ \underline{300} \\ 34 \\ \underline{30} \\ \text{Remainder} \to 4 \end{array}$$

Check:
$$\begin{array}{r} 15 \leftarrow \text{Divisor} \\ \underline{22} \leftarrow \text{Quotient} \\ 30 \\ \underline{300} \\ 330 \\ \underline{+4} \leftarrow \text{Remainder} \\ 334 \leftarrow \text{Dividend} \end{array}$$

In general form, these statements are:

$$\textbf{Dividend} - \textbf{Quotient} \times \textbf{Divisor} = \textbf{Remainder}$$

and

$$\textbf{Dividend} = \textbf{Quotient} \times \textbf{Divisor} + \textbf{Remainder}$$

Each of these equations is equivalent to a third,

$$\frac{\textbf{Dividend}}{\textbf{Divisor}} = \textbf{Quotient} + \frac{\textbf{Remainder}}{\textbf{Divisor}},$$

which for the example shown is $\frac{334}{15} = 22\frac{4}{15}$.

It is this last form which gives the **complete quotient**, $22\frac{4}{15}$. For contrast, 22 is sometimes called the "partial quotient."

When dividing polynomials, write the terms of the divisor and the dividend in order of decreasing degree in a chosen variable and follow essentially the same algorithm as was used in arithmetic.

The following three examples illustrate how to divide polynomials under various circumstances.

OPERATIONS WITH POLYNOMIALS 155

Example 1. $y + 3 \overline{) 2y^2 + 10y + 12}$

Solution:

$$2y^2 \div y = 2y$$

$$\begin{array}{r} 2y \\ y+3 \overline{) 2y^2 + 10y + 12} \\ \underline{2y^2 + 6y} \\ 4y + 12 \end{array}$$ ← Subtract $2y(y+3)$

$4y \div y = 4$

$$\begin{array}{r} 2y + 4 \\ y+3 \overline{) 2y^2 + 10y + 12} \\ \underline{2y^2 + 6y} \\ 4y + 12 \\ \underline{4y + 12} \\ 0 \end{array}$$ ← Subtract $4(y+3)$

$$\frac{2y^2 + 10y + 12}{y + 3} = 2y + 4$$

Check:
$$(2y + 4)(y + 3) + 0 \stackrel{?}{=} 2y^2 + 10y + 12$$
$$2y^2 + 10y + 12 = 2y^2 + 10y + 12$$

In Example 2, the steps in the division are shown compactly. This example also shows how to insert missing terms in a dividend, using 0 as a coefficient.

Example 2. $x^2 + x - 3 \overline{) x^3 - 3}$

Solution:

$$\begin{array}{r} x - 1 \\ x^2 + x - 3 \overline{) x^3 + 0x^2 + 0x - 3} \\ \underline{x^3 + x^2 - 3x} \\ -x^2 + 3x - 3 \\ \underline{-x^2 - x + 3} \\ 4x - 6 \end{array}$$

$$\frac{x^3 - 3}{x^2 + x - 3} = x - 1 + \frac{4x - 6}{x^2 + x - 3}$$

Check:
$$(x^2 + x - 3)(x - 1) + (4x - 6) \stackrel{?}{=} x^3 - 3$$
$$x^3 - 4x + 3 + (4x - 6) \stackrel{?}{=} x^3 - 3$$
$$x^3 - 3 = x^3 - 3$$

When does the division algorithm end? It ends when *the remainder is zero* or *the degree of the remainder in the chosen variable is less than that of the divisor.*

Example 3. $\dfrac{16a^2 + 46ab + 10b^2}{2a + 5b}$

Solution:

$$\begin{array}{r} 8a + 3b \\ 2a + 5b \overline{\smash{\big)}\,16a^2 + 46ab + 10b^2} \\ \underline{16a^2 + 40ab} \\ 6ab + 10b^2 \\ \underline{6ab + 15b^2} \\ -\ 5b^2 \end{array}$$

$$\dfrac{16a^2 + 46ab + 10b^2}{2a + 5b} = (8a + 3b) - \dfrac{5b^2}{2a + 5b}$$

EXERCISES

Corresponding to each quotient, write an equation of the form

$$\dfrac{\text{Dividend}}{\text{Divisor}} = \text{Quotient} + \dfrac{\text{Remainder}}{\text{Divisor}}$$

and check your results.

1. $\dfrac{y^2 + 3y + 2}{y + 1}$
2. $\dfrac{n^2 + 6n + 5}{n + 1}$
3. $\dfrac{x^2 - 5x + 6}{x - 2}$
4. $\dfrac{z^2 + 3z - 4}{z + 4}$
5. $\dfrac{a^2 - 9a + 20}{a - 4}$
6. $\dfrac{b^2 + 8b - 20}{b - 2}$
7. $\dfrac{x^2 + 2x + 3}{x + 1}$
8. $\dfrac{t^2 - 2t + 2}{t - 2}$
9. $\dfrac{-28 - 3x + x^2}{x - 7}$
10. $\dfrac{-24 - 5r + r^2}{r + 3}$
11. $\dfrac{m^2 - 4}{m + 2}$
12. $\dfrac{q^2 - 36}{q - 6}$
13. $\dfrac{2x^2 + 3x - 2}{2x - 1}$
14. $\dfrac{3y^2 + y - 2}{3y - 2}$
15. $\dfrac{6p^2 - 3p + 2}{2p - 3}$
16. $\dfrac{8d^2 - 3d + 2}{2d - 1}$
17. $\dfrac{x^3 - 8y^3}{x - 2y}$
18. $\dfrac{8a^3 + 27b^3}{2a + 3b}$
19. $\dfrac{8t^2 + 2ts - 15s^2}{4t - 5s}$
20. $\dfrac{18k^2 - 3rk - 10r^2}{6k - 5r}$

21. $\dfrac{64n^3 - 125m^3}{4n - 5m}$

22. $\dfrac{x^4 - 16y^4}{x - 2y}$

23. $x + 4 \overline{\smash{)}x^3 + 2x^2 - 5x + 12}$
24. $y - 3 \overline{\smash{)}2y^3 - 9y^2 - 8y - 3}$
25. $2r + 3 \overline{\smash{)}6r^3 + r^2 - 2r + 17}$
26. $3k - 2 \overline{\smash{)}12k^3 - 17k^2 + 21k - 8}$
27. $x^2 + 2x - 3 \overline{\smash{)}x^4 + 7x^3 + 9x^2 - 11x - 6}$
28. $z^2 - 2z + 2 \overline{\smash{)}2z^4 - z^3 - 8z^2 + 18z - 12}$
29. $x^2 - 1 \overline{\smash{)}x^8 - 1}$
30. $y^2 - 2 \overline{\smash{)}y^{10} + 20}$
31. One factor of $x^3 - 6x^2 + 11x - 12$ is $x - 4$. Find the other factor.
32. One factor of $x^6 + 1$ is $x^2 + 1$. Find the other factor.
33. Is $3x + 2$ a factor of $3x^3 - 4x^2 - x + 4$? Justify your answer.
34. Is $2y + 3$ a factor of $12y^3 + 18y^2 - 10y - 15$? Justify your answer.
★35. Find the number c for which $x + 3$ is a factor of $3x^2 + 2x + c$. *Hint:* The remainder when $3x^2 + 2x + c$ is divided by $x + 3$ must be 0.
★36. Find the number c for which $3y - 5$ is a factor of $6y^2 + 11y + c$.

chapter summary

1. To **add** polynomials, write the sum and add the similar terms.
   To **subtract** one polynomial from another, add to the minuend the opposite of each term of the subtrahend.
   Check your work with polynomials by substituting a particular value (except 0 and 1) for each variable and evaluating each expression.

2. To **multiply two powers with the same base**, use the rule: For all positive integers m and n, $b^m \cdot b^n = b^{m+n}$.
   To **multiply monomials**, multiply the numerical factors and the variable factors, applying the preceding rule together with the commutative and associative axioms of multiplication.

3. When the base of a power is a product, the rule $(ab)^m = a^m b^m$ applies for every positive integer m.
   When the base is itself a power, then for all positive integers m and n, $(b^m)^n = b^{mn}$.

4. To **multiply a polynomial by a monomial**, apply the distributive axiom, multiplying each term of the polynomial by the monomial and then adding the products.
To **multiply a polynomial by a polynomial**, use the distributive axiom to multiply one polynomial by each term of the other, and then add the products.

5. **Expand a power** of a polynomial by using the polynomial as a factor as many times as shown by the exponent and by performing the indicated multiplications.

6. To **divide two powers with the same base**, use the rules: For positive integers m and n and nonzero real numbers b, if $m > n$, $\dfrac{b^m}{b^n} = b^{m-n}$, and if $m < n$, $\dfrac{b^m}{b^n} = \dfrac{1}{b^{n-m}}$.

The **Property of Quotients** is: For all real numbers x and y, and nonzero real numbers c and d: $\dfrac{xy}{cd} = \dfrac{x}{c} \cdot \dfrac{y}{d}$

To **divide monomials**, divide the numerical factors and the variable factors, applying the preceding property together with the rules of exponents.

7. To **divide a polynomial by a monomial**, divide each term of the polynomial by the monomial, and then add the quotients.
To **divide a polynomial by a polynomial**, arrange the terms of the divisor and the dividend in order of decreasing degree in one variable, and then proceed as in arithmetic division. The process stops when the remainder is 0, or when its degree is *less* than that of the divisor. In general:

$$\text{Complete Quotient} = \frac{\text{Dividend}}{\text{Divisor}} = \text{Quotient} + \frac{\text{Remainder}}{\text{Divisor}}$$

8. By definition, for any nonzero real number b, $b^0 = 1$, $b^{-n} = \dfrac{1}{b^n}$, and $b^n = \dfrac{1}{b^{-n}}$.

CHAPTER REVIEW

6–1 Simplify each expression for a sum. Check your work by using the value 2 for the variable.

1. $(17y + 100) + (21y - 30)$

2. $\begin{array}{r} 5x^2 + 3x - 7 \\ -7x^2 + 2x - 8 \\ \hline \end{array}$

6–2
Subtract the lower polynomial from the one above it, and check your work by using the values 2, 3, and 5 for a, b, and c.

3. $4a - 2b + c$
   $\underline{a - 3b + 2c}$

4. $a^2 - ab + b^2$
   $\underline{2a^2 - ab - b^2}$

Solve each equation over $\mathcal{R}$.

5. $7x - (3 - x) = 29$

6. $3t - [4 + (t - 2)] = 0$

Simplify each expression for a product.

6–3 7. $(-3x^2y^3z)(-x^4z^2)$

8. $(2a^3b^2)(-3a^2)(-b)^3$

6–4 9. $(-3r^2t^3s)^3$

10. $(-2k)^3(kr^2)^2$

6–5 11. $-3s(5s - 6)$

12. $2x^2(x^2 - 3x - 2)$

6–6 13. $(t - 4)(t + 1)$

14. $(3x + 5y)(4x - y)$

6–7 15. A rectangular picture is 3 inches longer than it is wide. It has a frame 2 inches wide whose area is 92 square inches. Find the dimensions of the picture.

6–8 16. One number is 4 greater than another, and the difference of their squares is 64. Find the numbers.

6–9 17. Simplify: $\dfrac{-16x^3y^2z}{4x^2yz^3}$

6–10 Write each expression using positive exponents only.

18. $\dfrac{t^0 s^{-2}}{k}$

19. $\dfrac{(-3)^0 x^4 y^{-1}}{x^{-1} y z^{-2}}$

Simplify each expression for a quotient.

6–11 20. $\dfrac{26t^3 + 8t}{2t}$

21. $\dfrac{-8x^3 + 12x^2 - 16x}{4x}$

6–12 22. $\dfrac{6x^2 - 7x - 3}{2x - 3}$

23. $\dfrac{x^3 - 5x^2 + 10x - 12}{x - 3}$

7 special products and factoring

7-1 Factoring in Algebra

It is often necessary to know whether a number can be expressed as a product of two or more *factors*. Because

$$18 = 2 \times 9 \quad \text{and} \quad 18 = 3 \times 6,$$

2 and 9, and 3 and 6 are factors of 18. Of course, $18 = \frac{1}{2} \times 36$, so that $\frac{1}{2}$ and 36 might also be called factors of 18, and 5 could be called a factor of 18 since $18 = 5 \times \frac{18}{5}$. However, if no restrictions are placed on the kind of numbers allowed as factors, every nonzero real number is a factor of every real number. Therefore, a particular set of numbers to be used as factors is normally specified.

Finding numbers belonging to a given set of numbers and having their product equal to a given number is called **factoring the number over the given set**. In later work, *integers will be factored over the set of integers, unless another set is specified.*

However, for some purposes, an important subset of the integers is chosen as a set of possible factors, namely, the set of *prime numbers*. A **prime number** is an integer *greater than one*, having no positive integral factor other than itself and one. The first prime numbers are

$$2, 3, 5, 7, 11, 13, 17, 19, 23, \ldots.$$

To factor 18 over the set of primes, write

$$18 = 2 \cdot 3 \cdot 3 = 2 \cdot 3^2.$$

The **prime factors** of 18 are 2 and 3, with 3 occurring twice.

SPECIAL PRODUCTS AND FACTORING 161

To express a positive integer as a product of primes, continue factoring until each factor cannot be factored further. For example:

$$
\begin{aligned}
196 &= 7 \cdot 28 & \text{or} \quad 196 &= 2 \cdot 98 \\
&= 7 \cdot 7 \cdot 4 & &= 2 \cdot 2 \cdot 49 \\
&= 7 \cdot 7 \cdot 2 \cdot 2 & &= 2 \cdot 2 \cdot 7 \cdot 7 \\
&= 7^2 \cdot 2^2 & &= 2^2 \cdot 7^2
\end{aligned}
$$

In the second method (shown at the right above), look systematically for the smallest prime factor of the nonprime factor at each stage. That is, first try 2, and try it again and again until it no longer can be used. Then try 3, then 5, then 7, and so on until all the factors shown are prime numbers.

It can be proved that the prime factorization of a positive integer is *unique*, except for the order in which the factors may be written. If 1 were considered as a prime factor, the factorization would not be unique, because any number of factors 1 could be written:

$$196 = 2^2 \cdot 7^2 \cdot 1 = 2^2 \cdot 7^2 \cdot 1^2 = 2^2 \cdot 7^2 \cdot 1^3 = \cdots$$

Once we know the prime factors of an integer, it is easy to list all its positive integral factors. The (positive integral) factors of 196, for example, are 1 and all possible products of one or both of the primes 2 and 7, each with an exponent less than or equal to its exponent in 196. These factors are

or

$$
\begin{array}{lllllllll}
1, & 2, & 2^2, & 7, & 2 \cdot 7, & 2^2 \cdot 7, & 7^2, & 2 \cdot 7^2, & 2^2 \cdot 7^2, \\
1, & 2, & 4, & 7, & 14, & 28, & 49, & 98, & 196.
\end{array}
$$

By factoring integers into products of primes, the greatest integral factor of both of two integers can be determined. To find the **greatest common factor** of 196 and 1260, notice that

$$196 = 2^2 \cdot 7^2 \quad \text{and} \quad 1260 = 2^2 \cdot 3^2 \cdot 5 \cdot 7.$$

Thus, the *prime* factors common to 196 and 1260 are 2 and 7. The greatest power of 2 common to 196 and 1260 is 2^2, and the greatest common power of 7 is 7. The greatest common factor is, therefore $2^2 \cdot 7$, or 28.

In algebra we often need to express a polynomial as a product of **polynomial factors**. Expressing a given polynomial in this way is called **polynomial factoring**. For example, each term of the polynomial $5x + 5y$ contains 5 as a factor. Therefore, by the distributive axiom,

$$5x + 5y = 5(x + y).$$

Both 5 and $x + y$ are factors of $5x + 5y$.

When factoring polynomials whose numerical coefficients are integers, we look for factors that are either integers or polynomials with integral coefficients. Some of the factors of $6x^2y$ are 1, 2, 3, 6, $2x^2$, and $3xy$; however, $2x^3$ is not a factor of $6x^2y$, since there is no polynomial by which we can multiply $2x^3$ to obtain $6x^2y$. (Recall that $3x^{-1}y$ is not a monomial.)

EXERCISES

Factor each integer over the set of primes.

1. 34
2. 52
3. 63
4. 95
5. 144
6. 242
7. 19
8. 1024

Find all the positive integral factors of each number.

9. 15
10. 19
11. 24
12. 25
13. 30
14. 108
15. 444
16. 1000

Find the greatest common factor of each pair of integers.

17. 24; 60
18. 45; 105
19. 315; 350
20. 252; 288
21. 693; 882
22. 1925; 2100

Find the monomial with the greatest numerical coefficient and the greatest degree in each variable that is a factor of both monomials in each pair.

23. $154a^2b$; $132ab^4$
24. $78x^3y^2$; $52x^2y^4$
25. $126x^2yz^3$; $105x^3yz^2$
26. $108a^3b^2c$; $72a^2bc^3$
27. $176r^2s^3$; $208s^2t^3$
28. $132m^4n^3$; $156n^2p^3$

Give the second factor for each monomial.

29. $6a^2b = 6a(\underline{\ ?\ })$
30. $24r^2s = (12rs)(\underline{\ ?\ })$
31. $-15x^3y^2 = (-5x^2y^2)(\underline{\ ?\ })$
32. $-32p^3q^4 = (16p^3q^2)(\underline{\ ?\ })$
33. $42u^3v^2w = (-6u^2vw)(\underline{\ ?\ })$
34. $51x^4y^2z^3 = (-17x^3y^2z^2)(\underline{\ ?\ })$
35. $72r^3s^5t^2 = (18r^3s^4t^2)(\underline{\ ?\ })$
36. $102a^3b^2c = (-17a^3b)(\underline{\ ?\ })$

For each pair of monomials, find the highest power of the first monomial that is a factor of the second monomial.

Example. $3c$; $27c^2d$

Solution: $(3c)^1 = 3c$, and $27c^2d = (3c)(9cd)$
$(3c)^2 = 9c^2$, and $27c^2d = (9c^2)(3d)$
$(3c)^3 = 27c^3$, but c^3 is *not* a factor of c^2.
∴ the highest power is $(3c)^2$.

37. z; $8z^4t$
38. x^2; $7x^5n$
39. $2t$; $24t^3v^2$
40. $3n$; $81n^4p^2$
41. $5xy$; $75x^3y^2$
42. $3m^2n$; $54m^7n^4$
43. $4a^3b^2$; $128a^{15}b^9$
44. $6cd^2$; $108c^2d^8$

7-2 Identifying Monomial Factors

Because each term in $6xy + 9x$ has $3x$ as a factor, we can use the distributive axiom to write

$$6xy + 9x = 3x(2y + 3).$$

Since $3x$ is a monomial, we say that $3x$ is a *monomial factor* of the polynomial $6xy + 9x$. When we factor a polynomial, we first see whether each term has the same monomial as a factor. A monomial is a **monomial factor** of a polynomial if it is a factor of *every* term of the polynomial.

Notice that 3 and x are also monomial factors of $6xy + 9x$. We should be sure to continue factoring until we find the *greatest monomial factor*. The **greatest monomial factor** of a polynomial is the monomial factor having the greatest numerical coefficient and the greatest degree.

Examine this chart, observing how each polynomial is factored.

Given Polynomial	Factors		Factored Expression
$2a^2 + 5a$	a	$2a + 5$	$a(2a + 5)$
$7t^2 + 28t$	$7t$	$t + 4$	$7t(t + 4)$
$5x^4 - 15x^3 + 35x^2$	$5x^2$	$x^2 - 3x + 7$	$5x^2(x^2 - 3x + 7)$
$12c^3d^2 + 36c^2d$	$12c^2d$	$cd + 3$	$12c^2d(cd + 3)$

The associative and commutative axioms for addition, together with the distributive axiom, may enable us to factor a polynomial by *grouping terms*:

$$ax + by + ay + bx = (ax + bx) + (ay + by)$$
$$= (a + b)x + (a + b)y$$
$$= (a + b)(x + y)$$

In the last step, $(a + b)$ is treated as a single term in applying the distributive axiom.

Here is another polynomial that can be factored readily when we group the terms appropriately:

$$2ax - 3by - 6ay + bx = (2ax - 6ay) + (bx - 3by)$$
$$= 2a(x - 3y) + b(x - 3y)$$
$$= (2a + b)(x - 3y)$$

Of course, there may be more than one convenient way to group terms:

$$2ax - 3by - 6ay + bx = (2ax + bx) + (-6ay - 3by)$$
$$= x(2a + b) + (-3y)(2a + b)$$
$$= [x + (-3y)](2a + b)$$
$$= (x - 3y)(2a + b)$$

Because multiplication is commutative, this result is the same as the preceding one.

EXERCISES

Write in factored form. Check by multiplying the factors.

1. $4x^2 - 8$
2. $6 + 9y$
3. $4a^2 + 5a$
4. $6b^2 + 7b$
5. $10z^3 - 5z^2$
6. $12n^2 + 24n^3$
7. $6a^2b + 3ab^2$
8. $7rs^3 - 14r^2s$
9. $3p^2 + 3p - 9$
10. $4 - 6q + 10q^2$
11. $5n^2 - 15n + 20$
12. $6 - 3p + 18p^2$
13. $a^3x^3 + a^2x^2 - ax$
14. $b^4y^2 + b^3y^3 - b^2y^4$
15. $4b^2y^2 - 8b^2y + 24b^2$
16. $35k^2 - 42k^2t + 14k^2t^2$
17. $15x^2y - 30xy + 35xy^2$
18. $-18u^3v^2 + 12u^2v^2 - 6uv^2$
19. $9x^5y^2 - 6x^4y^3 + 3x^3y^4$
20. $-12x^3y^3 + 18x^2y^4 + 27xy^5$

Write each expression in factored form, and check.

21. $n(n - 1) + 3(n - 1)$
22. $x(2x + 3) + 2(2x + 3)$
23. $(3a - b)b + (3a - b)a$
24. $(c + 3d)(2c) + (c + 3d)(3d)$
25. $t^2(y + 5) - 5(y + 5)$
26. $k^2(t + 1) + 2k(t + 1)$
27. $n^2(2n + 1) + (2n + 1)$
28. $2m^2(3m + 1) + (3m + 1)$
29. $a^2(a - b) + b(a - b)$
30. $2n(n^2 + 1) + 3(n^2 + 1)$
31. $5c(a^3 + b) - (a^3 + b)$
32. $m(m + 2n) - n(m + 2n)$
★33. $n^2 + 2n + np + 2p$
★34. $k^2 + 3k + 2k + 6$
★35. $3ab - b^2 + 3a^2 - ab$
★36. $6y^2 - 3y + 2py - p$
★37. $n^2m + 2nm + 2n + n^2$
★38. $4x + 8x^3 + 1 + 2x^2$

SPECIAL PRODUCTS AND FACTORING 165

PROBLEMS

Write an algebraic expression in factored form for the area A of the shaded region shown.

Example.

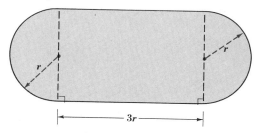

Solution:

A = Area of rectangle + area of circle
$ = (3r)(2r) + \pi r^2$
$ = 6r^2 + \pi r^2$
$ = (6 + \pi)r^2$

1.

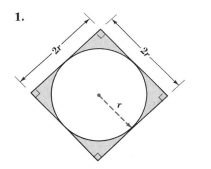

2.

3.

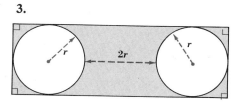

4.

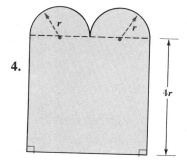

5.

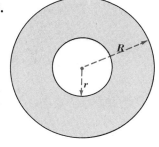

6.

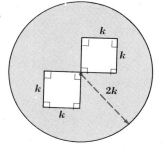

7. 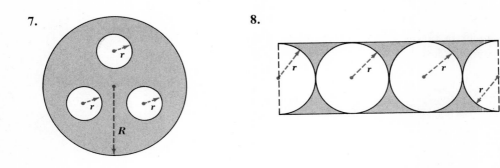 8.

7-3 Multiplying the Sum and Difference of Two Numbers

Certain products occur so often that they should be recognized at sight. Study each of the three examples below:

$$\begin{array}{ccc}
y + 2 & 3a - 2b & a + b \\
y - 2 & 3a + 2b & a - b \\
\hline
y^2 + 2y & 9a^2 - 6ab & a^2 + ab \\
 - 2y - 4 & 6ab - 4b^2 & - ab - b^2 \\
\hline
y^2 - 4 & 9a^2 - 4b^2 & a^2 - b^2
\end{array}$$

These examples illustrate this rule:

The product of the sum and difference of two numbers is the square of the first number minus the square of the second number; that is,

$$(a + b)(a - b) = a^2 - b^2.$$

Figure 7-1 shows that the area represented by the product $(a + b)(a - b)$ is the same as the area represented by $a^2 - b^2$.

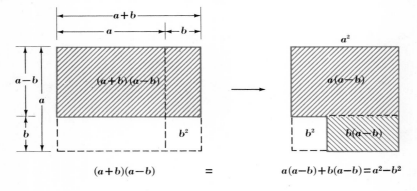

$(a+b)(a-b)$ = $a(a-b)+b(a-b)=a^2-b^2$

FIGURE 7-1

SPECIAL PRODUCTS AND FACTORING 167

EXERCISES

Multiply: **a.** by using the form $(a + b)(a - b)$;
b. by using the distributive axiom.

Example. $(58)(62)$

Solution: **a.** $(58)(62) = (60 - 2)(60 + 2) = 3600 - 4 = 3596$
b. $(58)(62) = 50(62) + 8(62) = 3100 + 496 = 3596$

Ordinary multiplication, which uses the distributive axiom, can also be used for part **b**.

1. $(8)(12)$
2. $(7)(13)$
3. $(17)(23)$
4. $(31)(29)$
5. $(15)(25)$
6. $(25)(35)$
7. $(41)(39)$
8. $(49)(51)$
9. $(34)(46)$
10. $(22)(38)$
11. $(88)(92)$
12. $(75)(85)$
13. $(95)(105)$
14. $(505)(495)$
15. $(6\frac{1}{3})(5\frac{2}{3})$
16. $(7\frac{1}{2})(6\frac{1}{2})$
17. $(1020)(980)$
18. $(1200)(800)$
19. $(0.7)(1.3)$
20. $(1.8)(2.2)$

Express each product as a polynomial.

21. $(y + 2)(y - 2)$
22. $(z + 3)(z - 3)$
23. $(x - y)(x + y)$
24. $(p - q)(p + q)$
25. $(t + 6)(t - 6)$
26. $(n - 8)(n + 8)$
27. $(2a - 1)(2a + 1)$
28. $(3b - 1)(3b + 1)$
29. $(y^2 - 5)(y^2 + 5)$
30. $(z^3 + 9)(z^3 - 9)$
31. $(3r + \frac{1}{2})(3r - \frac{1}{2})$
32. $(5k + \frac{2}{3})(5k - \frac{2}{3})$

7–4 Factoring the Difference of Two Squares

By the symmetric property of equality, the relation $(a + b)(a - b) = a^2 - b^2$ is reversible:

$$a^2 - b^2 = (a + b)(a - b)$$

We use this to factor an algebraic expression consisting of the difference of two squares:

$$x^2 - 9 = (x + 3)(x - 3)$$
$$16y^2 - x^2 = (4y + x)(4y - x)$$
$$25c^2 - 49d^2 = (5c + 7d)(5c - 7d)$$
$$-u^2v^2 + x^2y^2 = x^2y^2 - u^2v^2 = (xy + uv)(xy - uv)$$

If, as in $49x^2y^4$, the degree in *each* variable is even and the numerical coefficient is the square of an integer, then the monomial is a square,

$$49x^2y^4 = (7xy^2)^2.$$

Sometimes it is difficult to tell at sight whether or not a numerical coefficient is a square. However, Table 3 on page 358 enables us to tell whether or not each integer from 1 to 10,000 is the square of an integer.

EXERCISES

Factor and check by multiplication. Use Table 3 on page 358 as needed.

1. $t^2 - 9$
2. $k^2 - 4$
3. $4m^2 - 1$
4. $9r^2 - 1$
5. $t^2 - u^2$
6. $x^2 - 16y^2$
7. $25v^2 - 49$
8. $64 - 25r^2$
9. $225a^2 - b^2$
10. $4x^2 - 625$
11. $a^2b^4 - c^2$
12. $d^2 - f^2g^4$
13. $4y^2 - 1$
14. $16t^2 - 9$
15. $25x^2 - 36y^2$
16. $49m^2 - 64$
17. $4r^2 - 64$
18. $9p^2 - 81$
19. $x^2 - 169$
20. $144 - n^2$
21. $81t^2 - 100s^2$
22. $121a^2 - 144b^2$

Example. $4ax^{2n} - 36a$. Assume that $n \in \{\text{the positive integers}\}$.

Solution: $4ax^{2n} - 36a = 4a(x^{2n} - 9) = 4a(x^n - 3)(x^n + 3)$

23. $3t^2 - 27$
24. $5p^2 - 80t^2$
25. $x^4 - 25x^2$
26. $y^6 - 36y^4$
27. $-b^2 + c^2$
28. $-m + m^7$
29. $147z^2 - 75$
30. $112 - 63q^2$
31. $t^4 - k^4$
32. $r^4 - 1$
33. $-n^{2n} + 1$
34. $-x^{2y} + x^2$
35. $x^{2n} - 9$
36. $-t^{2n} + 25$
37. $k^2 - a^{2n}$
38. $z^{4n} - k^2$
39. $l^6 - w^4h^2$
40. $x^{2k} - x^2$
41. $x^{4n} - y^{4n}$
42. $81a^{4n} - 1$
43. $(x + 3)^2 - x^2$
44. $y^2 - (y - 2)^2$
45. $(x + 1)^2 - (x - 1)^2$
46. $9(x - 1)^2 - 25(x + 1)^2$

★47. Show that the difference between the squares
    a. of two consecutive integers equals the sum of the integers;
    b. of two consecutive odd integers equals twice the sum of the integers.

7-5 Squaring a Binomial

The binomial $a + b$ is squared at the right by the usual method of multiplication. Notice how each term in the product is obtained.

$$\begin{array}{r} a + b \\ a + b \\ \hline a^2 + ab \\ ab + b^2 \\ \hline a^2 + 2ab + b^2 \end{array}$$

1. Square the first term in the binomial.
2. Double the product of the two terms.
3. Square the second term in the binomial.

Now examine the square of a binomial difference. The binomial $a - b$ is squared by multiplication at the right. Again, notice how each term is obtained.

$$\begin{array}{r} a - b \\ a - b \\ \hline a^2 - ab \\ - ab + b^2 \\ \hline a^2 - 2ab + b^2 \end{array}$$

1. Square the first term in the binomial.
2. Double the product of the two terms.
3. Square the second term in the binomial.

Whenever we square a binomial, the product is a **trinomial square**, whose terms show this pattern:

$$(a + b)^2 = a^2 + 2ab + b^2$$
$$(a - b)^2 = a^2 - 2ab + b^2$$

Each side of the large square in Figure 7–2 is $(a + b)$ units in length. We can consider this square as being made up of four regions, as shown. The total area can be expressed as a square of a binomial, $(a + b)^2$, or as a trinomial square, $a^2 + 2ab + b^2$.

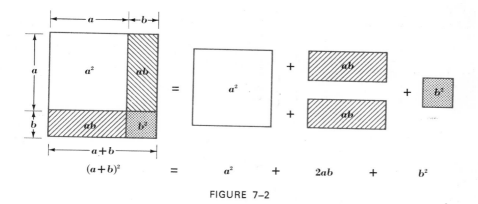

FIGURE 7–2

Knowing these relationships, we can write the square of a binomial without performing long multiplication.

Examples:
1. $(c + 1)^2 = c^2 + 2c + 1$
2. $(d - 1)^2 = d^2 - 2d + 1$
3. $(2x + 5)^2 = 4x^2 + 20x + 25$
4. $(6z^2 - w^3)^2 = 36z^4 - 12z^2w^3 + w^6$
5. $(-r^2s + t^3)^2 = (t^3 - r^2s)^2 = t^6 - 2t^3r^2s + r^4s^2$

EXERCISES

Write each power as a trinomial.

1. $(p + 7)^2$
2. $(q - 8)^2$
3. $(2x - 1)^2$
4. $(3y + 1)^2$
5. $(4t + 3)^2$
6. $(3s - 2)^2$
7. $(6r + 5)^2$
8. $(7k - 3)^2$
9. $(2x - 3y)^2$
10. $(5z + 2u)^2$
11. $(xy - 1)^2$
12. $(2 + rs)^2$
13. $(x^2 + 2)^2$
14. $(y^2 - 5)^2$
15. $(u^2v^2 + 7)^2$
16. $(2p^2 - 5q^2)^2$
17. $(x - \frac{2}{3})^2$
18. $(y + \frac{5}{2})^2$
19. $(\frac{1}{2} - m)^2$
20. $(\frac{3}{4} + x)^2$
21. $(n - 0.3)^2$
22. $(p + 2.4)^2$
23. $(\frac{3}{4}y + \frac{4}{3})^2$
24. $(\frac{1}{2}x - \frac{8}{5})^2$
25. $(0.6x + 1.5)^2$
26. $(1.1y - 3.2)^2$
27. $(-x^3 - y^2z)^2$
28. $(-m^2n^2 + p^4)^2$

29. Show that $(a + b + c)^2 = a^2 + b^2 + c^2 + 2ab + 2ac + 2bc$ by considering the figure at the left below.

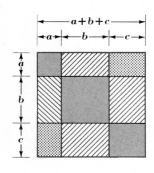
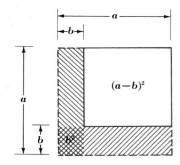

EX. 29   EX. 30

30. Show that $(a - b)^2 = a^2 - 2ab + b^2$ by considering the figure at the right above.

31. Show that
$$(a + b)^2 - (a - b)^2 = 4ab$$
by considering the figure at the left below.

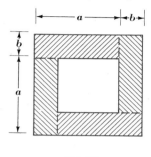

EX. 31   EX. 32

32. Show that
$$(a + b)^2 + (a - b)^2 = 2(a^2 + b^2)$$
by considering the figure at the right above.

★**33.** Use the figure at the left below to show that when n has the value 9,
$$1 + 2 + 3 + \cdots + n = \frac{n(n + 1)}{2}.$$

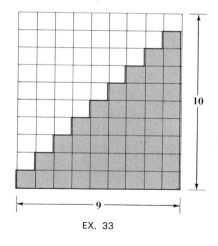
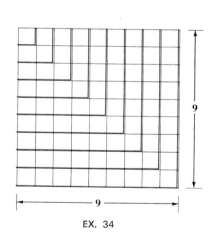

EX. 33   EX. 34

★**34.** Use the figure at the right above to show that when n has the value 9,
$$1 + 3 + 5 + \cdots + (2n - 1) = n^2.$$

7–6 Factoring a Trinomial Square

To factor a trinomial square, reverse the equations used in squaring a binomial. Thus:

$$a^2 + 2ab + b^2 = (a + b)^2$$
$$a^2 - 2ab + b^2 = (a - b)^2$$

Before one of these equations is used as a rule for factoring, we must be sure that the expression to be factored is a trinomial square. Arrange the terms of the trinomial with exponents in descending order, and then examine each term to see how it may have been obtained. Consider

$$y^2 + 10y + 25.$$

Is the first term a square? Yes, y^2 is the square of y. Is the third term a square? Yes, 25 is 5^2. Is the middle term (neglecting the sign) double the product of 5 and y? Yes, $10y = 2(y)(5)$. Therefore, this trinomial is a square. Since all its terms are **positive**, it is the square of a **sum**; thus,

$$y^2 + 10y + 25 = (y + 5)^2.$$

Example 1. Factor $81r^2 - 198rs + 121s^2$.

Solution: $81r^2 - 198rs + 121s^2 = (9r)^2 - 2(9r)(11s) + (11s)^2$
Thus, $81r^2 - 198rs + 121s^2 = (9r - 11s)^2$.

Check: $(9r - 11s)^2 = 81r^2 - 198rs + 121s^2$

Example 2. Factor $18y^2 - 12y + 2$.

Solution: $18y^2 - 12y + 2 = 2(9y^2 - 6y + 1) = 2(3y - 1)^2$

EXERCISES

Factor and check.

1. $n^2 - 2n + 1$
2. $m^2 + 2m + 1$
3. $k^2 - 6k + 9$
4. $t^2 - 8t + 16$
5. $r^2 + 10r + 25$
6. $s^2 + 12s + 36$
7. $4p^2 - 4p + 1$
8. $9q^2 + 6q + 1$
9. $16c^2 + 8c + 1$
10. $25d^2 - 10d + 1$
11. $4x^2 - 4xy + y^2$
12. $9u^2 + 6uv + v^2$

13. $1 - 4t + 4t^2$
14. $25 + 10k + k^2$
15. $64y^2 - 16yz + z^2$
16. $81k^2 + 18kt + t^2$
17. $9 + 12p + 4p^2$
18. $36 - 60q + 25q^2$
19. $4x^2y^2 - 12xyz + 9z^2$
20. $16t^2 + 24tuv + 9u^2v^2$
21. $x^4 + 2x^2 + 1$
22. $y^4 + 10y^2 + 25$
23. $25y^4 - 10y^2x + x^2$
24. $z^2 + 18zab + 81a^2b^2$
25. $7x^3 + 14x^2 + 7x$
26. $20ay^2 - 60ay + 45a$
27. $x^3 + 25x - 10x^2$
28. $4p^2q + pq^2 + 4p^3$
29. $24x + 24x^2 + 6x^3$
30. $3z + 42z^2 + 147z^3$
31. $4y^4 - 8y^2x^2 + 4x^4$
32. $6a^4 - 12a^2b^2 + 6b^4$
★33. $x^2 + 2x + 1 - y^2$
★34. $t^2 - 4t + 4 - s^2$
★35. $a^2 - b^2 + 2b - 1$
★36. $4p^2 - q^2 - 6q - 9$

Find k so that each trinomial will be a square.

37. $y^2 - 6y + k$
38. $b^2 + kb + 25$
39. $kx^2 - 12x + 9$
40. $y^2 - 2ky + 81$

7–7 Multiplying Binomials at Sight

To learn to write the product of two binomials of the form $(ax + b)(cx + d)$ at sight, study these examples which apply the distributive axiom.

Example 1. $(2y + 3)(7y - 5)$

Solution:

$$\begin{array}{r} 7y - 5 \\ 2y + 3 \\ \hline 14y^2 - 10y \\ 21y - 15 \\ \hline 14y^2 + 11y - 15 \end{array}$$

Example 2. $(ax + b)(cx + d)$

Solution:

$$\begin{array}{r} cx + d \\ ax + b \\ \hline acx^2 + adx \\ bcx + bd \\ \hline acx^2 + (ad + bc)x + bd \end{array}$$

To write the terms in the trinomial product of two binomials, $(ax + b)(cx + d)$, at sight:

1. Multiply the first terms of the binomials.
2. Multiply the first term of each binomial by the last term of the other, and add these products.
3. Multiply the last terms of the binomials.

Each term of a trinomial like $14y^2 + 11y - 15$ has a special name. The first, $14y^2$, is the *quadratic term* in the variable y: the second, $11y$, is the *linear term* in the variable y; and the third, -15, is the *constant term*. A **quadratic term** is a term of degree two in the variable. A **linear term** is a term of degree one in the variable. A **constant term** is a numerical term with no variable factor. The trinomial is itself called a **quadratic polynomial** because the term of the highest degree in it is a quadratic term.

EXERCISES

Write each product as a trinomial.

1. $(n + 6)(n + 3)$
2. $(t - 5)(t - 2)$
3. $(x + 10)(x - 5)$
4. $(y - 9)(y + 3)$
5. $(2x + 1)(x + 2)$
6. $(z + 3)(2z + 5)$
7. $(3t - 1)(t + 4)$
8. $(p - 3)(4p + 3)$
9. $(3x - 2)(2x - 3)$
10. $(5y + 2)(2y + 5)$
11. $(3r - 5)(2r + 1)$
12. $(5s + 2)(2s - 1)$
13. $(3 - x)(2 + x)$
14. $(5 - z)(4 + z)$
15. $(1 - 2s)(3 + 4s)$
16. $(2 + 5t)(3 - 2t)$
17. $(x - 2)(-x + 3)$
18. $(-y + 5)(y - 2)$
19. $(-z - 1)(-2z - 3)$
20. $(-r - 2)(-2r - 5)$
21. $x(x + 2)(x - 1)$
22. $z(3z + 2)(z - 1)$
23. $2y(3y - 1)(y + 5)$
24. $5k^2(k + 2)(2k - 1)$

Solve each equation and check.

25. $(x - 3)(x + 2) = (x - 6)(x - 1)$
26. $(x + 1)(x - 1) = (x + 5)(x - 2)$
27. $(4x + 3)(x + 3) = (2x + 5)(2x + 2)$
28. $(4x + 3)(x - 2) = (2x - 1)(2x - 3)$

Write each product as a trinomial.

29. $(x + \frac{2}{3})(x - \frac{1}{3})$
30. $(z - \frac{2}{7})(z - \frac{3}{7})$
31. $(y + \frac{4}{5})(y - \frac{2}{5})$
32. $(\frac{3}{7}x + 1)(\frac{2}{7}x - 1)$
33. $(a + 0.2)(a - 3.1)$
34. $(b - 2.3)(b + 5.1)$
35. $(2.1z + 0.2)(3z - 1.7)$
36. $(4.2y - 3.1)(2.3y + 1.6)$

Solve and check.

★37. $(3x + 1)(x - 1) = (2x - 3)(x + 5) + (x - 2)(x + 2)$
★38. $(2x - 5)(3x + 1) = (2x + 3)(x - 4) + (2x - 1)^2 - 4$
★39. $(2x + 3)(x + 4) = (x - 6)(x + 2) + (x + 3)^2 - 21$
★40. $(3x - 5)(2x + 7) - 7 = (2x - 1)^2 + (x - 3)(2x + 1) + 5$

7–8 Factoring the Product of Binomial Sums or Differences

We shall now need to consider expressing an integer as a product of two integers. For example,

$$30 = 1 \cdot 30 \quad = 2 \cdot 15 \quad = 3 \cdot 10 \quad = 5 \cdot 6 \qquad (1)$$
$$= (-1)(-30) = (-2)(-15) = (-3)(-10) = (-5)(-6) \qquad (2)$$

Notice that 30 may be expressed in several ways, either as a product of *two positive* integers (line 1 above) or as a product of *two negative* integers (line 2). In the work in this section we need to consider such pairs of factors of positive integers.

To factor the trinomial $z^2 + 6z + 8$, notice that

$$(z + r)(z + s) = z^2 + (r + s)z + rs$$

and compare the two: $z^2 + 6z + 8$

The two trinomials would be exactly alike if

$$rs = 8 \quad \text{and} \quad r + s = 6.$$

With these clues, we can find the two integers r and s and then write $z^2 + 6z + 8$ as a product of factors $(z + r)(z + s)$.

Observe first that the product, rs, of the desired integers is positive in this case, indicating that r and s are *both positive* or *both negative*. Observe next that their sum, $r + s$, is positive. Therefore, r and s cannot both be negative, and so in this case r and s must both be positive. There are two ways to express 8 as the product of two positive integers:

$$8 = 1 \cdot 8 \qquad \text{and} \qquad 8 = 2 \cdot 4$$
$$ \uparrow \ \uparrow \qquad\qquad\qquad\qquad \uparrow \ \uparrow$$
$$ r \ \ s \qquad\qquad\qquad\qquad r \ \ s$$
$$ \downarrow \searrow \qquad\qquad\qquad\qquad \downarrow \searrow$$
$$r + s = 1 + 8 = 9 \qquad\qquad r + s = 2 + 4 = 6$$

The second set of factors satisfies both clues, and so we conclude that

$$z^2 + 6z + 8 = (z + 2)(z + 4).$$

We can check this by multiplying the factors.

Of course, if we see the right pair of factors at the outset, there is no need to write the remaining possibilities.

Example. Factor $y^2 - 9y + 18$.

Solution:
$$\begin{aligned} y^2 - 9y + 18 &= (y \quad)(y \quad) \\ &= (y - \quad)(y - \quad) \\ &= (y - 3)(y - 6) \end{aligned} \quad \begin{cases} \text{Since } r + s \text{ is negative,} \\ \text{both } r \text{ and } s \text{ are negative.} \end{cases}$$

Check: $(y - 3)(y - 6) = y^2 - 9y + 18$

Each of the quadratic polynomials we have factored has a *positive constant term*. Then because $z^2 + 6z + 8$ has a *positive linear term*, it is factored as a product of two binomial sums, $(z + 2)(z + 4)$. On the other hand, $y^2 - 9y + 18$ has a *negative linear term*, and so it is factored as a product of two binomial differences, $(y - 3)(y - 6)$.

Not every quadratic trinomial can be written as a product of binomials having integral coefficients. To factor $x^2 + 2x + 6$, we would have to find positive integers r and s such that

$$rs = 6 \quad \text{and} \quad r + s = 2.$$

The two ways of writing 6 as a product of positive integers are:

$$6 = 1 \cdot 6 \qquad \text{and} \qquad 6 = 2 \cdot 3$$
$$ \uparrow \ \uparrow \qquad\qquad\qquad\qquad \uparrow \ \uparrow$$
$$ r \ \ s \qquad\qquad\qquad\qquad r \ \ s$$
$$r + s = 7 \qquad\qquad\qquad r + s = 5$$

In each case, $r + s \neq 2$; therefore, $x^2 + 2x + 6$ cannot be factored over the set of polynomials with integral coefficients. Such a polynomial is said to be *irreducible over this set of polynomials*. A polynomial which cannot be factored into polynomials of lower degree belonging to a designated set is said to be **irreducible** over that set of polynomials.

An irreducible polynomial whose greatest monomial factor is 1 is called a **prime polynomial**. Thus, $2x + 11$ is a prime polynomial. However, $8x + 44 = 4(2x + 11)$, and so $8x + 44$ is irreducible but not prime.

EXERCISES

Factor each trinomial and check by multiplication.

1. $n^2 + 8n + 7$
2. $t^2 + 7t + 6$
3. $y^2 + 5y + 4$
4. $y^2 + 9y + 8$
5. $k^2 - 7k + 6$
6. $r^2 - 6r + 8$
7. $a^2 - 9a + 8$
8. $b^2 - 7b + 12$
9. $x^2 + 7x + 10$
10. $x^2 + 11x + 18$
11. $n^2 - 15n + 26$
12. $p^2 - 23p + 60$
13. $14 + 9k + k^2$
14. $21 + 10d + d^2$
15. $42 - 13u + u^2$
16. $48 - 14v + v^2$
17. $x^2 + 8xy + 7y^2$
18. $r^2 + 8rs + 15s^2$
19. $m^2 - 11mn + 28n^2$
20. $p^2 + 14pq + 24q^2$
21. $r^2 - 23rt + 76t^2$
22. $x^2 - 21xy + 20y^2$
23. $a^2 - 26ab + 48b^2$
24. $m^2 - 29mn + 120n^2$

Determine all integral values for *k* for which the trinomial can be factored over the set of binomials with integral coefficients. In Exercises 29–32, list positive values only.

Example. $x^2 + kx + 15$

Solution: 15 can be factored into a product of two integers as follows:

$$1 \cdot 15, \quad 3 \cdot 5, \quad (-1)(-15), \quad (-3)(-5).$$

The corresponding values of *k* are 16, 8, -16, -8.

25. $y^2 + ky + 10$
26. $z^2 + kz + 12$
27. $x^2 + kx + 20$
28. $y^2 + ky + 4$
29. $z^2 + 3z + k$
30. $x^2 + 5x + k$
31. $y^2 + 6y + k$
32. $z^2 + 7z + k$

Factor each expression.

★33. $(x + y)^2 - 4(x + y) + 3$ ★34. $(a + b)^2 + 5(a + b) + 6$
★35. $(x - 3)^2 + 6(x - 3) + 8$ ★36. $(5 - y)^2 - 3(5 - y) + 2$

Show that each polynomial is prime over the set of polynomials with integral coefficients.

★37. $x^2 + 3x + 5$ ★38. $y^2 - 2y + 2$ ★39. $x^2 + 1$

7–9 Factoring the Product of a Binomial Sum and a Binomial Difference

In the preceding section we factored trinomials in which the constant term was positive. We will now learn how to factor trinomials in which the constant term is negative.

When a negative integer is written as a product of two integers, one factor must be a positive integer and the other, a negative integer. For example,

$$-30 = (-1)(30) = (-2)(15) = (-3)(10) = (-5)(6)$$
$$= (1)(-30) = (2)(-15) = (3)(-10) = (5)(-6)$$

To factor $a^2 + 3a - 10$, proceed as before to look for r and s such that

$$a^2 + 3a - 10 = (a + r)(a + s) = a^2 + (r + s)a + rs.$$

Thus
$$rs = -10 \quad \text{and} \quad r + s = 3.$$

Here, the product, rs, is negative, indicating that one integer, say r, must be positive, while the other, s, must be negative. But, $r + s$ is *positive*, which means that r, the positive member of the pair, must have the greater absolute value. On the basis of these conclusions, consider:

$$10(-1) \quad \text{and} \quad 5(-2)$$
$$\uparrow \uparrow \qquad\qquad \uparrow \uparrow$$
$$r \ \ s \qquad\qquad\quad r \ \ s$$
$$r + s = 9 \qquad\quad r + s = 3$$
$$\therefore a^2 + 3a - 10 = (a + 5)(a - 2).$$

On the other hand, in factoring $a^2 - 3a - 10$, search for two integers of opposite direction, but with a *negative* sum. Therefore, the negative integer would have to have the greater absolute value. This trinomial is factored:

$$a^2 - 3a - 10 = (a - 5)(a + 2)$$

EXERCISES

Factor each trinomial and check by multiplication.

1. $a^2 + a - 6$
2. $b^2 + 5b - 6$
3. $x^2 - 2x - 3$
4. $y^2 - 4y - 5$
5. $c^2 - 3c - 10$
6. $k^2 + k - 12$
7. $u^2 + 7u - 18$
8. $v^2 - 3v - 18$
9. $z^2 - 4z - 21$
10. $t^2 + 5t - 14$
11. $x^2 + x - 56$
12. $y^2 - 2y - 63$
13. $a^2 + a - 20$
14. $c^2 - c - 42$
15. $x^2 - 4x - 60$
16. $z^2 + 5z - 50$
17. $a^2 - 2ab - 8b^2$
18. $c^2 + 3cd - 10d^2$
19. $p^2 - 5pq - 24q^2$
20. $u^2 + 6uv - 55v^2$
21. $r^2 - 10rs - 24s^2$
22. $m^2 + 16mn - 36n^2$
23. $x^2 - 9xy - 36y^2$
24. $s^2 - 5st - 24t^2$

Determine all integral values of k for which the given trinomial can be factored over the set of binomials with integral coefficients.

25. $x^2 + kx - 10$
26. $y^2 + ky - 8$
27. $z^2 + kz - 12$
28. $t^2 + kt - 15$
29. $r^2 + kr - 24$
30. $p^2 + kp - 36$

Find the two negative integers k of least absolute value for which each trinomial can be factored.

31. $x^2 + x + k$
32. $y^2 - 2y + k$
33. $z^2 - 3z + k$
34. $u^2 + u + k$
35. $v^2 + 2v + k$
36. $t^2 + 3t + k$

Factor each expression.

★37. $(s + t)^2 - 5(s + t) - 66$
★38. $(x + y)^2 + 3(x + y) - 70$
★39. $(y - 2)^2 + 4(y - 2) - 45$
★40. $(3 - z)^2 - 2(3 - z) - 35$

7–10 General Method of Factoring Quadratic Trinomials

To factor a quadratic trinomial product whose quadratic term has a coefficient other than 1, we can use inspection and trial, as in the following examples on pages 180–181.

Example 1.   Factor $15x^2 - 31x + 10$.

Solution:

First clue: The constant term is positive, and the linear term is negative.
∴ both binomial factors are *differences*.

Second clue: The product of the linear terms of the binomial factors is $15x^2$, and the product of the constant terms of the binomial factors is 10.

The possible pairs of factors of $15x^2$ are

x and $15x$,
$3x$ and $5x$.

The possible pairs of factors of 10 are

-1 and -10,
-2 and -5.

∴ the possibilities to consider are as follows:

Possible Factors	Corresponding Linear Terms
$(x - 1)(15x - 10)$	$-10x - 15x = -25x$
$(x - 10)(15x - 1)$	$-x - 150x = -151x$
$(x - 2)(15x - 5)$	$-5x - 30x = -35x$
$(x - 5)(15x - 2)$	$-2x - 75x = -77x$
$(3x - 1)(5x - 10)$	$-30x - 5x = -35x$
$(3x - 10)(5x - 1)$	$-3x - 50x = -53x$
$(3x - 2)(5x - 5)$	$-15x - 10x = -25x$
$(3x - 5)(5x - 2)$	$-6x - 25x = -31x$

Third clue: The linear term of the trinomial is $-31x$.
Only the last possibility satisfies all three clues.

∴ $15x^2 - 31x + 10 = (3x - 5)(5x - 2)$.

Check: $(3x - 5)(5x - 2) = 15x^2 - 31x + 10$

As we gain experience, often we will not need to write down all the possibilities before discovering the factors.

Example 2. Factor $6x^2 - 7x - 3$.

Solution:
$$6x^2 - 7x - 3 = (\quad - \quad)(\quad + \quad)$$
$$= (\quad - 3)(\quad + 1)$$
$$= (2x - 3)(3x + 1)$$

Check: $(2x - 3)(3x + 1) = 6x^2 - 7x - 3$

EXERCISES

Factor each trinomial and check.

1. $2x^2 + 3x + 1$
2. $2z^2 + 5z + 3$
3. $3t^2 + 7t + 2$
4. $3k^2 + 8k + 5$
5. $5r^2 - 7r + 2$
6. $6s^2 - 11s + 3$
7. $3y^2 + 7y - 6$
8. $6x^2 - 13x - 5$
9. $4t^2 - 11t - 3$
10. $4k^2 + 4k - 15$
11. $5n^2 - 3n - 2$
12. $7x^2 + 9x + 2$
13. $7x^2 - 10x + 3$
14. $2y^2 - 9y - 5$
15. $5k^2 - 2k - 7$
16. $8r^2 + 2r - 3$
17. $2y^2 + xy - 6x^2$
18. $3b^2 - 17ab - 6a^2$
19. $12x^2 + 11x - 15$
20. $8y^2 - 27y - 20$
21. $6t^2 + 25t + 14$
22. $18z^2 - 19z - 12$
23. $6u^2 - u - 12$
24. $24v^2 + 5v - 36$
25. $10y^2 + 11yz - 18z^2$
26. $6a^2 - 47ab - 63b^2$
★27. $8(x + y)^2 + 14(x + y) - 15$
★28. $10(a - b)^2 - 11(a - b) - 6$
★29. $24(x - 1)^2 - 14(x - 1) - 3$
★30. $14(2 - x)^2 - 15(2 - x) - 11$

7–11 Combining Several Types of Factoring

In factoring a polynomial, first look for the greatest monomial factor. Sometimes a monomial factor conceals:

The difference of two squares	A trinomial square	A trinomial product
$8y^3 - 18y$	$-t^2 + 8t - 16$	$bz^3 - 6bz^2 + 5bz$
$= 2y(4y^2 - 9)$	$= -1(t^2 - 8t + 16)$	$= bz(z^2 - 6z + 5)$
$= 2y(2y - 3)(2y + 3)$	$= -1(t - 4)^2$	$= bz(z - 5)(z - 1)$

To factor a polynomial product:

1. Find the greatest monomial factor, and then consider the remaining polynomial factor.
2. If a factor is a binomial, is it the difference of two squares? Such a binomial can be factored.
3. If a factor is a trinomial, is it a square? A trinomial square can be factored.
4. If a factor is a trinomial that is not a square, assume that it is the product of two binomials, and search for them. Of course, a prime trinomial cannot be factored, but never decide that an expression is prime until all the ways of factoring it have been tried.
5. If a factor is neither a binomial nor a trinomial, can a common polynomial factor be found by grouping? If so, try to factor each of the resulting factors as in Steps 2, 3, and 4.
6. Write *all* the factors, including any monomial factor. The monomial factor may be a product, but all other factors should be prime; the *factoring should be complete.*
7. As a check, always multiply the factors to see that the product is the original expression; the *factoring should be correct.*

EXERCISES

Factor if possible. Check by multiplication.

1. $4x^2 - 4$
2. $3z^2 - 12$
3. $2y^2 + 6y + 4$
4. $3t^2 - 6t + 3$
5. $7x^2 + 13x - 2$
6. $5y^2 + 13y + 6$
7. $-3a^3 - 3ab^2$
8. $2cx^2 + 2c$
9. $ar^2 - 3ar - 4a$
10. $16b^2z^2 + 8bz + 1$
11. $6t^2 - 11t + 5$
12. $7k^2 + 21k - 28$
13. $50r^2 - 20r + 2$
14. $p^3 + 5p^2 + 4p$
15. $3y^2 - 2y - 5$
16. $12c^2 - 24c - 15$
17. $144y^2 - 49x^4$
18. $169a^4 - 36b^2$
19. $-15 + n + 6n^2$
20. $-5 - 28k + 12k^2$
21. $-4s - s^2 + 21$
22. $2t - 48 + t^2$
23. $28x^2 + 87x + 54$
24. $6y^2 - 28y - 480$
25. $6a^3b - 26a^2b^2 - 20ab^3$
26. $30c^2 + 4cd - 2d^2$
27. $42m^2n - 24mn^2 - 18n^3$
28. $12a^2 - 2ab - 24b^2$

29. $6s^4 - 19s^3 + 10s^2$
30. $2p^3 + 11p^2q + 12pq^2$
31. $z^4 - r^4$
32. $t^4 - 2t^2 + 1$
33. $z^4 - 10z^2 + 9$
34. $n^4 - 17n^2 + 16$
35. $-42a - 27az - 3az^2$
36. $-16a^2b - 10a^2br - a^2br^2$
37. $u^2(u - 3) - 10u(u - 3) - 24(u - 3)$
38. $n^2(n + 2) + 2n(n + 2) - 15(n + 2)$
39. $2x^2(x + 5) - 3x(x + 5) - 9(x + 5)$
40. $3y^2(y - 1) + 10y(y - 1) - 8(y - 1)$
41. $4x^2(x^2 - 9) - 16x(x^2 - 9) + 15(x^2 - 9)$
42. $1 - z^2 - 5z(1 - z^2) + 6z^2(1 - z^2)$
43. $(3x^2 - 2)^2 - x^2$
44. $(x^2 - 6)^2 - 6x(x^2 - 6) + 5x^2$
45. $5bz^2 - bw^2 + bz^2 - 5bw^2$
46. $k^2t^2 - 9s^2 - 9t^2 + k^2s^2$
47. $-20x^2 + 43xy - 14y^2$
48. $c^4 - 7c^2 + 6$
49. $x^4 + x^2 - 42$
50. $x^3 - x^2 + x - 1$
51. $n^3 + n^2 - n - 1$
52. $3x^4 - 288$

In Exercises 53–56, the given binomial is a factor of the trinomial over the set of polynomials with integral coefficients. Determine c in each case.

★53. $2x - 3$; $10x^2 - 3x + c$
★54. $3y + 2$; $21y^2 - y + c$
★55. $5w + 1$; $cw^2 + 3w - 1$
★56. $n - 3$; $cn^2 - 5n - 21$

7–12 Working with Factors Whose Product Is Zero

If we know that the product of two numbers is zero, we can say the following about the numbers.

First, if $ab = 0$ and $a \neq 0$, we can show that $b = 0$ as follows:

Since $a \neq 0$, the reciprocal of a, $\frac{1}{a}$, exists and is not zero. Using the multiplication property of equality (page 83), multiply each of the terms ab and 0 by $\frac{1}{a}$:

On the left, use the associative axiom, axiom of reciprocals (page 67), and the multiplicative axiom of 1 (page 63).

$$\frac{1}{a}(ab) = \frac{1}{a}(0)$$
$$\left(\frac{1}{a} \cdot a\right)b = 0$$
$$1 \cdot b = 0$$
$$b = 0$$

On the right, use the multiplicative property of 0 (page 64).

Similarly, if $ab = 0$ and $b \neq 0$, we can show that $a = 0$.

On the other hand, if either $a = 0$ or $b = 0$, then $ab = 0$ by the multiplicative property of 0. Thus, we have:

Zero-Product Property of Real Numbers

For all real numbers a and b, $ab = 0$ if and only if $a = 0$ or $b = 0$.

This property can be used to find solutions of equations in which one member is 0 and the other member is in factored form. For example, the solutions of the equation

$$(x - 3)(x + 4) = 0$$

are those real numbers for which one or the other factor in the left-hand member is 0. These values are, by inspection, 3 and -4. If we could not determine these values by inspection, we could have written the equivalent compound sentence

$$x - 3 = 0 \quad \text{or} \quad x + 4 = 0$$

and solved each equation for x. The solution set in either case is $\{3, -4\}$.

Example. $\left(\dfrac{1}{x} - \dfrac{2}{3}\right)\left(\dfrac{3}{x} + \dfrac{1}{2}\right) = 0$

Solution: For the given equation to become a true statement either $\dfrac{1}{x} - \dfrac{2}{3} = 0$ or $\dfrac{3}{x} + \dfrac{1}{2} = 0$. Thus:

$$\dfrac{1}{x} - \dfrac{2}{3} = 0 \quad \text{or} \quad \dfrac{3}{x} + \dfrac{1}{2} = 0$$

$$\dfrac{1}{x} = \dfrac{2}{3} \qquad\qquad \dfrac{3}{x} = -\dfrac{1}{2}$$

Taking reciprocals, $x = \tfrac{3}{2}$ 　　Taking reciprocals, $\dfrac{x}{3} = -2$

$$x = -6$$

Check: $\dfrac{1}{\tfrac{3}{2}} - \dfrac{2}{3} = \dfrac{2}{3} - \dfrac{2}{3} = 0;$ 　 $\dfrac{3}{-6} + \dfrac{1}{2} = -\dfrac{1}{2} + \dfrac{1}{2} = 0.$

$\therefore$ the solution set is $\{\tfrac{3}{2}, -6\}$.

EXERCISES

Solve each equation.

1. $x(x - 5) = 0$
2. $3z(z + 7) = 0$
3. $(t - 6)(t - 8) = 0$
4. $(p - 10)(p - 3) = 0$

5. $(k + 2)(k - 7) = 0$
6. $(y + 6)(y - 8) = 0$
7. $(n + 4)(2n - 3) = 0$
8. $(m - 6)(3m - 1) = 0$
9. $(2r - 1)(3r - 7) = 0$
10. $(4a - 3)(7a - 2) = 0$
11. $(2b + 7)(2b + 5) = 0$
12. $(9d + 2)(6d + 1) = 0$
13. $2x(x - 1)(x + 3) = 0$
14. $3r(r + 6)(r - 5) = 0$
15. $0(p + 7)(3p - 2) = 0$
16. $0(m + 6)(5m - 8) = 0$
17. $4\left(\dfrac{1}{3} - \dfrac{2}{t}\right) = 0$
18. $-3\left(\dfrac{5}{4} + \dfrac{3}{k}\right) = 0$
19. $\left(\dfrac{2}{v} - 3\right)\left(\dfrac{1}{v} + 4\right) = 0$
20. $\left(\dfrac{3}{x} - 7\right)\left(\dfrac{1}{x} + 6\right) = 0$
21. $\left(\dfrac{2}{y} - \dfrac{1}{7}\right)\left(-\dfrac{1}{y} + \dfrac{3}{8}\right) = 0$
22. $\left(\dfrac{5}{a} + \dfrac{3}{7}\right)\left(\dfrac{1}{a} - \dfrac{5}{8}\right) = 0$
23. $\left(\dfrac{3}{5} - \dfrac{7}{n}\right)\left(\dfrac{5}{3} + \dfrac{2}{n}\right) = 0$
24. $\left(6 - \dfrac{2}{3x}\right)\left(5 + \dfrac{3}{2x}\right) = 0$

7–13 Solving Polynomial Equations by Factoring

A **polynomial equation** is an equation whose left and right members are polynomials. A polynomial equation is in **standard form** when one of its members is zero and the other is a polynomial in simple form (page 130) in which all similar terms have been combined and the terms have been arranged in descending powers of the variable. Thus, the standard form for

$$y^2 - 10y = 24$$

is

$$y^2 - 10y - 24 = 0.$$

This equation is of degree two and is called a **quadratic equation**. The **degree of a polynomial equation** is the greatest of the degrees of the terms of the equation when written in standard form. An equation of degree one, like

$$3x + 2 = x - 5,$$

is called a **linear equation**, while an equation like

$$z^3 = 2z - 3z$$

whose degree is three, is a **cubic equation**. If we transform a polynomial equation into standard form, and if we can factor the left member, then we can obtain its roots by finding the numbers for which at least one of those factors is zero.

Example 1. Solve $y^2 - 10y = 24$.

Solution:

1. Transform the equation into standard form. $y^2 - 10y - 24 = 0$
2. Factor the left-hand member. $(y - 12)(y + 2) = 0$
3. Determine solutions by inspection (12 and -2) or
3a. Set each factor equal to zero and solve the resulting linear equations. $y - 12 = 0 \;\vert\; y + 2 = 0$
   $y = 12 \;\vert\; y = -2$
4. Check in original equation: $y^2 - 10y = 24$

$$(12)^2 - 10(12) \stackrel{?}{=} 24 \qquad (-2)^2 - 10(-2) \stackrel{?}{=} 24$$
$$144 - 120 \stackrel{?}{=} 24 \qquad 4 + 20 \stackrel{?}{=} 24$$
$$24 = 24 \qquad 24 = 24$$

∴ the solution set is $\{12, -2\}$.

Several situations may arise when we try to solve a polynomial equation by factoring. First, the polynomial may have a common numerical factor. Since such a factor would be a nonzero number, we should eliminate it by transforming the equation by division (page 88). Second, two or more factors may be identical. Such factors will yield a *double* or *multiple root*, which should be written only once in the roster of the solution set.

Of course, not all polynomial equations can be solved in this way, because it may not be possible to factor the nonzero polynomial member.

Example 2. Solve $5r^2 - 30r + 45 = 0$.

Solution:

The equation is in standard form. $5r^2 - 30r + 45 = 0$
1. Divide each member by 5. $r^2 - 6r + 9 = 0$
2. Factor the left-hand member. $(r - 3)(r - 3) = 0$
3. Determine solution by inspection or
3a. Set each factor equal to zero and solve the resulting linear equations. $r - 3 = 0 \;\vert\; r - 3 = 0$
   $r = 3 \;\vert\; r = 3$
4. Check in original equation: $5(3)^2 - 30(3) + 45 \stackrel{?}{=} 0$
   $45 - 90 + 45 \stackrel{?}{=} 0$
   $0 = 0$

∴ the solution set is $\{3\}$.

If we have a common monomial factor which contains a variable, we should not eliminate it by division. Such a factor may be zero and give us a root.

Example 3. Solve $x^3 = 2x^2 + 8x$.

Solution: 1. Transform the equation into standard form. $x^3 - 2x^2 - 8x = 0$

2. Factor the left-hand member.

$$x(x^2 - 2x - 8) = 0$$
$$x(x - 4)(x + 2) = 0$$

3. Determine solutions by inspection $(0, 4, -2)$, or

3a. Set each factor equal to 0 and solve the resulting equations.

$x = 0 \quad | \quad x - 4 = 0 \quad | \quad x + 2 = 0$
$ x = 4 \quad x = -2$

4. Check in the original equation: $x^3 = 2x^2 + 8x$

$(0)^3 \stackrel{?}{=} 2(0)^2 + 8(0)$	$(4)^3 \stackrel{?}{=} 2(4)^2 + 8(4)$	$(-2)^3 \stackrel{?}{=} 2(-2)^2 + 8(-2)$
$0 \stackrel{?}{=} 0 + 0$	$64 \stackrel{?}{=} 32 + 32$	$-8 \stackrel{?}{=} 8 - 16$
$0 = 0$	$64 = 64$	$-8 = -8$

∴ the solution set is $\{0, 4, -2\}$.

EXERCISES

Find the solution set of each equation.

1. $x^2 + x - 6 = 0$
2. $y^2 - 2y + 1 = 0$
3. $r^2 + 2r - 3 = 0$
4. $t^2 - 3t - 10 = 0$
5. $k^2 - 16 = 0$
6. $p^2 - 25 = 0$
7. $x^2 - 5x = 0$
8. $s^2 + 3s = 0$
9. $n^2 = 6n + 7$
10. $q^2 - 8 = 2q$
11. $h^2 + h = 56$
12. $w^2 - 2w = 15$
13. $r^2 + 15 = -8r$
14. $z^2 + 18 = -11z$
15. $4n^2 + 4n = -1$
16. $9t^2 - 6t + 1 = 0$
17. $2s^2 = s + 6$
18. $3x^2 + 5x = 2$
19. $2n^2 + 9n + 10 = 0$
20. $3t^2 + 13t = -14$
21. $3v^2 = 198 + 15v$
22. $2t^2 = 216 + 6t$
23. $6z^2 = 34z - 20$
24. $20x^2 = 22x - 6$
25. $10 = 17n - 3n^2$
26. $11q = 3 + 10q^2$
27. $10y^2 = 75 - 35y$
28. $8c^2 = 22c + 6$
29. $4y = y^2$
30. $6x^2 = -x$
31. $p^4 - 5p^2 + 4 = 0$
32. $t^4 - 13t^2 + 36 = 0$
33. $2d^3 - d^2 = 10d$
34. $10c^3 = 18c - 3c^2$
35. $4k^3 - 12k^2 + 9k = 0$
36. $3a^3 - 6a^2 + 3a = 0$
37. $(z - 4)(z + 3) = -10$
38. $(b - 5)(b + 2) = -12$
39. $(r - 2)(r + 1) = r(2 - r)$
40. $p(3p + 2) = (p + 2)^2$

Find an equation of the lowest degree having the given solution set.

Example. $z \in \{2, -1\}$.

Solution: Since $z = 2$ or $z = -1$, we have $z - 2 = 0$ or $z + 1 = 0$.

∴ $(z - 2)(z + 1) = 0$ and $z^2 - z - 2 = 0$.

41. $x \in \{3, 6\}$
42. $t \in \{5, -7\}$
43. $p \in \{0, 2, 5\}$
44. $q \in \{-3, 0, 3\}$
45. $p \in \{-1, 1, 3\}$
46. $y \in \{-3, -2, 2, 3\}$

7-14 Using Factoring in Problem Solving

With the ability to solve polynomial equations by factoring, we can solve a wider variety of problems. However, because polynomial equations usually have more than one solution, we must exercise judgment and reject answers (even when they are solutions of the equation we have set up) which are not sensible in the light of the conditions of the problem.

Example 1. Mr. Carlton wishes to make a pan by cutting squares from each corner of a 16-inch by 26-inch rectangular sheet of tin and folding up the sides. What should be the length of the side of each square if the base of the pan is to have an area of 200 square inches?

Solution:

1. Let $x =$ length (in inches) of edge of square.
   Then $26 - 2x =$ length (in inches) of base of pan,
   and $16 - 2x =$ width (in inches) of base of pan.

2. $(26 - 2x)(16 - 2x) = 200$

3. $416 - 84x + 4x^2 = 200$
   $4x^2 - 84x + 216 = 0$
   $x^2 - 21x + 54 = 0$
   $(x - 18)(x - 3) = 0$
   $x - 18 = 0 \quad | \quad x - 3 = 0$
   $x = 18 \quad | \quad x = 3$
   (Rejected)

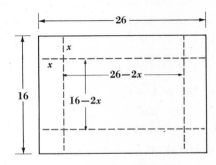

Although 18 is a solution of the equation, reject it because we cannot cut an 18-inch square from a rectangle only 16 inches wide.

Check: 4. If 3-inch squares are cut from each corner and the sides folded up, will the base have an area of 200 square inches?

$$(26 - 6)(16 - 6) \stackrel{?}{=} 200$$
$$(20)(10) \stackrel{?}{=} 200$$
$$200 = 200$$

∴ the length of the sides of each square should be 3 inches.

The next problem also has only one solution. The problem is especially interesting because it employs a most important rule:

$$d = rt + 16t^2$$

This rule gives a good approximation to the distance (in feet) covered in t seconds by an object falling freely toward the ground with an initial velocity of r feet per second.

Example 2.   From the top of a 200-foot building, an object is thrown toward the ground with an initial velocity of 40 feet per second. After how many seconds will the object hit the ground?

Solution: 1. Let t = number of seconds it takes the object to reach the ground,

d = distance fallen = 200 (feet),

r = rate at which object starts to fall

= 40 (feet per second).

2. $d = rt + 16t^2$
$200 = 40t + 16t^2$

3. $16t^2 + 40t - 200 = 0$ $\quad 2t - 5 = 0 \mid t + 5 = 0$
$2t^2 + 5t - 25 = 0$ $\quad\quad 2t = 5 \quad\mid\quad t = -5$
$(2t - 5)(t + 5) = 0$ $\quad t = \frac{5}{2}$ | (Rejected)

The solution -5 is rejected because the object could not have hit the ground before it was thrown.

4. Does an object take $\frac{5}{2}$ seconds to fall 200 feet when it starts falling at the rate of 40 feet per second?

$$200 \stackrel{?}{=} 40(\tfrac{5}{2}) + 16(\tfrac{5}{2})^2$$
$$200 \stackrel{?}{=} 100 + 16(\tfrac{25}{4})$$
$$200 \stackrel{?}{=} 100 + 100$$
$$200 = 200$$

∴ the object strikes the ground after $2\tfrac{1}{2}$ seconds.

Remember that we solve a problem by reasoning that if a number satisfies the requirements stated in the problem, then that number must satisfy the equation obtained in Step 2. On the other hand, just because a number satisfies the equation, we cannot conclude that the number will satisfy the problem. The solution set of the equation gives the *possible solutions* of the problem. By checking these possibilities in the statement of the problem, we find the *actual solutions* of the problem.

PROBLEMS

Solve each problem, rejecting solutions that do not meet the conditions of the problem.

1. Find two consecutive positive integers whose product is 56.
2. Find two consecutive positive integers whose product is 132.
3. The sum of the squares of two consecutive negative integers is 113. Find the integers.
4. The sum of the squares of two consecutive negative odd integers is 130. Find the integers.
5. If the length of a rectangle is 3 feet greater than the width and it has an area of 28 square feet, find its dimensions.
6. The length of a rectangle is 2 inches greater than twice its width, and its area is 60 square inches. Find its dimensions.
7. An object is thrown downward from the top of a 280-foot tower at a rate of 24 feet per second. In how many seconds does it hit the ground?
8. From an airplane flying at an altitude of 2560 feet, an object is thrown downward at a rate of 96 feet per second. In how many seconds will it strike the ground?
9. The front of a house is in the shape of a triangle on top of a rectangle. If the rectangle is twice as long as it is tall, the altitude of the triangular part is 5 feet less than the length of its base, and the total area of the front is 1500 square feet, find the width of the house.
10. The rectangular base of the Mark VII computer is 3 times as long as it is wide. The old Mark VI was 1 foot longer and 2 feet wider than the Mark VII. If the base area of the Mark VII is 13 square feet less than that of the Mark VI, find the dimensions of the Mark VII.
11. The perimeter of a rectangle is 60 inches and the area is 161 square inches. Find the dimensions of the rectangle.
12. The sum of the squares of three consecutive integers is 77. Find the integers.

The equation $h = rt - 16t^2$ is needed to solve the next four problems. It gives the height h, in feet, that an object will reach in t seconds when it is projected upward with a starting speed of r feet per second.

13. An object is projected upward at 160 feet per second. In how many seconds will the object be 400 feet above the ground?

14. A ball is thrown upward at 80 feet per second. In how many seconds will the object be 100 feet above the ground?

15. A ball is thrown upward at 64 feet per second. John is on top of a building 48 feet in height and catches the ball on its way down. How many seconds had the ball been in the air when John caught it?

16. A projectile is fired upward at 1600 feet per second. In how many seconds will the projectile hit the ground?

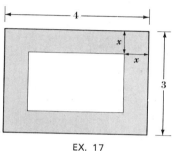

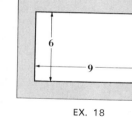

EX. 17 EX. 18

17. A strip of masking tape is placed around the edges of a rectangular window prior to painting its frame. If the window measures 3 feet by 4 feet, and the masking tape covers $\frac{1}{2}$ the area of the window, how wide is the tape?

18. The area of a concrete walk around a rectangular fishpond is equal to the area of the pond. If the pond measures 6 yards by 9 yards, find the width of the walk.

19. Let f be a function with domain $\mathcal{R}$ such that $f: x \to 2x^2 + 5x + 8$. Determine the least value of x such that $f(x) = 11$.

20. Let g be a function whose domain is the set of positive numbers. Given that $g: t \to 6t^2 + t - 2$, determine the value of t such that $g(t) = 0$.

21. The sum S of the first n consecutive natural numbers is given by the formula $S = \frac{1}{2}n(n + 1)$. How many such natural numbers must be added to give a sum of 28?

22. In a plane, consider a set of p points no three of which lie on a line. The number N of segments that can be drawn connecting all possible pairs of these points is given by the formula $N = \frac{1}{2}p(p - 1)$. How many points are there in the particular set for which the total number of segments joining pairs of points is 21?

23. What is the error in the following argument? The equation

$$x^2 + 4x + 3 = 1$$

is equivalent to

$$(x + 1)(x + 3) = 1 \cdot 1$$

Hence, the given equation is equivalent to:

$$x + 1 = 1 \quad \text{or} \quad x + 3 = 1$$
$$x = 0 \qquad\qquad x = -2$$

∴ its solution set is $\{0, -2\}$.

24. A picture 15 inches wide and 20 inches long is enclosed in a frame of uniform width and whose area is two-thirds that of the picture. Find the width of the frame.

25. The sides of 2 cubes differ by 2 inches and their volumes differ by 152 cubic inches. Find the length of a side of the smaller cube.

26. A farmer makes a rectangular enclosure using a stone wall for one side and 100 feet of fence for the other three sides. Find the dimensions of the enclosure if the area enclosed is 1200 square feet.

★**27.** Show that the sum of the squares of any two consecutive integers is 1 more than a multiple of 4. (*Hint:* Let n and $n + 1$ denote the integers. Also use the fact that one of every two consecutive integers is even.)

★**28.** Show that the square of an odd integer is 1 more than a multiple of 8. (*Hint:* If n denotes an integer, then $2n + 1$ denotes an odd integer. Square $2n + 1$ and use the fact that one of every two consecutive integers is even.)

chapter summary

1. **To find the greatest common factor** of a number of integers, factor each as a product of prime numbers. The factors of a polynomial with integral coefficients usually are limited to integers and polynomials with integral coefficients.

2. **To factor a polynomial**, use the distributive axiom to form a product of the greatest monomial factor, if any, and the remaining polynomial factor. Next, consider the possibilities of factoring this polynomial.

3. Certain **special products** should be read and factored at sight:

   The sum of two numbers times their difference:
   $$(a + b)(a - b) = a^2 - b^2$$
   The square of a binomial sum: $\quad (a + b)^2 = (a^2 + 2ab + b^2)$
   The square of a binomial difference: $\quad (a - b)^2 = (a^2 - 2ab + b^2)$

4. **To factor trinomial products** such as $ax^2 + bx + c, a > 0$:
   If b and c are positive, both binomial factors are sums; if b is negative and c is positive, both binomial factors are differences; if c is negative, the binomial factors are a sum and a difference. By inspection and trial, find factors of the quadratic and constant terms which produce binomials whose product contains a linear term with coefficient b.

5. **Factoring** must be complete; each polynomial factor must be prime over the set of polynomials with appropriate coefficients. The correctness of factoring should be checked by multiplication.

6. **To solve a polynomial equation by factoring**: Transform the equation into standard form with the right member zero and the left member a polynomial in descending powers of the variable. Factor the left member. Set each factor equal to zero, applying the principle that a product is zero if and only if at least one of its factors is zero. Solve the resulting linear equations. Check each possible root in the original equation. Write the solution set, listing multiple roots only once.

7. Problems leading to quadratic equations may have two answers. However, some problems have only one answer even though the equation has two roots. Therefore, all possible answers must be checked against the wording of the problem.

CHAPTER REVIEW

7–1   **1.** Find the greatest common factor of 60 and 450.
        2. $27a^2bc^2 = (-3abc^2)(\underline{\ ?\ })$
7–2   **3.** Write $45t^3 - 15t^2$ in factored form.
        4. Write $z(z + 2) + 3(z + 2)$ in factored form.
        5. Factor $x^3 - 15 - 3x^2 + 5x$.
7–3   **6.** $(-7y^4)^2 = \underline{\ ?\ }$ **7.** $(xy - 3)(xy + 3) = \underline{\ ?\ }$

        Factor each expression.

7–4   **8.** $27x^2 - 48$ **9.** $-121 + z^2$
7–5   **10.** $(2z + 3u)^2 = \underline{\ ?\ }$ **11.** $(4a - 3b)^2 = \underline{\ ?\ }$

7–6   **12.** Factor $9c^2 + 12cd + 4d^2$.
7–7   **13.** Solve and check: $(4x + 3)(x - 2) = (2x + 1)^2 + 2$.

Factor each expression.

7–8   **14.** $x^2 + 5x + 6$ **15.** $x^2 - 8xy + 7y^2$
7–9   **16.** $t^2 - 4t - 21$ **17.** $p^2 + 4p - 12$
7–10 **18.** $3r^2 + 20r + 12$ **19.** $6a^2 + ab - 12b^2$
7–11 **20.** $20n^2 - 22n - 12$

Solve each equation and check.

7–12 **21.** $(y + 3)(y - 9) = 0$ **22.** $3z(3z + 8) = 0$
7–13 **23.** $t^2 = 88 - 3t$ **24.** $64k^2 - 25 = 0$
7–14 **25.** Find the dimensions of a rectangle with area 24 square feet if its length is 2 feet greater than twice its width.

8
operations with fractions

8–1 Defining Fractions

Any indicated quotient of two mathematical expressions, such as $\frac{3}{1}$, $\frac{7}{13}$, $-\frac{11}{3}$, $\frac{t}{2}$, $\frac{5}{z}$, and $\frac{n^2 + 2n - 3}{2n + 1}$, is called a **fraction**. In a fraction $\frac{r}{s}$, r is called the **numerator**, and s the **denominator**. Since division by zero is not permitted, a *fraction is defined only when its denominator is not zero*. In the fractions below, the indicated numbers must be excluded from the replacement set of x. To find such excluded values, set the denominator of each fraction equal to zero, and solve the resulting equation.

Fraction	$\frac{5}{x}$	$\frac{2}{x-1}$	$\frac{3x+5}{x+7}$	$\frac{2}{x^2-1}$	$\frac{5x-2}{3}$
Excluded values of the variable	0	1	−7	1 and −1	No exclusions

EXERCISES

Express as a fraction. State any values of the variable for which the fraction is not defined.

1. $6 \div x$
2. $-3 \div b$
3. 0.23
4. 0.7
5. $0.5y$
6. $7z$
7. $(x - 1) \div x$
8. $t \div (t - 2)$

9. $h \div (3h - 6)$
10. $n \div (5n + 15)$
11. $(7z^2 + 2) \div 5$
12. $(8r^2 - 16) \div 4$
13. $(p - 3) \div (7p + 14)$
14. $(t - 8) \div (10t - 30)$
15. $1 \div z(z - 7)$
16. $1 \div k(k + 1)$

Give the set of excluded values of the variable.

17. $\dfrac{2z + 3}{z^2 - 8z + 15}$
18. $\dfrac{3x - 8}{x^2 + 8x + 12}$
19. $\dfrac{-2}{t^2 - 3t - 28}$
20. $\dfrac{1}{r^2 + 5r - 14}$
21. $\dfrac{2n}{3n^2 - 5n - 2}$
22. $\dfrac{-2r + 4}{5r^2 - 11r + 6}$
23. $\dfrac{k + 7}{k^2 - 49}$
24. $\dfrac{a - 8}{a^2 - 64}$
25. $\dfrac{-2}{c(c - d)}$
26. $\dfrac{4t}{(t - s)(t - 2s)}$
27. $\dfrac{x - 2y}{x^2 - 2xy + y^2}$
28. $\dfrac{3u + v}{u^2 - 3uv - 18v^2}$
29. $\dfrac{p - 3r}{2p^2 + pr - r^2}$
30. $\dfrac{a^2 - 2a + 4}{3a^2 - 11ab + 6b^2}$

8–2 Reducing Fractions to Lowest Terms

Using the property of quotients (page 147), we can show that the fractions $\frac{4}{3}$ and $\frac{12}{9}$ name the same number:

$$\frac{12}{9} = \frac{4 \cdot 3}{3 \cdot 3} = \frac{4}{3} \cdot \frac{3}{3} = \frac{4}{3} \cdot 1 = \frac{4}{3}$$

Similarly, any fraction of the form $\dfrac{4c}{3c}$ ($c \neq 0$) equals $\dfrac{4}{3}$ because if $c \neq 0$, then $\dfrac{c}{c}$ equals 1.

This example illustrates the following property:

Multiplication Property of Fractions

Dividing or multiplying the numerator and denominator of a fraction by the same nonzero number produces a fraction equal to the given one. That is:

$$\frac{ac}{bc} = \frac{a}{b}, \text{ provided } c \neq 0$$

Thus, $-\dfrac{24}{56} = -\dfrac{3 \cdot 8}{7 \cdot 8} = -\dfrac{3}{7}$ and $\dfrac{15z}{20z^2} = \dfrac{3 \cdot 5z}{4z \cdot 5z} = \dfrac{3}{4z}$ if $z \neq 0$.

A fraction whose numerator and denominator are polynomials is said to be **in lowest terms** when the numerator and denominator have no common factor other than 1 and -1. **Reducing a fraction to lowest terms** is the process of dividing the numerator and denominator by their *greatest common factor* (page 161).

Example 1. Reduce $\dfrac{4x - 12}{2x^2 - 5x - 3}$ to lowest terms.

Solution: Factor numerator and denominator. $\rightarrow \dfrac{4x - 12}{2x^2 - 5x - 3} = \dfrac{4(x - 3)}{(2x + 1)(x - 3)}$

What values of x are excluded? $\rightarrow -\tfrac{1}{2}$ and 3

Divide numerator and denominator by their greatest common factor, $x - 3$. $\rightarrow \dfrac{4(x - 3) \div (x - 3)}{(2x + 1)(x - 3) \div (x - 3)}$

Simplify the result. $\rightarrow = \dfrac{4}{2x + 1}$ if $x \notin \{-\tfrac{1}{2}, 3\}$.

Example 2. Simplify $\dfrac{5 - y}{y^2 - 25}$.

Solution: Factor numerator and denominator. $\rightarrow \dfrac{5 - y}{y^2 - 25} = \dfrac{5 - y}{(y + 5)(y - 5)}$

Note that $y \notin \{-5, 5\}$.

To show the common factor, express the numerator as a product having -1 as a factor. $\rightarrow = \dfrac{-1(y - 5)}{(y + 5)(y - 5)}$

Simplify. $\rightarrow = \dfrac{-1}{y + 5}$ if $y \notin \{-5, 5\}$.

The fraction $\dfrac{-1}{y + 5}$ in Example 2 above also can be written in the form $-\dfrac{1}{y + 5}$, because

$$\dfrac{-1}{y + 5} = (-1) \cdot \dfrac{1}{y + 5} = -\dfrac{1}{y + 5}.$$

Note that a fraction can be reduced only when the numerator and denominator have a common *factor*. Compare the fractions below:

$$\frac{5 \cdot 7}{5} = \frac{7}{1}$$

$$\frac{ab}{a} = \frac{b}{1} = b, \text{ if } a \neq 0.$$

a and b are factors of the numerator. This fraction can be reduced because a is a common factor of the numerator and denominator.

$$\frac{5 + 7}{5}$$

$$\frac{a + b}{a}, \text{ if } a \neq 0.$$

a and b are *not* factors of the numerator. This fraction can *not* be reduced, for no factor (other than 1 or -1) is common to both numerator and denominator.

EXERCISES

Write each fraction in lowest terms, noting all necessary restrictions on values of the variables.

1. $\dfrac{15a}{3}$
2. $\dfrac{28b^2}{7}$
3. $\dfrac{14}{21x}$
4. $\dfrac{15}{10y}$

5. $\dfrac{24a^2b}{-24ab}$
6. $\dfrac{-8xy^2}{24xy}$
7. $\dfrac{2c + 2d}{5c + 5d}$
8. $\dfrac{3x - 3y}{3x + 3y}$

9. $\dfrac{7a - 7b}{a^2 - b^2}$
10. $\dfrac{z^2 - 9}{z - 3}$
11. $\dfrac{k^2 - 9}{3 - k}$
12. $\dfrac{8rt - 2r}{2rt^2}$

13. $\dfrac{9p^2q}{3pq^2 + 6p^2q}$
14. $\dfrac{4 - r^2}{r - 2}$

15. $\dfrac{x^2 - 9}{x^2 + 6x + 9}$
16. $\dfrac{y^2 + 8y + 16}{y^2 - 16}$

17. $\dfrac{p - p^2q}{q - pq^2}$
18. $\dfrac{7t^2 - 28}{t + 2}$

19. $\dfrac{2z^2 + 4}{2z + 4}$
20. $\dfrac{kr^2 + k}{kr + k}$

21. $\dfrac{x^2 - 3x}{x^2 - 2x - 3}$
22. $\dfrac{y^2 + y - 6}{3y^2 - 27}$

23. $\dfrac{3x^2 + 15x + 18}{3x^2 - 12}$
24. $\dfrac{(x - 3)^2}{x^2 - 9}$

25. $\dfrac{x^2 + 6xy + 9y^2}{x^2 - 9y^2}$
26. $\dfrac{r^2 + rs - 6s^2}{r^2 - 4s^2}$

27. $\dfrac{y^2 - 2y - 8}{y^2 - 4y}$
28. $\dfrac{3x^2 - 9x}{x^2 - 5x + 6}$

29. $\dfrac{n^2 - 5n + 6}{n^2 - 4n + 4}$ 30. $\dfrac{y^2 - 8y + 15}{y^2 + 4y - 21}$

31. $\dfrac{n^2 + n - 6}{n^2 - 7n + 10}$ 32. $\dfrac{z^2 - 2z - 15}{z^2 + 3z - 40}$

Explain why each of the following fractions cannot be reduced.

33. $\dfrac{3 + y}{y}$ 34. $\dfrac{3 - n}{2 - n}$ 35. $\dfrac{t^2 - 1}{t + 3}$ 36. $\dfrac{z^2}{z + 4}$

Write each fraction in lowest terms, noting all necessary restrictions on the values of the variables.

37. $\dfrac{2y^3 - y^2 - 10y}{y^3 - 2y^2 - 8y}$ 38. $\dfrac{2z^3 + z^2 - 3z}{6z^3 + 5z^2 - 6z}$

39. $\dfrac{4r^4 + 2r^3 - 6r^2}{4r^4 + 26r^3 + 30r^2}$ 40. $\dfrac{3x^3 + 2x^2 - 8x}{3x^4 - x^3 - 4x^2}$

41. $\dfrac{18x^3 + 3x^2 - 36x}{12x^3 - 31x^2 + 20x}$ 42. $\dfrac{3n^4 + 27n^3 + 60n^2}{6n^2 + 6n - 72}$

43. $\dfrac{x^2 - x - 6}{x^2 + 2a + ax + 2x}$ 44. $\dfrac{rt - us - ur + st}{rt - 2us + 2st - ur}$

8–3 Ratio

To compare the enrollments at two schools, one having 8000 students and the other 2000 students, we can say that the first school has four times as many students as the second. This comparison is made by computing the quotient $\frac{8000}{2000} = 4$. We can also say that the enrollments are in the *ratio* of 4 to 1.

A **ratio** of one number to another one (not zero) is the quotient of the first number divided by the second. The ratio 7 to 5 can be expressed by:

1. An indicated quotient using the division sign $\div$ $\longrightarrow$ $7 \div 5$
2. An indicated quotient using the ratio sign $\;:\;$ $\longrightarrow$ $7 : 5$
3. A fraction $\longrightarrow$ $\frac{7}{5}$
4. A fraction in decimal notation $\longrightarrow$ 1.4

By the multiplication property of fractions (page 196), the ratio $7:5$ compares not only the numbers 7 and 5, but also 14 and 10, 21 and 15, -28 and -20, and $7n$ and $5n$, where $n \neq 0$. However, to compare a 7-pound weight to a 5-ounce weight, we must change the 7 pounds to 112 ounces and then use the ratio $\frac{112}{5}$ or $112 : 5$.

To find the ratio of two quantities of the same kind:

1. Find the measures in the same unit.

2. Then divide these measures.

Example. In an alloy, the ratio of copper to tin is 2 : 5. How many pounds of each metal are in 140 pounds of the alloy?

Solution:
1. Let $2n$ = the number of pounds of copper.
   Then $5n$ = the number of pounds of tin.
2. $2n + 5n = 140$
3. $7n = 140$
   $n = 20$
   $\therefore 2n = 40,\ 5n = 100$

Check:
4. Are the weights in the ratio of 2 to 5? $\quad \frac{40}{100} \stackrel{?}{=} \frac{2}{5},\ \frac{2}{5} = \frac{2}{5}$
   Do the weights of copper and tin total 140 pounds? $\quad 40 + 100 \stackrel{?}{=} 140,\ 140 = 140$
   $\therefore$ there are 40 pounds of copper and 100 pounds of tin in the alloy.

Any real number that is the ratio of two integers (the second integer not zero) is called a **rational number**. Thus, the following are rational numbers:

$$\frac{3}{4}, \quad 3 = \frac{3}{1}, \quad 2\frac{1}{2} = \frac{5}{2}, \quad 0.7 = \frac{7}{10}, \quad -\frac{2}{3} = \frac{-2}{3} = \frac{2}{-3}$$

Polynomials or fractions whose numerators and denominators are polynomials are often called **rational expressions**. Thus, $\frac{x}{2}$ might be called a "rational expression" although it may not represent a rational number.

EXERCISES

Give each ratio in its lowest terms.

1. The area of a 3-inch by 4-inch rectangle to that of one 2 by 3 inches.
2. The area of a 7-inch by 10-inch rectangle to that of one 5 by 7 inches.
3. The area of a 2-inch by 3-foot rectangle to that of one 10 by 20 inches.

4. The area of a 6-inch square to that of a 0.75-foot square.
5. The cost per ounce of flower seed to the cost of $32 per pound.
6. 384 students to 12 teachers.
7. The ratio of miles to hours on a trip of 496 miles completed in 8 hours.
8. Wins to losses in a season of 38 games with 8 losses.

In Exercises 9–12, use the rule that in a triangle

$$\text{Area} = \tfrac{1}{2}(\text{Base} \times \text{Altitude}).$$

9. The area of a triangle with a 10-yard base and a 12-yard height to that of one with a 16-yard base and an 8-yard height.
10. The area of a triangle with a 12-meter base and a 20-meter height to that of one with a 30-meter base and a 4-meter height.
11. The area of a triangle with a 3-foot base and an 8-inch height to that of a rectangle measuring 6 inches by 18 inches.
12. The area of a triangle with a 24-inch base and a $1\tfrac{2}{3}$-foot height to that of a square with sides each measuring 1 yard.

Find the ratio $x:y$ in each case.

Example. $5x = 7y$

Solution: $5x = 7y$

$$\frac{5x}{5y} = \frac{7y}{5y}$$

$$\frac{x}{y} = \frac{7}{5}$$

Thus, $x:y = 7:5$.

13. $2x = 3y$
14. $9x = 7y$
15. $5x = 9y$
16. $3x = 3y$
17. $7y = 14x$
18. $3y = 12x$
19. $4x - 3y = 0$
20. $-7x + 2y = 0$
21. $\dfrac{2x + 3y}{3y} = \dfrac{3}{4}$
22. $\dfrac{5x - y}{y} = \dfrac{4}{3}$
23. $\dfrac{3x + 2y}{2y} = \dfrac{7}{2}$
24. $\dfrac{5x - 3y}{3y} = -\dfrac{12}{7}$
★25. $\dfrac{x^2 + 2y^2}{y^2} = \dfrac{2x + y}{y}$
★26. $\dfrac{x^2 + 5y^2}{y^2} = \dfrac{y - 4x}{y}$

PROBLEMS

1. Find the greater of two numbers in the ratio of 7 to 2, whose difference is 25.
2. Find the lesser of two numbers in the ratio of 2 to 5, whose sum is -14.
3. How many of the 280 members of a golf club are men if the ratio of women to men is 2 to 5?
4. How many pages of a 34-page newspaper are classified advertisements if the ratio of classified advertisements to other material in the newspaper is 3 to 14?
5. If a car goes 52 miles on 4 gallons of gasoline, how far will it travel on 7 gallons?
6. If 38 ounces of a salt solution contain 15 ounces of salt, how much salt will 30 ounces of the solution contain?
7. A mutual fund invests in bonds and stocks in the ratio of 4 to 5. How much of $27,000 invested will go into bonds?
8. In the Acme Manufacturing Company, 4 out of every 5 dollars expended by the company goes to labor costs. How much of an expenditure of $100,000 will go to labor costs?
9. A 90-foot rope is cut into two pieces whose lengths are in the ratio of 26 to 19. How long is the shorter piece of rope?
10. Mr. Fisher plants wheat and alfalfa in the ratio of 5 to 3. How many acres of a 160-acre plot are devoted to alfalfa?
11. Which is a better buy on a canned vegetable, a 16-ounce can at 72 cents or a 12-ounce can at 57 cents?
12. Miss Avery types 2000 words in 50 minutes, while Miss Benson types 738 words in 19 minutes. Which one is the faster typist?
13. If 2 of every 9 trees in a forest containing 831,762 trees are cedars, how many trees in the forest are not cedars?
14. If 7 out of every 11 dollars collected by Central City in taxes are spent on education, how many of the $24,026,343 collected goes to items other than education?
15. In a dry concrete mixture the ratio of sand to cement is $4:1$ and the ratio of gravel to sand is $5:4$. How many pounds of each are in 1100 pounds of the concrete?
16. A casting is made from an alloy containing 4 parts lead, 3 parts copper, and 2 parts tin. How many pounds of each does a 117-pound casting contain?
17. A profit of $1800 is divided among 3 persons in the ratio of $8:5:2$. How much does each person receive?
18. Three house lots have frontages of 70, 80, and 100 feet. How much of an assessment of $2000 should be assigned the owner of each lot?

8-4 Percent and Percentage Problems

The ratio of one number to another is often expressed as a *percent*. The word **percent** (denoted by %) stands for "divided by 100" or "hundredths." Hence:

$$7\% \text{ is another way of writing } \frac{7}{100} \text{ or } 0.07$$
$$150\% \text{ is another name for } \frac{150}{100} \text{ or } 1.5$$
$$1\% = \frac{1}{100} \quad \text{and} \quad 100\% = \frac{100}{100} = 1$$

To write a ratio as a percent:

1. **Write the ratio as a fraction with denominator 100.**
2. **Then write the numerator followed by a percent sign.**

Example 1. Express each number as a percent: $\frac{3}{8}$, 3.4

Solution:
$$\frac{3}{8} = \left(\frac{3}{8} \cdot 100\right)\frac{1}{100} = \frac{37\frac{1}{2}}{100} = 37\frac{1}{2}\%$$
$$3.4 = (3.4)(100)\frac{1}{100} = \frac{340}{100} = 340\%$$

A **percentage** is a number equal to the product of the percent and another number, called the **base**. Since percent is the ratio of the percentage to the base, it is often called the **rate** to avoid confusion with *percentage*. The key to percent and percentage problems is this basic relationship:

$$\textbf{Percentage} = \textbf{Rate} \times \textbf{Base} \quad \text{or} \quad p = rb, \quad b \neq 0.$$
$$\frac{\textbf{Percentage}}{\textbf{Base}} = \textbf{Rate} \quad \text{or} \quad \frac{p}{b} = r, \quad b \neq 0.$$

Example 2. How much is a 15% discount on an item whose price is $40?

Solution:
1. Let p = discount (percentage)
   $r = 15\%$
   $b = 40$

2. $p = rb$ or $\frac{p}{b} = r$
   $p = (0.15)(40)$
   $\frac{p}{40} = \frac{15}{100}$

3. $p = 6.00$

4. Check that the discount (percentage) is $6.

EXERCISES

Determine each of the following.

1. 12% of 180
2. 3% of 3.2
3. 42% of 18.5
4. 80% of 1000
5. 25% of 16
6. 100% of 83
7. 200% of 15
8. 150% of 44
9. 400% of 1
10. 3.25% of 48
11. $\frac{3}{5}$% of 820
12. 0.02% of 1000

Find the number.

13. 18 is 60% of the number.
14. 23 is 25% of the number.
15. 4% of the number is 3.4.
16. 75% of the number is 9.75.
17. 100% of the number is 218.
18. 150% of the number is 84.
19. $\frac{1}{5}$% of the number is 1.23.
20. $1\frac{1}{2}$% of the number is 30.2.

Determine each rate.

21. What % of 64 is 48?
22. What % of 56 is 24?
23. What % of 8 is 24?
24. What % of 12 is 40?
25. 2 is what % of 400?
26. 5 is what % of 900?
27. 80 is what % of 5?
28. 270 is what % of 30?

PROBLEMS

1. If 4% of 2400 persons polled expressed no opinion, how many of the persons did express an opinion?
2. If Jupiter Airlines Flight 203 was filled to 65% of capacity, how many of the 120 seats were occupied?
3. How many minutes out of an hour's TV program are taken up by commercials if 15% of the program is alloted to commercials?
4. How much zinc is in 30 pounds of an alloy containing 28% zinc?
5. Mr. Tamura paid $124 in sales tax on his new car. If this represents 4% of the price of the car, what was the price of the car?
6. Jack received 462 votes for president of student body. If this represented 55% of the votes cast for president, how many votes were cast?
7. Mr. McGee receives a commission of 5% for selling a house. If he received an $1860 commission, what was the price of the house he sold?
8. During a sale, a coat sells for $77.50. If this is 62% of the usual price, what is the usual price of the coat?
9. The sales tax on $140 is $6.30. What is the rate of sales tax?

10. Out of a shipment of 12,000 bolts, 360 are defective. What percent of the shipment are defective?
11. The salesman's commission on a $22,000 home is $1540. What is his rate of commission?
12. The list price of a camera is $150. If a discount of $7.50 is given for cash, what percent of the list price is the discount?
13. The price of one share of a certain stock rose from $64.20 to $67.41. By what percent of the first price did the price of the stock increase?
14. For tax purposes, a house worth $28,000 one year was worth $26,880 the next year. By what percent of the initial value had the value of the house depreciated?
15. Mr. Fisher received $1136 as trade-in for his old car. What was the original price of the old car if it had depreciated by 60% of its original purchase price?
16. If an ore contains 15% copper, how many tons of ore are necessary to obtain 18 tons of copper?
17. A dealer pays $190 for a TV set. He has overhead (expenses) totaling 15% of the selling price and wishes to make 40% profit on the selling price. What price should he charge for the set?
18. An article is marked $24 and an 18% discount on that price is given. What profit is made if the article cost $12?

Example. Find the number of degrees in each central angle of this graph, called a **circle graph**.

Solution: The sum of all the adjacent angles around a point is 360°. Thus,

Oxygen, $\frac{48}{100} \times 360° = 172.8° \doteq 173°$

Silicon, $\frac{28}{100} \times 360° = 100.8° \doteq 101°$

Aluminum, $\frac{8}{100} \times 360° = 28.8° \doteq 29°$

Other, $\frac{16}{100} \times 360° = 57.6° \doteq 58°$

Check: $173° + 101° + 29° + 58° = 361°$. (We have 1° too many, because we rounded each measurement to a whole number of degrees. To compensate, we can replace 58° by 57°.

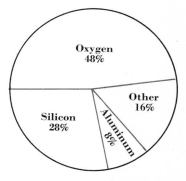

Average Chemical Composition of the Earth's Crust

19. Of the registered voters in Valley City, 43% are registered Democratic, 29% Republican, 15% Independent; the rest decline to state a party.
    a. Make a circle graph, using a protractor to draw the central angles.
    b. If 540 voters decline to state a party, how many registered voters are there in Valley City?

20. A survey found that at a given hour in Central City, of the TV sets in operation, 38% were tuned to channel 2, 35% to channel 5, 16% to channel 7, 5% to channel 9, and the rest to channel 11.

   a. Make a circle graph.

   b. If 980 sets were tuned to channel 5, how many sets in all were in operation?

Make a circle graph from the data given in each table. Label each graph.

21.

Ages of Licensed Drivers in U.S.	Under 20	20–25	26–35	36–45	46–65	Over 65
% of Licensed Drivers in U.S.	10	10	25	25	25	5

22.

Ages of U.S. population in 1967	Under 5	5–19	20–44	45–64	65 and over
% of population of given age to nearest 1%	10	29	31	20	10

8–5 Multiplying Fractions

When the property of quotients (see page 147)

$$\frac{xy}{cd} = \frac{x}{c} \cdot \frac{y}{d}$$

is read from right to left, we have the following:

Rule for Multiplying Fractions

For any real numbers x, y, c, and d, if $c \neq 0$ and $d \neq 0$, then:

$$\frac{x}{c} \cdot \frac{y}{d} = \frac{xy}{cd}$$

That is, when fractions are multiplied, the product is a fraction whose numerator is the product of the numerators and whose denominator is the product of the denominators of the given fractions.

OPERATIONS WITH FRACTIONS 207

Thus, by the rule for multiplying fractions, we know that:

$$\frac{(x-4)}{(x-3)} \cdot \frac{(x+5)}{(x+3)} = \frac{(x-4)(x+5)}{(x-3)(x+3)} = \frac{x^2+x-20}{x^2-9}, \quad x \notin \{3, -3\}$$

$$7t \cdot \frac{5t}{r} = \frac{7t}{1} \cdot \frac{5t}{r} = \frac{35t^2}{r}, \quad r \neq 0$$

A product that is not in lowest terms should be reduced. Thus:

$$\frac{a}{b} \cdot \frac{b}{a} = \frac{ab}{ba} = \frac{ab}{ab} = 1, \quad a \neq 0, \; b \neq 0$$

$$\frac{x+y}{x-y} \cdot \frac{x-y}{5} = \frac{(x+y)(x-y)}{5(x-y)} = \frac{x+y}{5}, \quad x \neq y$$

The multiplication of fractions can be simplified by first factoring where possible:

$$\frac{4y-6}{3y+5} \cdot \frac{9y^2-25}{2y^2+y-6} = \frac{2(2y-3)}{(3y+5)} \cdot \frac{(3y+5)(3y-5)}{(2y-3)(y+2)}, \quad y \notin \{-\tfrac{5}{3}, \tfrac{3}{2}, -2\}$$

$$= \frac{2(3y-5)(2y-3)(3y+5)}{(y+2)(2y-3)(3y+5)}$$

$$= \frac{2(3y-5)}{y+2}, \text{ or } \frac{6y-10}{y+2}, \quad y \notin \{-\tfrac{5}{3}, \tfrac{3}{2}, -2\}$$

Hereafter, in the sets of exercises it will be assumed that the replacement sets of the variables include no value for which the denominator is zero.

EXERCISES

Express each product as a single fraction in lowest terms.

1. $\frac{3}{5} \cdot \frac{2}{11} \cdot 3$
2. $\frac{1}{3} \cdot \frac{5}{4} \cdot (-5)$
3. $\frac{2}{9} \cdot \frac{3}{5}$
4. $\frac{4}{9} \cdot \frac{3}{8}$
5. $\frac{2}{3} \cdot \frac{3}{5} \cdot \frac{5}{4}$
6. $\frac{3}{8} \cdot \frac{4}{9} \cdot \frac{12}{5}$
7. $\frac{24}{30} \cdot \frac{20}{36} \cdot \frac{3}{4}$
8. $\frac{9}{15} \cdot \frac{6}{8} \cdot \frac{2}{10}$
9. $\frac{8a}{13b} \cdot \frac{26ab}{4a}$
10. $\frac{-6rs}{5} \cdot \frac{10r}{3r^2s}$
11. $\frac{2}{3} \cdot \frac{9z^2}{4}$
12. $6a^2b \cdot \frac{2}{3a^2}$
13. $\frac{-16xy^2}{8x} \cdot \frac{14x^2y}{14y}$
14. $\frac{21a^2b}{8c} \cdot \frac{-3c^2}{7ab}$
15. $\frac{3x+15}{2x} \cdot \frac{4x}{4x+40}$
16. $\frac{2t+14}{6t} \cdot \frac{9t^2}{t^2+7t}$
17. $\frac{y-3}{8y-4} \cdot \frac{10y-5}{5y-15}$
18. $\frac{2z-4}{3z+6} \cdot \frac{2z+3}{z-2}$

19. $\dfrac{a^2 - b^2}{a^2 - 16} \cdot \dfrac{a + 4}{a + b}$

20. $\dfrac{y^2 - 4}{y^2 - 1} \cdot \dfrac{y - 1}{y - 2}$

21. $\dfrac{z^2 - 2z - 3}{3z^2} \cdot \dfrac{6z}{z + 1}$

22. $\dfrac{t^2 - 2t + 1}{4t} \cdot \dfrac{8t^2}{t - 1}$

23. $\dfrac{x^2 + 5x + 6}{2x - 2} \cdot \dfrac{x^2 - x}{x + 3}$

24. $\dfrac{n^2 - 3n - 4}{n^2 - 2n} \cdot \dfrac{n - 2}{n + 1}$

25. $\dfrac{r^2 - r - 20}{r^2 + 7r + 12} \cdot \dfrac{r^2 + 9r + 18}{r^2 - 7r + 10}$

26. $\dfrac{p^2 + p - 2}{p^2 - 3p + 2} \cdot \dfrac{p^2 - p - 2}{p^2 + 5p + 6}$

27. $\dfrac{x - y}{x^2 + xy} \cdot \dfrac{x^2 - y^2}{x^2 - xy}$

28. $\dfrac{r^2 + s^2}{r^2 - s^2} \cdot \dfrac{r - s}{r + s}$

29. $\dfrac{n^2 - 11n + 30}{n^2 - 6n + 9} \cdot \dfrac{n^2 - 3n}{n^2 - 5n}$

30. $\dfrac{t^2 - 2t - 3}{t^2 - 9} \cdot \dfrac{t^2 + 5t + 6}{t^2 - 1}$

31. $\dfrac{a^2 - 4}{a^2 - 5a + 6} \cdot \dfrac{a^2 - 2a - 3}{a^2 + 3a + 2}$

32. $\dfrac{c^2 - d^2}{c^2 + 4cd + 3d^2} \cdot \dfrac{c^2 + cd - 6d^2}{c^2 + cd - 2d^2}$

33. $\dfrac{2a^2 - a - 3}{6a^2 - 13a + 6} \cdot \dfrac{3a^2 - 2a}{a + 1}$

34. $\dfrac{z^2 - z - 6}{z^3 - 9z} \cdot \dfrac{z + 3}{3z + 9}$

35. $\dfrac{u^2 + 3u + 2}{u^2 + u} \cdot \dfrac{u^2 + 3u}{u^2 + 5u + 6}$

36. $\dfrac{b^2 + 5bc + 4c^2}{bc + 4c^2} \cdot \dfrac{b^2 + 5bc}{b^2 + 6bc + 5c^2}$

37. $\dfrac{n^2 + 4n + 3}{n^2 - 1} \cdot \dfrac{n^2 - 2n + 1}{n + 3} \cdot \dfrac{n + 1}{n - 1}$

38. $\dfrac{3t^2 - 27}{t^2 + t - 6} \cdot \dfrac{t^2 + 3t}{6} \cdot \dfrac{2t - 4}{t - 3}$

39. $\dfrac{20 + y - y^2}{y^2 - 6y + 5} \cdot \dfrac{6 - 5y - y^2}{y^2 + 7y + 12} \cdot \dfrac{y^2 - 9}{36 - y^2}$

40. $\dfrac{12 + r - r^2}{9 - r^2} \cdot \dfrac{r + 2}{r^2 + r} \cdot \dfrac{3 + 2r - r^2}{8 + 2r - r^2}$

41. $\dfrac{k^2 + 4k + 3}{k^2 - 8k + 7} \cdot \dfrac{35 + 2k - k^2}{k^2 - 7k - 8} \cdot \dfrac{k^2 - 9k + 8}{k^2 + 8k + 15}$

42. $\dfrac{2b^2 - b - 3}{4b^2 - 5b + 1} \cdot \dfrac{b^2 - 1}{(b + 1)^2} \cdot \dfrac{1 - 3b - 4b^2}{3 - 5b + 2b^2}$

8–6 Dividing Fractions

A quotient can be expressed as the product of the dividend and the reciprocal of the divisor (page 86). Thus:

$$12 \div 4 = 12 \times \tfrac{1}{4}; \qquad 8 \div \tfrac{1}{5} = 8 \times 5; \qquad \tfrac{2}{3} \div \tfrac{4}{9} = \tfrac{2}{3} \times \tfrac{9}{4}$$

Since the reciprocal of $\dfrac{c}{d}$ is $\dfrac{d}{c}$, we have the following:

Rule for Dividing Fractions

For any real numbers a, b, c, and d, if $b \neq 0$, $c \neq 0$, and $d \neq 0$, then:

$$\frac{a}{b} \div \frac{c}{d} = \frac{a}{b} \cdot \frac{d}{c} = \frac{ad}{bc}$$

That is, to divide fractions, multiply the dividend by the reciprocal of the divisor.

Example. Simplify: $\dfrac{x^2 - 16}{x + 4} \div \dfrac{x^2 - 8x + 16}{4 - x}$

Solution:
$$\frac{x^2 - 16}{x + 4} \div \frac{x^2 - 8x + 16}{4 - x} = \frac{x^2 - 16}{x + 4} \cdot \frac{4 - x}{x^2 - 8x + 16}$$

$$= \frac{(x + 4)(x - 4)}{(x + 4)} \cdot \frac{4 - x}{(x - 4)(x - 4)}$$

$$= \frac{(x + 4)(x - 4)(-1)(x - 4)}{(x + 4)(x - 4)(x - 4)} = -1$$

EXERCISES

Simplify each expression.

1. $\frac{2}{5} \div \frac{7}{10}$
2. $\frac{4}{9} \div \frac{2}{15}$
3. $\dfrac{x}{y^2} \div \dfrac{x^2}{y}$
4. $\dfrac{r^2}{s^2} \div \dfrac{r}{s^3}$
5. $\dfrac{p^2}{2q} \div \dfrac{p^4}{4q^3}$
6. $\dfrac{n^2}{m^3} \div \dfrac{3n}{m^4}$
7. $\dfrac{3t}{8s^2} \div \dfrac{12t^2}{4s^3}$
8. $\dfrac{81k^2}{28k} \div \dfrac{9k}{7k^3}$
9. $(-16z^2) \div \dfrac{4z}{3}$
10. $\dfrac{3ab}{4} \div (-12b^2)$
11. $\dfrac{r + s}{18} \div \dfrac{r + s}{3}$
12. $\dfrac{3z - 1}{36} \div \dfrac{3z - 1}{9}$
13. $\dfrac{y^2 - 4}{2y} \div (y + 2)$
14. $\dfrac{t^2 - 2t + 1}{t^2} \div (t - 1)$
15. $\dfrac{z + 2}{z^2 - 9} \div \dfrac{1}{z - 3}$
16. $\dfrac{k^3}{k^2 + 4k + 4} \div \dfrac{k}{k + 2}$
17. $\dfrac{4n - 8}{3n + 9} \div \dfrac{2n - 4}{6n + 18}$
18. $\dfrac{r^2 + 2rs}{2rs + s^2} \div \dfrac{r^3 + 2r^2s}{rs + s^2}$
19. $\dfrac{3a + 3z}{4a^2} \div \dfrac{a^2 - z^2}{2a^2}$
20. $\dfrac{x^2 - 4}{x^3} \div \dfrac{x^2 - 4x + 4}{x^2}$

21. $\dfrac{n^2 - m^2}{n^2 - 3n - 4} \div \dfrac{n - m}{n^2 + n}$

22. $\dfrac{y^2 - 9}{y^2 - 6y + 9} \div \dfrac{3y + 9}{7y - 21}$

23. $\dfrac{9 - a^2}{3a - 3b} \div \dfrac{9 - 6a + a^2}{b^2 - a^2}$

24. $\dfrac{4z^2 + 8z + 3}{2z^2 - 5z + 3} \div \dfrac{1 - 4z^2}{6z^2 - 9z}$

25. $\dfrac{1 - 4t^2}{t^2 - 4} \div \dfrac{4t + 2}{t^2 + 2t}$

26. $\dfrac{c^2 + 2c^3}{9 - c^2} \div \dfrac{c - 4c^3}{3c + c^2}$

27. $\dfrac{2n^2 - 18}{n^2 + 6n - 7} \div \dfrac{8n^2 + 4n - 24}{n^2 - 1}$

28. $\dfrac{20 + r - r^2}{r^2 + 7r + 12} \div \dfrac{(r - 5)^2}{(r + 3)^2}$

29. $\dfrac{3s^2 - 14s + 8}{2s^2 - 3s - 20} \div \dfrac{6 - 25s + 24s^2}{15 - 34s - 16s^2}$

30. $\dfrac{2x^2 - 5x - 3}{3x^2 - 10x - 8} \div \dfrac{9 - x^2}{12 + x - x^2}$

★31. $\dfrac{ab - 3a + b - 3}{a - 2} \div \dfrac{a + 1}{a^2 - 4}$

★32. $\dfrac{2rs + 4r + 3s + 6}{2r + 3} \div \dfrac{s + 2}{r - 1}$

★33. $\dfrac{ac - bc + bt - at}{3x^3 - 3xy^2} \div \dfrac{ac - at}{y^2 - 2xy + x^2}$

★34. $\dfrac{ux + 2uy - vx - 2vy}{x^2 + xy - 2y^2} \div \dfrac{u^2 + uv - 2v^2}{x^2 - 2xy + y^2}$

8–7 Expressions Involving Multiplication and Division

In the absence of parentheses, the rule for order of performing multiplications and divisions (page 28) is applied to an expression containing fractions. Replace only the fraction immediately following a division sign by its reciprocal.

Example. Simplify: $\dfrac{x^2 - x - 2}{x^2 + 2x + 1} \div \dfrac{x - 2}{7} \cdot \dfrac{4}{x}$

Solution: $\dfrac{x^2 - x - 2}{x^2 + 2x + 1} \div \dfrac{x - 2}{7} \cdot \dfrac{4}{x} = \dfrac{x^2 - x - 2}{x^2 + 2x + 1} \cdot \dfrac{7}{x - 2} \cdot \dfrac{4}{x}$

$= \dfrac{(x + 1)(x - 2)(7)(4)}{(x + 1)(x + 1)(x - 2)x}$

$= \dfrac{28}{(x + 1)x}$, or $\dfrac{28}{x^2 + x}$

Check: Let x have the value 3. The check is left to you.

EXERCISES

Simplify each expression and check by substitution.

1. $\dfrac{2}{y^2} \cdot \dfrac{y}{6} \div \dfrac{x}{9}$

2. $\dfrac{z^3}{8} \cdot \dfrac{2}{z^4} \div \dfrac{z}{6}$

3. $\dfrac{rs^2}{t} \cdot \dfrac{st^2}{r} \div rst$

4. $\dfrac{4ab^3}{3c} \cdot \dfrac{6bc}{5a} \div \dfrac{abc}{10}$

5. $\dfrac{y^2 - 4}{y^2} \cdot \dfrac{y}{y + 2} \div \dfrac{y - 2}{2y}$

6. $\dfrac{4x}{4x - 3} \cdot \dfrac{8x - 6}{6x^2} \div \dfrac{x + 1}{3}$

7. $\dfrac{3t + 4}{8st} \div \dfrac{9t + 12}{12s^2} \cdot \dfrac{4t^2}{9}$

8. $\dfrac{x - 3y}{3x} \div \dfrac{8x - 24y}{9x^2} \cdot \dfrac{16y}{3x}$

9. $\dfrac{p^2}{p^2 - q^2} \cdot \dfrac{p + q}{p - q} \div \dfrac{p}{(p - q)^2}$

10. $\dfrac{k^2 + 4k + 4}{3k} \cdot \dfrac{k - 2}{k + 2} \div \dfrac{k^2 - 4}{4k^3}$

11. $\dfrac{x}{x + 3} \div \dfrac{3x^2}{3x + 9} \cdot \dfrac{x^2 + 4x + 3}{x^2 - 9}$

12. $\dfrac{2y - 1}{4y^2} \div \dfrac{4y + 2}{y^3} \cdot \dfrac{4y^2 + 4y + 1}{4y^2 - 1}$

13. $\dfrac{t^2}{4t} \div \dfrac{s^2t^2 - 4}{2t} \cdot \dfrac{st + 2}{st}$

14. $\dfrac{5a}{a^2b^2 - 9} \div \dfrac{20}{ab} \cdot \dfrac{3ab + 9}{a^2b}$

15. $\dfrac{x^2 + 9x + 14}{x^2 - 3x} \cdot \dfrac{2x^2 + 2x}{x^2 + 6x - 7} \div \dfrac{x + 2}{x - 3}$

16. $\dfrac{y^2 + 2y - 15}{6y^2} \cdot \dfrac{24y}{y^2 + 7y + 10} \div \dfrac{y - 3}{y + 2}$

17. $\dfrac{4n^2}{2n - m} \div \dfrac{12n^3}{4n^2 - m^2} \cdot \dfrac{2n^2}{6n^2 - 3nm}$

18. $\dfrac{a^2c^2}{a^3 - a^2c} \div \dfrac{4ac^3}{a^2 - c^2} \cdot \dfrac{2a + 2c}{ac}$

19. $\dfrac{y^2 - yc - 2c^2}{10y + 5c} \cdot \dfrac{4y^2 - c^2}{3y - 6c} \div \dfrac{y^2 - c^2}{15y - 15c}$

20. $\dfrac{r^2 - r}{r^2 - 2r - 3} \cdot \dfrac{r^2 + 2r + 1}{r^2 + 4r} \div \dfrac{r^2 - 3r - 4}{r^2 - 16}$

8-8 Sums and Differences of Fractions with Equal Denominators

Consider the sum $\frac{a}{b} + \frac{c}{b}$. Since $\frac{a}{b} = a\left(\frac{1}{b}\right)$ and $\frac{c}{b} = c\left(\frac{1}{b}\right)$ (page 86), by the distributive axiom we know that

$$\frac{a}{b} + \frac{c}{b} = a\left(\frac{1}{b}\right) + c\left(\frac{1}{b}\right) = (a+c)\left(\frac{1}{b}\right) = \frac{a+c}{b}.$$

Similarly,

$$\frac{a}{b} - \frac{c}{b} = a\left(\frac{1}{b}\right) - c\left(\frac{1}{b}\right) = (a-c)\frac{1}{b} = \frac{a-c}{b}.$$

These chains of equalities suggest the following rule:

Adding and Subtracting Fractions with Equal Denominators

For any real numbers *a*, *b*, and *c*, if $b \neq 0$, then

$$\frac{a}{b} + \frac{c}{b} = \frac{a+c}{b} \quad \text{and} \quad \frac{a}{b} - \frac{c}{b} = \frac{a-c}{b}.$$

That is, the sum of fractions with equal denominators is a fraction whose numerator is the sum of the numerators and whose denominator is the common denominator of the given fractions. The difference of two fractions with equal denominators is a fraction whose numerator is the difference of the numerators and whose denominator is the common denominator of the given fractions.

Example 1. $\dfrac{6}{11} + \dfrac{8}{11} - \dfrac{9}{11} = \dfrac{6 + 8 - 9}{11} = \dfrac{5}{11}$

Example 2. $\dfrac{n}{3n+1} + \dfrac{n+3}{3n+1} - \dfrac{5-n}{3n+1}$

$$= \frac{n + (n+3) - (5-n)}{3n+1}$$

$$= \frac{3n-2}{3n+1}$$

Example 3. $\dfrac{3x}{2x^2+3x} + \dfrac{5-x}{2x^2+3x} - \dfrac{2}{2x^2+3x}$
$= \dfrac{3x+(5-x)-2}{2x^2+3x} = \dfrac{2x+3}{2x^2+3x} = \dfrac{2x+3}{x(2x+3)} = \dfrac{1}{x}$

EXERCISES

Find a fraction in lowest terms equivalent to each expression.

1. $\dfrac{3}{17} + \dfrac{8}{17}$
2. $\dfrac{9}{23} + \dfrac{14}{23}$
3. $\dfrac{2}{7} + \dfrac{8}{7} - \dfrac{4}{7}$
4. $\dfrac{3}{8} - \dfrac{5}{8} + \dfrac{7}{8}$
5. $\dfrac{4}{3x} - \dfrac{5}{3x} + \dfrac{2}{3x}$
6. $\dfrac{7}{10z} + \dfrac{9}{10z} - \dfrac{19}{10z}$
7. $\dfrac{x+4}{2} + \dfrac{2x-1}{2}$
8. $\dfrac{3z}{5} + \dfrac{z+4}{5}$
9. $\dfrac{4x}{x+y} + \dfrac{4y}{x+y}$
10. $\dfrac{y}{y-7} - \dfrac{7}{y-7}$
11. $\dfrac{x^2}{x-y} - \dfrac{y^2}{x-y}$
12. $\dfrac{r^2}{r+3} - \dfrac{9}{r+3}$
13. $\dfrac{2a-3b}{3ab} + \dfrac{4a+2b}{3ab} + \dfrac{3a+b}{3ab}$
14. $\dfrac{4x+y}{4xy} - \dfrac{2x-3y}{4xy} + \dfrac{10y-2x}{4xy}$
15. $\dfrac{3ab}{a+2b} + \dfrac{a^2+2b^2}{a+2b}$
16. $\dfrac{r^2-3s^2}{r+s} - \dfrac{2rs}{r+s}$
17. $\dfrac{k^2+k}{k^2-9} + \dfrac{k-3}{k^2-9}$
18. $\dfrac{p+1}{p^2-3p-10} - \dfrac{6}{p^2-3p-10}$
19. $\dfrac{3z}{z^2-2z-15} - \dfrac{2z+5}{z^2-2z-15}$
20. $\dfrac{b^2+2b}{b^2+4b-12} - \dfrac{b+6}{b^2+4b-12}$

★21. Find a fraction whose sum with $\dfrac{a-a^2}{a^3+9a}$ is $\dfrac{1}{a^2+9}$.

★22. Find a fraction whose sum with $\dfrac{x^3-3x^2}{r^3+2r^2+3r}$ is $\dfrac{x}{r^2+2r+3}$.

8–9 Sums and Differences of Fractions with Unequal Denominators

To add $\frac{5}{4}$ and $\frac{7}{18}$, first express them as fractions with equal denominators. Then use the method developed in the preceding section. As a common denominator, we may use any positive integer having 4 and 18 as factors, but for convenience we usually seek the least common denominator (LCD). To find the LCD systematically, write 4 and 18 as products of primes (recall page 160) and take each prime factor the greatest number of times it appears in 4 or 18.

$$4 = 2 \cdot 2 = 2^2 \qquad 18 = 2 \cdot 3 \cdot 3 = 2 \cdot 3^2$$

$$\therefore \text{ the LCD} = 2^2 \cdot 3^2 = 36.$$

To convert $\frac{5}{4}$ and $\frac{7}{18}$ to fractions with denominator 36, note that

$$36 \div 4 = 9 \quad \text{and} \quad 36 \div 18 = 2.$$

Thus,

$$\frac{5}{4} = \frac{5 \cdot 9}{4 \cdot 9} = \frac{45}{36} \quad \text{and} \quad \frac{7}{18} = \frac{7 \cdot 2}{18 \cdot 2} = \frac{14}{36}.$$

We then have

$$\frac{5}{4} + \frac{7}{18} = \frac{45}{36} + \frac{14}{36} = \frac{45 + 14}{36} = \frac{59}{36}.$$

Example 1. Simplify: $\frac{5}{27} - \frac{11}{36} + \frac{7}{30}$

Solution:

1. To find the LCD, first factor each denominator. $\rightarrow$
$$27 = 3^3$$
$$36 = 2^2 \cdot 3^2$$
$$30 = 2 \cdot 3 \cdot 5$$

   Then take each prime factor the greatest number of times it appears in any denominator. $\rightarrow$ LCD $= 2^2 \cdot 3^3 \cdot 5$

2. Replace each fraction with an equal fraction having the LCD, apply the rule for adding and subtracting fractions, and simplify:

$$\frac{5}{27} - \frac{11}{36} + \frac{7}{30}$$
$$= \frac{5 \cdot 2^2 \cdot 5}{(3^3) \cdot 2^2 \cdot 5} - \frac{11 \cdot 3 \cdot 5}{(2^2 \cdot 3^2) \cdot 3 \cdot 5} + \frac{7 \cdot 2 \cdot 3^2}{(2 \cdot 3 \cdot 5) \cdot 2 \cdot 3^2}$$
$$= \frac{100 - 165 + 126}{4 \cdot 27 \cdot 5}$$
$$= \frac{61}{540}$$

Example 2. Simplify: $\dfrac{n+10}{n^2-2n} + \dfrac{2}{n} - \dfrac{6}{n-2}$

Solution:

1. To find the LCD, first factor each denominator.

   $n^2 - 2n = n(n-2)$
   $n = n$
   $n - 2 = n - 2$

   Then take each prime factor the greatest number of times it appears in any denominator: → LCD $= n(n-2)$

2. Replace each fraction with an equal fraction having the LCD:

   $$\dfrac{n+10}{n(n-2)} + \dfrac{2}{n} - \dfrac{6}{n-2}$$

   $$= \dfrac{n+10}{n(n-2)} + \dfrac{2(n-2)}{n(n-2)} - \dfrac{6n}{n(n-2)}$$

3. Apply the rule for adding and subtracting fractions, and simplify:

   $$\dfrac{n+10+2(n-2)-6n}{n(n-2)} = \dfrac{n+10+2n-4-6n}{n(n-2)}$$

   $$= \dfrac{-3n+6}{n(n-2)}$$

   $$= \dfrac{-3(n-2)}{n(n-2)} = -\dfrac{3}{n}$$

Check: Let n have the value 3. The check is left to you.

EXERCISES

Find a fraction in lowest terms equivalent to each expression.

1. $\dfrac{2}{a} + \dfrac{1}{3}$

2. $\dfrac{1}{4} - \dfrac{a}{x}$

3. $\dfrac{3}{2a} - \dfrac{1}{a}$

4. $\dfrac{5}{3b} + \dfrac{2}{b}$

5. $\dfrac{x+2}{6} + \dfrac{2}{3}$

6. $\dfrac{3}{5} - \dfrac{x+1}{10}$

7. $\dfrac{2z+1}{3} - \dfrac{z-1}{9}$

8. $\dfrac{3x-2}{4} + \dfrac{2x-1}{6}$

9. $\dfrac{3t+4}{2} + \dfrac{4t-1}{3}$

10. $\dfrac{2-x}{6} + \dfrac{3+x}{2}$

11. $\dfrac{2}{x} + \dfrac{3}{x^2} - \dfrac{1}{x^3}$

12. $\dfrac{3}{b^3} - \dfrac{1}{b^2} + \dfrac{2}{b^3}$

13. $\dfrac{5c+1}{6c} + \dfrac{3}{2c}$

14. $\dfrac{3n-2}{2n} - \dfrac{1}{n}$

15. $\dfrac{x+7}{ax} + \dfrac{3}{a}$

16. $\dfrac{2}{b^2} - \dfrac{6x+5}{b^2 x}$

17. $\dfrac{2x-y}{4y} - \dfrac{x-3y}{6x}$

18. $\dfrac{r-s}{rs} - \dfrac{s-t}{st}$

19. $\dfrac{2}{c^2 - d^2} - \dfrac{3}{c+d}$

20. $\dfrac{6}{p^2 - q^2} + \dfrac{p}{p-q}$

21. $\dfrac{5}{6r+6} - \dfrac{3}{2r+2}$

22. $\dfrac{6}{5x-10} + \dfrac{7}{3x-6}$

23. $\dfrac{y}{y+2} - \dfrac{y}{y-2}$

24. $\dfrac{3}{z+3} - \dfrac{3}{z-3}$

25. $\dfrac{2}{t+2} + \dfrac{3}{t+3}$

26. $\dfrac{3}{3b-4} - \dfrac{5}{5b+6}$

27. $\dfrac{y+1}{y+2} - \dfrac{y+2}{y+3}$

28. $\dfrac{z-1}{z+1} - \dfrac{z+1}{z-1}$

29. $\dfrac{3x}{x^2 - 4x + 3} + \dfrac{2}{x-3}$

30. $\dfrac{3z-4}{z^2 - z - 20} + \dfrac{2}{z-5}$

31. $\dfrac{2x-3}{16x^2} - \dfrac{2-x}{8x} + \dfrac{3}{4x}$

32. $\dfrac{2y+1}{3y} - \dfrac{y-5}{2y} + \dfrac{y+4}{18y^2}$

33. $\dfrac{1}{z^2 - z - 2} - \dfrac{3}{z^2 + 2z + 1}$

34. $\dfrac{3y}{y^2 + 3y - 10} - \dfrac{2y}{y^2 + y - 6}$

35. $\dfrac{3}{z+2} + \dfrac{5}{z-2} + \dfrac{2z-5}{4-z^2}$

36. $\dfrac{4}{k^2 - 25} - \dfrac{2}{k+5} - \dfrac{k+2}{5-k}$

★37. $\dfrac{2}{a^2 - 9} - \dfrac{3}{a^2 - 1} + \dfrac{1}{a^2 + 2a + 3}$

★38. $\dfrac{n+2}{(n-2)^2} - \dfrac{n}{n^2 - 4} + \dfrac{2}{n-2}$

★39. $\dfrac{1}{t^2 - 5t + 6} - \dfrac{1}{4 - t^2} + \dfrac{1}{6 + t - t^2}$

★40. $\dfrac{p+1}{p^2 - 2p - 3} - \dfrac{1}{p^2 + p} - \dfrac{3}{p^2 - 3p}$

8–10 Mixed Expressions

A **mixed numeral** like $3\tfrac{2}{5}$ denotes the sum of an integer and a fraction. To transform it into a fraction, write the integer as a fraction with denominator 1 and add the fractions:

$$3\tfrac{2}{5} = \tfrac{3}{1} + \tfrac{2}{5} = \tfrac{15}{5} + \tfrac{2}{5} = \tfrac{17}{5}$$

Similarly, $-3\frac{2}{5} = -(3 + \frac{2}{5}) = -3 - \frac{2}{5} = -\frac{17}{5}$.

The sum or difference of a polynomial and a fraction is called a **mixed expression**. A mixed expression can be written as a single fraction, as shown below:

$$y + \frac{5}{y} = \frac{y}{1} + \frac{5}{y} = \frac{y^2}{y} + \frac{5}{y} = \frac{y^2 + 5}{y}$$

$$3 - \frac{x - 2z}{x + z} = \frac{3}{1} - \frac{x - 2z}{x + z} = \frac{3(x + z)}{x + z} - \frac{x - 2z}{x + z} = \frac{2x + 5z}{x + z}$$

A rational expression can be changed from a fraction to a mixed expression by applying the division algorithm (page 154):

$$\frac{6n^2 + 5}{2n} = 3n + \frac{5}{2n}$$

$$\frac{x^2 - 3x - 10}{x + 1} = x - 4 - \frac{6}{x + 1}$$

EXERCISES

Express each mixed expression as a fraction in lowest terms.

1. $a + \dfrac{2}{a + 3}$
2. $t + \dfrac{2r}{t + r}$
3. $2 + \dfrac{x + 2y}{x - y}$
4. $3 + \dfrac{a - 4b}{a + b}$
5. $y - 2 + \dfrac{1}{y + 2}$
6. $z + 3 + \dfrac{1}{z - 3}$
7. $\dfrac{4}{n + 2} + 1$
8. $\dfrac{5}{n - 3} + 1$
9. $2 + \dfrac{a}{b} + \dfrac{b}{a}$
10. $\dfrac{p}{q} + 2 + \dfrac{q}{p}$
11. $y + 3 + \dfrac{2y - 1}{y - 2}$
12. $z + 2 - \dfrac{z + 1}{z - 1}$

Change each fraction to a mixed expression.

13. $\dfrac{27}{4}$
14. $\dfrac{231}{18}$
15. $\dfrac{12 + 18z^3}{6z}$
16. $\dfrac{12 - 6t^2}{2t}$
17. $\dfrac{14p^3 - 3}{7p^2}$
18. $\dfrac{20t^3 - 5}{4t^3}$
19. $\dfrac{16x^2y^2 + 12xy}{16xy}$
20. $\dfrac{12a^2b^2 - 4ab}{4a^2b^2}$
21. $\dfrac{y^2 - 3y + 2}{y + 3}$
22. $\dfrac{b^2 + 5b - 2}{b + 2}$
23. $\dfrac{9t^2 - 6t + 5}{3t - 2}$
24. $\dfrac{15n^2 - 2n + 8}{3n - 1}$

8–11 Complex Fractions

A **complex fraction** is a fraction whose numerator or denominator contains one or more fractions. Complex fractions may be changed to simple ones by two methods.

Method I: Multiply the numerator and denominator by the LCD of all fractions within them.

Method II: Express the fraction as a quotient, using the sign $\div$, and apply the rule for dividing fractions.

Example 1. Simplify: $\dfrac{\frac{2}{3}}{\frac{5}{7}}$

Solution:

Method I

$$\frac{\frac{2}{3}}{\frac{5}{7}} = \frac{\frac{2}{3}(21)}{\frac{5}{7}(21)}$$
$$= \frac{14}{15}$$

Method II

$$\frac{\frac{2}{3}}{\frac{5}{7}} = \frac{2}{3} \div \frac{5}{7}$$
$$= \frac{2}{3} \cdot \frac{7}{5}$$
$$= \frac{14}{15}$$

Example 2. Simplify: $\dfrac{\frac{x+3y}{2y}}{\frac{2x-y}{4y^2}}$

Solution:

Method I

$$\frac{\frac{x+3y}{2y}}{\frac{2x-y}{4y^2}} = \frac{\frac{x+3y}{2y}(4y^2)}{\frac{2x-y}{4y^2}(4y^2)}$$
$$= \frac{2y(x+3y)}{2x-y}$$

Method II

$$\frac{\frac{x+3y}{2y}}{\frac{2x-y}{4y^2}} = \frac{x+3y}{2y} \div \frac{2x-y}{4y^2}$$
$$= \frac{x+3y}{2y} \cdot \frac{4y^2}{2x-y}$$
$$= \frac{4y^2(x+3y)}{2y(2x-y)}$$
$$= \frac{2y \cdot 2y(x+3y)}{2y(2x-y)}$$
$$= \frac{2y(x+3y)}{2x-y}$$

EXERCISES

Simplify each fraction.

1. $\dfrac{\frac{3}{4}}{\frac{9}{8}}$

2. $\dfrac{\frac{12}{27}}{\frac{2}{9}}$

3. $\dfrac{\frac{x}{y}}{\frac{x}{y}}$

4. $\dfrac{\frac{a}{b^2}}{\frac{a}{b}}$

5. $\dfrac{\frac{10x^2y^2}{9z}}{\frac{5xy^2}{3z}}$

6. $\dfrac{\frac{18a^2}{5ab^2}}{\frac{9ab}{25b^4}}$

7. $\dfrac{\frac{x+y}{x}}{\frac{x-y}{y}}$

8. $\dfrac{\frac{a+3}{6a}}{\frac{a-2}{3a^2}}$

9. $\dfrac{y^2-9}{y+3}$

10. $\dfrac{\frac{t^2-4}{t-2}}{t}$

11. $\dfrac{\frac{2}{3}+\frac{1}{4}}{2-\frac{1}{6}}$

12. $\dfrac{\frac{7}{8}+\frac{5}{6}}{\frac{5}{12}-2}$

13. $\dfrac{\frac{x+3}{x-3}}{\frac{3x+9}{x^2-9}}$

14. $\dfrac{\frac{cy-cz}{y^2-z^2}}{\frac{y-c}{y+c}}$

15. $\dfrac{\frac{a}{b}+2}{1-\frac{a}{b}}$

16. $\dfrac{\frac{p}{q}-2}{1+\frac{p}{q}}$

17. $\dfrac{\frac{x^2+y^2}{xy}-2}{\frac{4x^2-4y^2}{2xy}}$

18. $\dfrac{\frac{4a^2-b^2}{3ab}}{\frac{2a^2-b^2}{ab}+1}$

19. $\dfrac{x+2-\frac{12}{x+3}}{x-5+\frac{16}{x+3}}$

20. $\dfrac{n-\frac{2}{n+1}}{n+\frac{n-3}{n+1}}$

21. $\dfrac{\frac{x^2+y^2}{x^2-y^2}}{\frac{x-y}{x+y}-\frac{x+y}{x-y}}$

22. $\dfrac{x+\frac{2x+1}{x-1}}{x+\frac{2}{x-1}}$

23. $1-\dfrac{1}{1-\frac{1}{x-2}}$

24. $2+\dfrac{1}{1+\frac{2}{x+\frac{1}{x}}}$

Determine each solution set.

★25. $1+\dfrac{2+\frac{12}{y}}{\frac{y}{2}+3}=\dfrac{2}{y}$

★26. $1+\dfrac{2-\frac{2}{x}}{x-1}=\dfrac{1}{2x}$

chapter summary

1. The **Multiplication Property of Fractions**: For each a, each b, and each c other than zero, $\dfrac{ac}{bc} = \dfrac{a}{b}$. This property permits us to **reduce a fraction to lowest terms** by dividing its numerator and denominator by their **greatest common factor**.

2. From the **Property of Quotients**, $\dfrac{xy}{ab} = \dfrac{x}{a} \cdot \dfrac{y}{b}$, comes the rule for **multiplying fractions**: The product is the fraction whose numerator is the product of the numerators and whose denominator is the product of the denominators of the fractions. The product should be expressed in lowest terms. If the numerators and denominators are first factored, the product of fractions may usually be computed more readily.

3. To **divide fractions**, multiply the dividend by the reciprocal of the divisor; thus, $\dfrac{a}{b} \div \dfrac{c}{d} = \dfrac{a}{b} \cdot \dfrac{d}{c} = \dfrac{ad}{bc}$.

4. To find the **least common denominator** of several fractions, factor each denominator, and find the product of the different prime factors, each taken the greatest number of times it appears in any denominator.

5. To **add and subtract fractions**, use the multiplication property of fractions to replace each fraction by one equal to it and having as its denominator the least common denominator of the given fractions. The **sum of fractions with equal denominators** is the fraction whose numerator is the sum of the numerators and whose denominator is the common denominator of the fractions.

6. To change a **mixed expression** to a fraction, write the polynomial as a fraction whose denominator is 1, and add this fraction to the fractional part of the expression. To change a **rational expression** to a mixed expression, divide the numerator by the denominator.

CHAPTER REVIEW

8–1 1. For what values of t is $\dfrac{t-3}{t^2-1}$ not defined?

8–2 Reduce to lowest terms.

2. $\dfrac{8x + 16y}{3x + 6y}$ 3. $\dfrac{n^2 - 7n}{n^2 - 8n + 7}$ 4. $\dfrac{r^2 - 4r + 4}{4 - r^2}$

8–3 5. Express the ratio of 3 feet to 8 inches in lowest terms.

6. If a 20-foot string is cut into two pieces whose lengths are in the ratio 3:2, how long is each of the two pieces?

8–4 7. What would be the sale price on an item regularly priced at $4.50 if a 30% discount is given?

8. Draw a circle graph, showing the following use of the Berry family's income: Housing 20%, Food 30%, Savings 10%, Other 40%.

Express as a single fraction in lowest terms.

8–5 9. $\dfrac{4z}{9z^2} \cdot \dfrac{3z}{2z}$ 10. $\dfrac{p^2 - q^2}{r^2 - s^2} \cdot \dfrac{r + s}{p - q}$

8–6 11. $\dfrac{6r - 3s}{4r^2 - s^2} \div \dfrac{3}{2r + s}$ 12. $\dfrac{n^2 + n}{n^2} \div \dfrac{n^2 - 1}{3n - 3}$

8–7 13. $\dfrac{3p - 1}{3p} \div \dfrac{3p + 1}{3p^2} \cdot \dfrac{p^2 + 2p - 3}{p^2 - p}$

8–8 14. $\dfrac{3a - 3b}{2ab} - \dfrac{2a + 5b}{2ab} + \dfrac{8b - a}{2ab}$

8–9 15. $\dfrac{2z + 1}{6} - \dfrac{3z - 5}{9}$ 16. $\dfrac{1}{x + y} + \dfrac{x}{x^2 - y^2}$

17. $\dfrac{r}{r^2 - 25} - \dfrac{1}{2r + 10}$

8–10 18. $t - 3 + \dfrac{5t - 6}{t - 2}$

19. Change to a mixed expression: $\dfrac{4c^2 + 6c - 8}{2c}$

8–11 Simplify.

20. $\dfrac{\frac{8x^2 y}{9z}}{\frac{4xy}{3z}}$ 21. $\dfrac{\frac{t + 1}{t - 1}}{\frac{2t + 2}{t^2 - 1}}$ 22. $\dfrac{\frac{x^2 + y^2}{xy} + 2}{\frac{x^2 - y^2}{3}}$

9 using fractions

9–1 Solving Equations and Inequalities

Transformations previously used to solve open sentences (pages 90 and 108) may be used also when the numerical coefficients are fractions. Two methods are shown in the following Example. We will need to find the least common denominator (LCD) of the fractional coefficients in applying either method.

Example. $\dfrac{3y}{2} + \dfrac{8 - 4y}{7} = 3$

Solution:

Method 1:

$$\dfrac{3y}{2} + \dfrac{8 - 4y}{7} = 3$$

Multiply each member by the LCD, 14

$$14\left(\dfrac{3y}{2} + \dfrac{8 - 4y}{7}\right) = 14(3)$$

$$14\left(\dfrac{3y}{2}\right) + 14\left(\dfrac{8 - 4y}{7}\right) = (3)$$

$$21y + 16 - 8y = 42$$

$$13y + 16 = 42$$

$$13y = 26$$

$$y = 2$$

The check is left to you.

Method 2:

$$\dfrac{3y}{2} + \dfrac{8 - 4y}{7} = 3$$

Replace each term in each member by an equivalent fraction having the LCD as denominator: LCD = 14

$$\dfrac{21y}{14} + \dfrac{2(8 - 4y)}{14} = 3$$

$$\dfrac{21y + 16 - 8y}{14} = 3$$

$$\dfrac{13y + 16}{14} = 3$$

EXERCISES

Solve. If the sentence is an inequality, graph its solution set.

1. $\frac{x}{2} + \frac{x}{5} = 7$
2. $\frac{y}{3} - \frac{y}{6} = 1$
3. $\frac{5z}{2} - z < \frac{3}{2}$
4. $\frac{4t}{5} - \frac{3t}{10} > \frac{3}{2}$
5. $\frac{3}{4}z - \frac{3}{2}z = 3$
6. $\frac{2}{3}n - \frac{5}{9}n = -1$
7. $0.03b - 0.01b \geq 0.2$
8. $0.5a - 0.2a \leq 1.5$
9. $\frac{3}{4}t - \frac{2}{5}t = \frac{7}{20}$
10. $\frac{1}{3}s + \frac{2}{5}s = \frac{11}{15}$
11. $\frac{3}{8}x - \frac{1}{4}x > \frac{3}{2}$
12. $\frac{1}{6}y \geq \frac{2}{3} + \frac{1}{3}y$
13. $\frac{t}{2} = \frac{3-t}{4}$
14. $\frac{x+1}{4} - \frac{3}{2} = \frac{2x-9}{10}$
15. $\frac{z}{6} - \frac{20-z}{8} = 1$
16. $\frac{n+1}{10} - \frac{n}{3} = \frac{19}{15}$
17. $\frac{3x-14}{6} \leq \frac{2x}{9} + \frac{x}{4}$
18. $\frac{5x-2}{2} + \frac{x+1}{4} \geq 2$
19. $\frac{1}{10}(10y - 3) - \frac{3}{5}(y + 1) = \frac{1}{10}$
20. $\frac{1}{3}(x - 3) + \frac{2}{5}(x + 10) = \frac{56}{15}$
21. $\frac{z+2}{2} + \frac{3(28-z)}{8} = 11$
22. $\frac{x+8}{16} - \frac{x-4}{12} = 1$
23. $0.08y + 0.12(10{,}000 - y) = 960$
24. $0.03n + 0.05(2000 - n) = 68$
25. $\frac{4x+3}{15} - \frac{2x-3}{9} = \frac{3x+2}{3} - x$
26. $5z + \frac{3z-4}{7} + \frac{5z+3}{3} = 43$
27. $\frac{2t}{3} - \frac{t+5}{2} = \frac{3t-3}{4}$
28. $\frac{3n}{2} - \frac{n+3}{3} = 8 - \frac{n+2}{4}$
29. $0.12x - 0.02(x - 3) = 0.03x - 0.03(3x - 9)$
30. $0.06t - 0.03(3 - t) = 0.06 - 0.03(t + 2)$

9–2 Percent Mixture Problems

An interesting type of problem involving percent is the percent mixture problem. Recall (page 203) that the basic percentage equation is

$$p = rb$$

and that percents are equal to fractions with denominator 100.

Example 1. How many ounces of water must be added to 12 ounces of an 8% salt solution to produce a 6% solution?

Solution: 1. Let x = number of ounces of water to be added.
Amount of original solution: 12 ounces
Amount of final solution: $(12 + x)$ ounces
Salt in original solution: 8% of $12 = 0.08(12)$
Salt in final solution: 6% of $(12 + x) = 0.06(12 + x)$

2. $\underbrace{\text{Amount of salt in original solution}}_{0.08(12)} = \underbrace{\text{Amount of salt in final solution}}_{0.06(12 + x)}$ LCD = 100

3. $100(0.08)(12) = 100(0.06)(12 + x)$
$8(12) = 6(12 + x)$
$96 = 72 + 6x$
$24 = 6x$
$4 = x$

4. The check is left for you.

Example 2. How many ounces of a 75% acid solution must be added to 30 ounces of a 15% acid solution to produce a 50% acid solution?

Solution: 1. Let x = number of ounces of 75% acid solution to be added.
Amount of original 15% acid solution: 30 ounces
Amount of final 50% acid solution: $(30 + x)$ ounces
Acid in 75% solution: 75% of x
Acid in *original* 15% solution: 15% of 30
Acid in *final* 50% solution: 50% of $(30 + x)$

2. $\underbrace{\text{Amount of acid in 75% acid solution}}_{0.75x} + \underbrace{\text{Amount of acid in } original\text{ 15% acid solution}}_{0.15(30)} = \underbrace{\text{Amount of acid in final 50% acid solution}}_{0.50(30 + x)}$

LCD = 100

3. $100(0.75)x + 100(0.15)(30) = 100(0.50)(30 + x)$
$75x + 450 = 1500 + 50x$
$25x = 1050$
$x = 42$

4. The check is left for you.

PROBLEMS

1. How many pounds of water must be added to 10 pounds of a 10% salt solution to produce a 4% solution?
2. How many ounces of water must be added to 4 ounces of a 50% antiseptic solution to produce a 40% solution?
3. How many pounds of water must be added to 50 pounds of a 90% acid solution to produce an 80% acid solution?
4. How many ounces of water must be added to 2 ounces of a 30% antifreeze solution to produce a 20% solution?
5. How many pounds of water must be evaporated from 40 pounds of a 10% salt solution to produce a 25% solution?
6. How many pounds of water must be evaporated from a barrel containing 80 pounds of a 6% brine (salt and water) to obtain a 10% brine?
7. How many pounds of pure alcohol must be added to 20 pounds of an 80% pure alcohol to produce an 85% pure alcohol?
8. How many ounces of pure acid must be added to 20 ounces of a 20% acid solution to produce a 50% solution?
9. How many pounds of a 30% solution of acid must be added to 40 pounds of a 12% solution to produce a 20% solution?
10. How many ounces of a 5% antiseptic solution must be added to 30 ounces of a 10% solution to produce an 8% solution?
11. How many pounds of a 35% silver alloy must be melted with how many pounds of a 65% silver alloy to obtain 20 pounds of a 41% silver alloy?
12. How many pounds of a 70% copper alloy must be melted with how many pounds of a 90% copper alloy to obtain 100 pounds of an 81% copper alloy?
★13. At most, how many quarts of an antifreeze solution containing 40% glycerine should be drawn from a radiator containing 32 quarts and replaced with water in order that the radiator be filled with at least a 20% glycerine solution?
★14. An automobile radiator contains 20 quarts of a 25% antifreeze solution. At least, how many quarts of this solution should be drawn off and replaced with pure antifreeze to fill the radiator with at least a 40% antifreeze solution?

9–3 Investment Problems

The amount of interest, i dollars, paid on an investment of P dollars at the interest rate r per year for 1 year is found by applying the percentage equation,

$$\text{Percentage} = \text{Rate} \times \text{Base}.$$

Here the percentage is i, the rate of interest is r, and the base is P:

$$i = rP$$

To find the **simple interest** for t years, multiply rP by t. Thus:

Simple interest, i dollars, paid on P dollars for t years at rate r per year is given by

$$i = Prt.$$

Thus, the simple interest on $1000 invested at 4% per year for 3 months is $10, since

$$i = Prt = 1000(\tfrac{4}{100})(\tfrac{3}{12}) = 10.$$

Investment problems may concern money invested at different rates of interest.

Example. Mr. Ford invested $8000, part at 3% and the remainder at 4% annual interest. If his yearly return on these investments is $275, how much does he have invested at each rate?

Solution: 1. Let x = number of dollars invested at 3%.
Then $8000 - x$ = number of dollars invested at 4%.

2. $\underbrace{\text{Income from}}_{(0.03)(x)}\text{3% investment} + \underbrace{\text{Income from}}_{(0.04)(8000-x)}\text{4% investment} = \underbrace{\text{Total}}_{275}\text{income}$

LCD = 100

3. $100(0.03)(x) + 100(0.04)(8000 - x) = 100(275)$
$$3x + 4(8000 - x) = 27{,}500$$
$$3x + 32{,}000 - 4x = 27{,}500$$
$$4500 = x$$
$$8000 - x = 3500$$

4. The check is left for you.

PROBLEMS

1. If $700 is invested at 4% per year, how much simple interest is earned in 2 years?
2. What is the simple interest for 3 years on $500 invested at 6% per year?
3. Mr. James received $120 as interest for 1 year on a sum of money he invested at 6% per year. How much had he invested?
4. An investment at 4% simple annual interest returned $576 in 4 years. How much was invested?

5. Mr. Stevens borrowed $700 at simple interest of 3% per year. Three years later he paid back the $700, together with the amount of interest. What was the total amount he paid?

6. If $800 is borrowed at simple interest of 5% per year, how much money must be paid all together after 5 years?

7. Mr. Lopez has some money invested at 4% annual interest. If he adds $2000 more to his investment, his annual interest will amount to $260. How much did Mr. Lopez originally invest?

8. Mrs. Lee had $5200 in an account at 3% annual interest. After withdrawing some money to buy a car, what remained in her account drew $72 annual interest. How much did she withdraw?

9. Mr. Ulrich invested twice as much money at 5% per year as he did at 4% per year. How much did he invest at each rate if his annual return totaled $168?

10. Mr. Mooney invested a sum of money at 6% per year, and 3 times as much at 4% per year. How much did he invest at each rate if his annual return totaled $144?

11. Mr. Daniels has some money invested at 5% per year. If he adds $200 to what he has, the new total will return the same amount each year at 4% as his original investment does at 5%. How much does he have invested at 5%?

12. Mrs. Cohen withdrew all of her savings from an account. The savings had been earning 3% per year. After spending $400 of the money, she reinvested what remained at 5% per year. She then earned $4 more in interest each year than she had on the original account. How much did she have in her original account?

13. Mr. Kirchner invested $8000, part at 4% and the rest at 5% per year. How much did he invest at each rate if his annual interest was $380?

14. A total of $9000 is invested, part at 6% and the rest at 3% per year. If the annual interest is $396, how much is invested at each rate?

15. An investment fund has $3000 more invested at 4% per year than it does at 5%, and the interest on the amount invested at 4% is $30 per year greater than the annual interest on the amount invested at 5%. How much does the fund have invested at each rate?

16. Mr. Conte invests $\frac{1}{2}$ of his money at 5% per year and $\frac{1}{5}$ of it at 4%. The rest, invested at 3%, is $1000 more than that invested at 4%. If his annual interest is $420, how much is invested at each rate?

17. Mrs. Danton invests some money at 7% per year, $\frac{3}{4}$ as much at 6% as she does at 7%, and $150 less at 5% than she does at 6%. If her yearly interest is $633, how much has she invested at each rate?

★18. The Omer Fund invested part of $150,000 at 6% and the rest at 4%. If it had invested twice as much at 6% and the rest at 4%, it would have increased its income by $1200. How much was invested at 6%?

9–4 Solving Fractional Equations

An equation which has a variable in the denominator of one or more terms is called a **fractional equation**.

Example. Solve $\dfrac{30}{x^2 - 9} + 2 = \dfrac{5}{x - 3}$.

Solution: Factor the denominators: $\dfrac{30}{(x - 3)(x + 3)} + 2 = \dfrac{5}{x - 3}$

LCD $= (x - 3)(x + 3)$, $x \notin \{3, -3\}$ (Recall page 87.)

$$(x - 3)(x + 3)\dfrac{30}{(x - 3)(x + 3)} + (x - 3)(x + 3)(2)$$
$$= (x - 3)(x + 3)\dfrac{5}{x - 3}$$

$30 + 2(x^2 - 9) = 5(x + 3)$
$30 + 2x^2 - 18 = 5x + 15$
$2x^2 - 5x - 3 = 0$
$(2x + 1)(x - 3) = 0$

$x = -\tfrac{1}{2}$ | $x = 3$ Excluded above, this cannot be a root of the original equation.

Check:
$$\dfrac{30}{x^2 - 9} + 2 = \dfrac{5}{x - 3}$$
$$\dfrac{30}{(-\tfrac{1}{2})^2 - 9} + 2 \stackrel{?}{=} \dfrac{5}{(-\tfrac{1}{2}) - 3}$$
$$\dfrac{30}{-\tfrac{35}{4}} + 2 \stackrel{?}{=} \dfrac{5}{-\tfrac{7}{2}}$$
$$-\tfrac{120}{35} + 2 \stackrel{?}{=} -\tfrac{10}{7}$$
$$-\tfrac{24}{7} + \tfrac{14}{7} \stackrel{?}{=} -\tfrac{10}{7}$$
$$-\tfrac{10}{7} = -\tfrac{10}{7}$$

$\therefore$ the solution set is $\{-\tfrac{1}{2}\}$.

In the Example above, notice that multiplying the given equation by $(x - 3)(x + 3)$ led to an equation that was *not equivalent* to the given one. This new equation had the extra root 3, a number for which the multiplier $(x - 3)(x + 3)$ represents zero. Observe the following caution:

Multiplying an equation by a variable expression which can represent zero may produce an equation having roots not satisfying the original equation. Only values producing true statements when substituted in the original equation belong to the solution set.

EXERCISES

Solve each equation. Check your solutions.

1. $\dfrac{12}{z} = \dfrac{4 + 4z}{z}$

2. $\dfrac{3}{n} - 2 = \dfrac{5}{2n} - \dfrac{3}{2}$

3. $\dfrac{t-2}{t} = \dfrac{14}{3t} - \dfrac{1}{3}$

4. $\dfrac{y-2}{5y} = \dfrac{1}{6} - \dfrac{4}{15y}$

5. $\dfrac{n}{n+2} = \dfrac{3}{5}$

6. $\dfrac{2}{3n} = \dfrac{2}{n+4}$

7. $\dfrac{6-x}{6x} = \dfrac{1}{x+1}$

8. $\dfrac{r+1}{r-1} = \dfrac{2}{r(r-1)}$

9. $\dfrac{y-3}{2} = \dfrac{4y}{y+3}$

10. $\dfrac{x-5}{8x} = \dfrac{3}{x+5}$

11. $\dfrac{4}{s} - 3 = \dfrac{5}{2s+3}$

12. $\dfrac{4}{3a} + \dfrac{3}{3a+1} = -2$

13. $1 + \dfrac{2}{b-1} = \dfrac{2}{b^2-b}$

14. $\dfrac{1}{k^2-k} = \dfrac{3}{k} - 1$

15. $\dfrac{14}{y-6} = \dfrac{1}{2} + \dfrac{6}{y-8}$

16. $\dfrac{2}{r-3} + 1 = \dfrac{6}{r-8}$

17. $\dfrac{7}{p-3} - \dfrac{1}{2} = \dfrac{3}{p-4}$

18. $\dfrac{4}{q-2} - \dfrac{2}{15} = \dfrac{7}{q-3}$

19. $\dfrac{3z-1}{z+3} + 3 = \dfrac{4z}{z-3}$

20. $\dfrac{k+2}{k-2} - \dfrac{2}{k+2} = -\dfrac{7}{3}$

★21. $\dfrac{2x+11}{x+4} + \dfrac{x-2}{x-4} - \dfrac{12}{x^2-16} = \dfrac{7}{2}$

★22. $\dfrac{2c}{2c-3} = \dfrac{15-32c^2}{4c^2-9} + \dfrac{3c}{2c+3}$

★23. $\dfrac{z-2}{z^2-z-6} = \dfrac{1}{z^2-4} + \dfrac{3}{2z+4}$

★24. $\dfrac{y-4}{2y^2+5y-3} = \dfrac{4y-1}{4y^2+13y+3} - \dfrac{2y+7}{8y^2-2y-1}$

9–5 Rate-of-Work Problems

To solve problems about work, we use the formula

$$w = rt,$$

where w is the amount of work done, r is the rate of doing work, and t is the time worked. When two or more persons work on and complete a job, each does a fractional part of the job. Since together they complete a job, the sum of their fractional parts is 1.

Example 1. Janet can key-punch a number of computer cards in 6 hours, while Martha requires 8 hours to punch the same number. How long will it take both girls to punch that number of cards if they work together at the same rates at which they work individually?

Solution:

1. Let x = number of hours for both to complete the work together.

    $\frac{1}{6}x$ = fractional part of the job done by Janet ($\frac{1}{6}$ of the job in 1 hour)

    $\frac{1}{8}x$ = fractional part of the job done by Martha ($\frac{1}{8}$ of the job in 1 hour)

2. $\underbrace{\text{Part of work done by Janet}}_{\frac{1}{6}x} + \underbrace{\text{Part of work done by Martha}}_{\frac{1}{8}x} = \underbrace{\text{Whole job done together}}_{1}$

    LCD = 24

3. $24(\frac{1}{6}x + \frac{1}{8}x) = 24(1)$

    $4x + 3x = 24$

    $7x = 24$

    $x = \frac{24}{7}$

Check:

4. In $\frac{24}{7}$ hours, Janet finishes $\frac{1}{6}(\frac{24}{7})$, or $\frac{4}{7}$, of the job.

    In $\frac{24}{7}$ hours, Martha finishes $\frac{1}{8}(\frac{24}{7})$, or $\frac{3}{7}$, of the job.

    Does $\frac{4}{7} + \frac{3}{7} = 1$? Yes.

    ∴ it takes $3\frac{3}{7}$ hours for both girls working together to complete the job.

Sometimes those doing a job may not work for the same length of time. In such cases, the values substituted for time in the equation may differ.

Example 2. How long would it take to punch all the cards in Example 1, if they started at the same time, but Janet stopped work after 2 hours?

Solution:
1. Let t = number of hours Martha works. Janet works 2 hours.
2. $\frac{1}{6}(2) + \frac{1}{8}t = 1$

Steps 3 and 4 are left for you.

PROBLEMS

1. The main engine alone on a rocket can consume the fuel supply in 60 seconds, while the auxiliary engine alone can consume the fuel in 80 seconds. How long can both engines be fired together?

2. One spillway can empty the reservoir behind Beaver Creek Dam in 5 days and a second spillway can empty it in 7 days. How long will it take to empty the reservoir if both spillways are used at the same time?

3. Working alone, Mr. Golden can paint his house in 3 days, while his son can paint it in 5 days. How long will it take them to paint the house if they work together?

4. One pipe can fill a tank in 12 hours and another can fill the tank in 8 hours. How long will it take both pipes together to fill the tank?

5. A heater raises the temperature 15 degrees in 24 minutes. With a second heater also in operation, the change takes place in 6 minutes. How long would it take the second heater working alone to produce the same temperature change?

6. One pipe can fill a small reservoir in 15 hours, while with a second pipe also in operation, the reservoir can be filled in 6 hours. How long would it take the second pipe alone to fill the reservoir?

7. One electronic reader can read a deck of punched cards in $\frac{1}{2}$ the time of another reader. Together, they can read the deck in 8 minutes. How long would it take each reader alone to read the deck?

8. A battery can operate one radio receiver for 4 times as long as it can operate another radio receiver. If both receivers are operated together, the battery will last 16 hours. How long would the battery operate each receiver alone?

9. A filler pipe can fill a tank in 10 hours, while an outlet pipe can empty the tank in 15 hours. How long will it take to fill the empty tank with both pipes operating?

10. One pipe can fill a tank in 6 hours while another can empty it in 2 hours. How long will it take to empty the full tank if both pipes are open at once?

11. Mr. Johnson paints $\frac{5}{9}$ of his house in 10 hours, and is then joined on the job by his neighbor, Mr. Prima. Together, they finish painting the house in 3 hours. How long would it have taken Mr. Prima to do the entire job by himself?

12. Doris and Julie addressed all the invitations to a wedding in 2 hours. Doris completed 3 invitations in the time it took Julie to do 2. How long would it take Julie to address all the invitations?

13. The reservoir behind Buford Dam can be emptied in 15 days by opening both of its spillways. With the reservoir full, one spillway is opened. Then after 5 days, the second spillway is opened, and the reservoir is then emptied by both spillways in 13 days more. How long would it take each spillway alone to empty the reservoir?

14. Jason can paint the fence around his house in 15 hours, while his brother Amos can do it in 18 hours. If Jason paints for 7 hours alone and then turns the rest of the job over to Amos, how long will it take Amos to finish painting the fence?

★ 15. An outlet pipe can empty a cistern in 18 hours. With the cistern full, this pipe and a second pipe are opened, and together they empty $\frac{3}{5}$ of the cistern in 6 hours. If the first pipe is then shut off, how long will it take the second pipe to empty the cistern?

★ 16. Joe started out to paint the family garage alone. After $1\frac{1}{3}$ hours he coaxes his younger brother into helping him and they finished in $2\frac{2}{3}$ hours more. If Joe had worked 2 hours before calling in his brother, it would have taken them $2\frac{2}{7}$ hours to complete the job. How long would it take each boy to paint the garage alone?

9–6 Motion Problems

Certain motion problems can be solved by using fractional equations. Recall (page 120) that the basic uniform motion equation is

$$d = rt,$$

where d is the distance traveled, r is the rate (speed), and t is the time.

Example. The riverboat Memphis Belle sailing at the rate of 18 miles per hour in still water can go 63 miles down the river in the same time it takes it to go 45 miles up the river. What is the speed of the current in the river?

Solution: 1. Let $r =$ the speed of the current.
Since the boat's speed in still water is 18 miles per hour:
$18 + r =$ speed downstream
$18 - r =$ speed upstream
Distance downstream $= 63$ miles
Distance upstream $= 45$ miles

2.

	r	d	$t = \dfrac{d}{r}$
Downstream	$18 + r$	63	$\dfrac{63}{18 + r}$
Upstream	$18 - r$	45	$\dfrac{45}{18 - r}$

$$\underbrace{\text{Time downstream}}_{\dfrac{63}{18+r}} = \underbrace{\text{Time upstream}}_{\dfrac{45}{18-r}}$$

$$\frac{63}{18+r} = \frac{45}{18-r}$$

LCD $= (18 + r)(18 - r)$, $r \notin \{18, -18\}$

3. $(18 - r)(18 + r)\dfrac{63}{18 + r} = (18 - r)(18 + r)\dfrac{45}{18 - r}$

$63(18 - r) = 45(18 + r)$
$7(18 - r) = 5(18 + r)$
$126 - 7r = 90 + 5r$
$36 = 12r$
$3 = r$

4. Is the speed of the current 3 miles per hour?
The speed of the boat downstream:
$18 + 3 = 21$ (miles per hour)
The speed of the boat upstream:
$18 - 3 = 15$ (miles per hour)

Since $t = \dfrac{d}{r}$: time downstream $= \dfrac{63}{21} = 3$ (hours)
time upstream $= \dfrac{45}{15} = 3$ (hours)

Time downstream $\stackrel{?}{=}$ time upstream; $3 = 3$

∴ the speed of the current is 3 miles per hour.

PROBLEMS

1. A motorboat goes 24 miles upstream in the same length of time it takes to go 36 miles downstream. If the current is flowing at 3 miles per hour, what is the boat's rate in still water?

2. Water flows down the Chino River at 2 miles per hour. Rowing steadily, Mr. McCoy can go 12 miles downstream in the same time it would take him to go 4 miles upstream. What is his rate of rowing in still water?

3. The riverboat Lucky Piece can travel 12 miles per hour in still water. It takes the Lucky Piece as long to move 27 miles upstream as it does to move 45 miles downstream. What is the rate of the current?

4. An airplane whose cruising speed in still air is 220 miles per hour can travel 520 miles with the wind in the same length of time it travels 360 miles against the wind. What is the speed of the wind?

5. Jeremy can row in still water at a rate twice that of the current in the Juta River. What is the rate of the current in the river if it takes Jeremy 2 hours less to row 2 miles up the river than it does to row 9 miles down the river?

6. One day Mr. Kimbel found that his light airplane cruised at 5 times the rate of the wind. If he flew 132 miles against the wind in one-half hour less time than he flew 396 miles with the wind, what was the speed of the wind?

7. Julie's motorboat cruises at 20 miles per hour in still water. It takes her twice as long to go 90 miles upstream as it does to return 75 miles downstream. What is the speed of the current?

8. Paul can row 3 miles per hour in still water. If it takes him twice as long to row 8 miles upstream as it does the same distance downstream, what is the rate of the current?

9. Mr. Sigura drove 90 miles to visit his brother. He averaged 15 miles per hour more on the return trip than he did on the trip going. If his total travel time was $3\frac{1}{2}$ hours, what was his average rate on the return trip?

10. It takes Jack's brother an hour longer to walk his 4-mile paper route than it does to ride his bicycle over the route. If he averages 2 miles per hour more riding than walking, what is his rate walking?

11. Mike sailed 6 miles across a lake in $1\frac{1}{2}$ hours less than it took James to walk 12 miles around the lake to meet him. If Mike's rate was 6 miles per hour greater than James' rate, what was the rate of each?

12. Mr. Egbert had driven 20 miles at a constant speed, but found that he had to increase his speed by 20 miles per hour in order to cover the last 30 miles in time to make an appointment. If his total traveling time was 1 hour, what was his original rate?

★13. Fred gave Bill a five-yard head start in a 100-yard dash, and Fred was beaten by $\frac{1}{4}$ yard. In how many yards more would Fred have overtaken Bill?

★14. Two cars race on a 4-mile oval track. The sum of the rates at which they travel is 200 miles per hour. Find the rate of each if the faster car gains one lap in 40 minutes.

9–7 Loops and Subscripts in Flow Charts (Optional)

The flow chart below pictures a program to compute and print the **compound amount** A (principal plus interest) of 1 unit of principal (say $1) at the end of n interest periods at interest rate x per period, where

$$A = (1 + x)^n.$$

In the program, x is taken to be 1%, and A is computed for values of n from 1 to 50 in steps of 1.

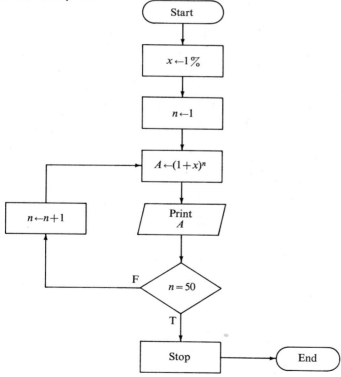

This flow chart as well as those on pages 37 and 69–72 involves a **loop**, which is a set of steps that are to be done over and over again.

Each loop has four parts:

1. One or more steps that set up the **initial conditions** to start the loop. In the compound amount flow chart, there are two such **initialization steps**:

and

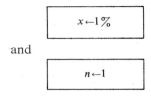

2. One or more steps that are to be repeatedly executed. These are called the **iteration steps**. In the preceding flow chart there are two iteration steps:

and

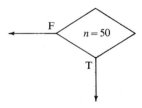

3. One or more **test steps** used to decide when to *exit from*, or break off, the loop. The flow chart on page 235 has one test step:

4. One or more steps that **update** the loop, that is, prepare the iteration steps for the next time around. The step

updates the loop on page 235.

Now look back at the flow charts on page 37 and 69–72 and identify the steps that form the four parts of the loops in those flow charts.

Often a program contains a loop inside another loop (a **nested loop**). For instance, the flow chart below shows the program to compute the compound amount for n varying from 1 to 50 in steps of 1, first with $x = 1\%$, then with $x = 1.5\%$, and so on, with x varying from 1% to 8% in steps of 0.5%.

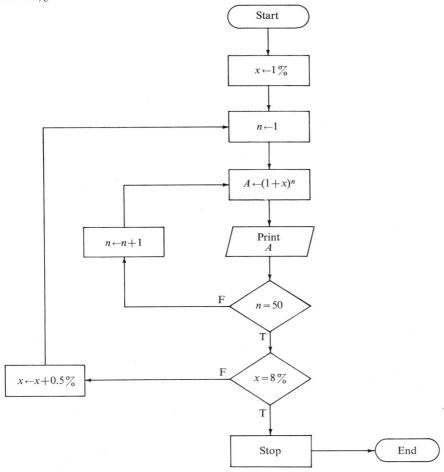

Many loops involve computations with long lists, or arrays, of data. Suppose, for instance, that we want to write a program to compute the sum of the squares of the ten numbers:

$$5.6,\ -2,\ 3,\ 1.25,\ -8,\ 6,\ 4.5,\ 0.4,\ -2.1,\ 15$$

For convenience in referring to this array, we might call it V. We can then refer to any number in the array V by writing the array name followed by a numeral in parentheses (called a **subscript**) that shows the position of the number in the list. For example, the first number in the array V is 5.6; therefore,

$$V(1) = 5.6,$$

read "the value of V-sub-one is five and six tenths." Similarly, since -8 is the fifth number in the array, $V(5) = -8$. In general, if I denotes any one of the integers from 1 to 10, inclusive, then the Ith number in array V is $V(I)$. We call $V(I)$ a *subscripted variable*. Notice that in the ordinary notation of algebra, V_1 is written in place of $V(1)$, V_5 in place of $V(5)$, and V_I in place of $V(I)$.

The array V is really a function whose domain is the set of natural numbers $\{1, 2, 3, 4, 5, 6, 7, 8, 9, 10\}$ and whose range is the numbers in the array?

Example. Draw a flow chart of a program to read an array V of ten numbers, and to compute and write the sum of the squares of the numbers in the array.

Solutions: Let I denote the subscript and let S denote the sum of the squares.

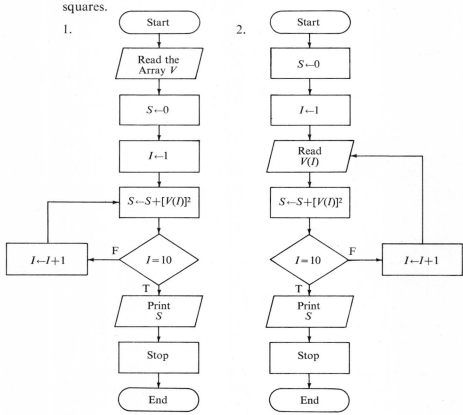

EXERCISES

Draw a flow chart of the program described in each of the following exercises.

1. Read an array R of 50 numbers and compute and write the arithmetic mean, or average, of the numbers in the array.
2. Read an array G of 20 numbers and compute and write the product of the numbers in the array.
3. Read an array V of 25 numbers and an array W of 25 numbers and compute and write the value of the sum of products:

$$V(1) \times W(1) + V(2) \times W(2) + \cdots + V(25) \times W(25)$$

4. Read two arrays C and D each containing 20 numbers, and compute and print the values of the sums:

$$C(1) + D(1), \quad C(2) + D(2), \quad \ldots, \quad C(20) + D(20)$$

5. Read an array X of 30 numbers and write an array Y of 30 numbers when the values in the array Y are computed by the following rule:

$$Y(I) = X(I) \text{ if } X(I) \geq 0, \qquad Y(I) = -X(I) \text{ if } X(I) < 0$$

6. Read an array X of 40 numbers and compute and write the values of $\overline{X}$ and V where

$$\overline{X} = \frac{X(1) + X(2) + \cdots + X(40)}{40},$$

$$V = \frac{(X(1) - \overline{X})^2 + (X(2) - \overline{X})^2 + \cdots + (X(40) - \overline{X})^2}{39}.$$

7. Read an array B containing 100 numbers and compute and write the sum P of the positive numbers in B and the sum N of the negative numbers.

chapter summary

1. Equations whose numerical coefficients are fractions and fractional equations having the variable in the denominator of a fraction may be solved by multiplying both members by the least common denominator (LCD) of the terms of the equation.
2. Whenever the LCD may represent zero, the roots of the resulting equation may not satisfy the original equation. Check all roots in the original equation.
3. Fractions often occur in the solution of problems involving percent mixtures, investments, work, or motion. We use these equations:

$$p = rb, \quad i = Prt, \quad w = rt, \quad d = rt$$

CHAPTER REVIEW

9–1 1. Solve and graph the solution set: $3x - \dfrac{7x}{3} \leq 1$

9–2 2. How many ounces of water must be added to 16 ounces of a 25% salt solution to obtain a 10% solution?

9–3 3. A sum of $10,000 is invested part at 4% per year and part at 6%. If the annual return from both investments is $500, how much is invested at each rate?

9–4 4. Solve: $\dfrac{4x}{x - 3} - 3 = \dfrac{3x - 1}{x + 3}$

9–5 5. A pipe can fill a tank in 5 hours while an outflow pipe can empty it in 8 hours. How long would it take to fill the tank if both pipes are opened together?

9–6 6. Tom can row 9 miles downstream in the same length of time it takes him to row 3 miles upstream. If the current in the river flows at 6 miles per hour, how fast does Tom row in still water?

10 functions, relations, and graphs

10–1 Functions Described by Tables

The table shown at the right reports the average daily circulation of the five newspapers in Central County. The contents of this table can also be presented as a *correspondence* or as a list of **ordered pairs**:

CENTRAL COUNTY

Newspaper	Average Daily Circulation
Times	17,500
Examiner	16,000
Courier	25,000
Journal	9,000
Blade	12,000

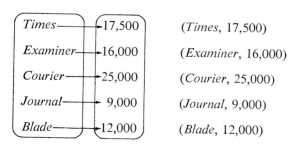

(Times, 17,500)
(Examiner, 16,000)
(Courier, 25,000)
(Journal, 9,000)
(Blade, 12,000)

Since corresponding to each newspaper, there is one and only one value, this table describes a *function* (page 101). The *domain* D of this function is the set of first elements in the ordered pairs above,

{Times, Examiner, Courier, Journal, Blade},

and the *range* R is the set of second elements in the ordered pairs above,

{17,500, 16,000, 25,000, 9,000, 12,000}.

The table above shows that the *Courier* has the greatest average daily circulation, while the *Journal* has the least. However, comparisons can be made more easily when the numbers in the range are represented by lengths in a diagram. One way of doing this is by drawing a *bar graph* as shown in Figure 10–1.

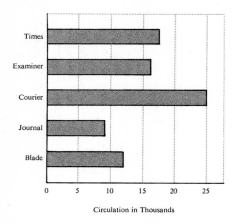

FIGURE 10–1

In this graph the elements of the domain are listed down the left-hand side of the graph. For each element of the domain, a horizontal bar is drawn to represent the corresponding element of the range. The length of each bar is determined by the horizontal scale at the bottom of the diagram. From this bar graph we can easily see how the circulations of the newspapers compare.

Pictographs or *pictorial graphs* like Figure 10–2 are special bar graphs in which rows of uniform symbols replace the bars.

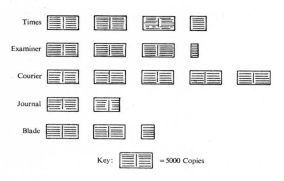

FIGURE 10–2

The adjoining table gives the numbers of college graduates in the United States at 5-year intervals from 1945 through 1970. The domain of this function is

{1945, 1950, 1955, 1960, 1965, 1970}.

The range (in hundred thousands) is

{2.5, 3, 4, 5.5, 7.5}.

This function is pictured as a bar graph in Figure 10–3. Note that in this figure the elements of the domain are listed across the bottom of the graph.

Year of Graduation	United States College Graduates (nearest 50,000)
1945	250,000
1950	400,000
1955	300,000
1960	400,000
1965	550,000
1970	750,000

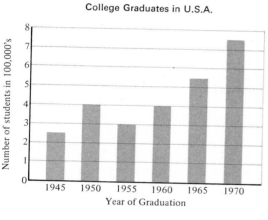

FIGURE 10–3

FIGURE 10–4

Sometimes when the elements of the domain of a function form a numerical succession, the bars of a vertical bar graph are replaced by dots located where the tops of the bars are. Then these dots are connected by line segments, giving a *broken-line graph* as shown in Figure 10–4.

Notice carefully that the line segments in Figure 10–4 do not show the number of graduates for "in-between" years. They do, however, help us to visualize the changes or trends over periods of 5 years.

In working with ordered pairs of numbers, such as (1960, 400,000), the first number, 1960, is called the **first component** or **first coordinate** and the second number, 400,000, is called the **second component** or **second coordinate**. Two ordered pairs of numbers are defined to be **equal** when *both* their first components *and* their second components are equal. Thus, (1960, 400,000) = (1960, 4×10^5), but (1960, 400,000) ≠ (1960, 400,001).

EXERCISES

Make a bar graph for the facts shown in each table and give the graph a title. Give the domain and the range of each function.

1.

Name of Satellite	Telstar	Tiros 5	Discoverer 18	Mariner 4
Satellite weight in hundreds of pounds	1.7	2.8	3.0	5.75

2.

Computer Program Language	Fortran IV	Fortran II	PL/1	Cobol F	RPG
Running time of Program T in seconds	70	80	115	85	110

Make a broken-line graph for the facts shown in each table, and give the graph a title. Give the domain and the range of each function.

3.

Year	1930	1940	1950	1960	1970
Population of Central City in thousands	30	32	35	46	52

4.

Year	1955	1960	1965	1970
Male High School Graduates in U.S. during year (nearest 100,000)	7	9	13	15

5–8. Make a pictograph for the facts shown in the table in Exercises 1–4 above. Give each pictograph a title.

Find all real values of a and b for which the given ordered pairs are equal.

Example. $(a + 2, 3b + 5)$; $(2a - 1, b - 3)$

Solution: $(a + 2, 3b + 5) = (2a - 1, b - 3)$ if and only if

$a + 2 = 2a - 1$ and $3b + 5 = b - 3$
$3 = a$ $\quad\quad\quad\quad\quad\quad\quad 2b = -8$, or $b = -4$

The check is left to you.

∴ the values of a and b are 3 and -4, respectively.

9. $(2a - 3, b); (9, 3)$
10. $(3a, b - 7); (15, -2)$
11. $(-a, 7b + 1); (3a - 6, 15)$
12. $(3a + 9, 4b); (2a + 11, -4)$
13. $(a, b + 5); (2a, 2b - 13)$
14. $(3a + 8, b + 2); (7a, 8 - b)$
15. $(1 - a, 4); (2 + a, |b|)$
16. $(|a| + 1, b); (4, 5b)$
17. $(2a + 5, b^2); (5a + 2, b)$
18. $(3 - a^2, b^2); (2a, 25)$
19. $(|a + 1|, b^2); (0, -1)$
20. $(3 - a, b^2 - 4); (1 - a, 0)$

10–2 Coordinates in a Plane

The functions that we deal with most often have sets of numbers as domains and ranges. To work with them, we need to know how to *graph ordered pairs of numbers* as points in a plane.

Draw a horizontal number line, called the *horizontal axis*, and draw a second number line intersecting it at right angles such that both number lines have the same zero point, called the *origin* (O). The second number line is called the *vertical axis* (Figure 10–5). The positive direction (indicated by the single arrowhead) is usually to the right on the horizontal axis and upward on the vertical axis.

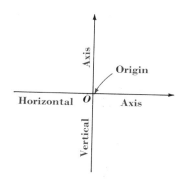

FIGURE 10–5

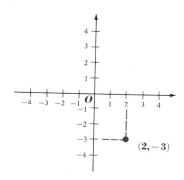

FIGURE 10–6

To locate the *graph of the ordered pair* $(2, -3)$, find the graph (page 6) of 2 on the horizontal axis (Figure 10–6), and draw a vertical line through it. Then find the graph of -3 on the vertical axis, and draw a horizontal line through that point. The point of intersection of these lines is the graph of $(2, -3)$. Mark the point with a dot. Marking a point in this way is called *plotting the point*.

It is convenient to draw the axes on squared paper, using the length of the side of a square as the unit on each axis (Figure 10–7).

The axes separate the plane into four regions, called **quadrants**, numbered as shown in Figure 10–7.

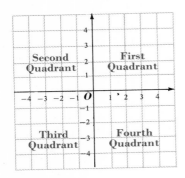
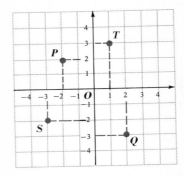

FIGURE 10–7   FIGURE 10–8

From point P in Figure 10–8, draw a vertical line to meet the horizontal axis. The coordinate (page 6) of the point where it meets this axis is called the **abscissa** of P, -2.

Draw a horizontal line from P to meet the vertical axis. The coordinate of this meeting point is called the **ordinate** of P, 2.

Together, the abscissa and ordinate of P are called the **coordinates** of P. The coordinates are written as an ordered pair, with the abscissa first, $(-2, 2)$.

Verify the coordinates of the other points graphed in Figure 10–8:

$$T(1, 3), \quad S(-3, -2), \quad Q(2, -3), \quad O(0, 0)$$

Notice that in the first quadrant, both coordinates are positive; in the second quadrant, the abscissa is negative but the ordinate is positive; in the third quadrant, both coordinates are negative; in the fourth quadrant, the abscissa is positive but the ordinate is negative.

When a coordinate system is set up on a plane as we have just done, the axes are called **coordinate axes** and the plane is called a **coordinate plane**. In working with a coordinate plane, the following facts are taken for granted:

1. **There is exactly one point in the coordinate plane paired with each ordered pair of real numbers.**
2. **There is exactly one ordered pair of real numbers paired with each point in the coordinate plane.**

Thus, there is a one-to-one correspondence (page 15) between ordered pairs of real numbers and points of a coordinate plane. This correspondence is called a **plane rectangular coordinate system**.

EXERCISES

Plot the graph of each ordered pair.

1. (2, 3)
2. (5, 0)
3. (1, −4)
4. (0, −3)
5. (−2, −3)
6. (−5, 0)
7. (−4, 2)
8. (0, 6)
9. (7, 7)
10. (−3, −3)
11. (5, −5)
12. (−6, 6)

Each of Exercises 13–18 lists three vertices of a rectangle. Graph these, determine the coordinates of the fourth vertex, and sketch the rectangle.

13. (1, 1), (1, −1), (−1, −1)
14. (2, 6), (0, 0), (2, 0)
15. (−2, −3), (5, 1), (−2, 1)
16. (1, 6), (−2, 6), (−2, −3)
17. (8, −2), (2, −2), (8, −5)
18. (−6, −1), (−6, −5), (−1, −1)

Plot three points in at least two quadrants whose coordinates are integers satisfying the given requirement.

19. The abscissa equals the ordinate.
20. The abscissa is half the ordinate.
21. The ordinate is one greater than the abscissa.
22. The ordinate is two greater than twice the abscissa.

10–3 Relations

The diagram at the right shows how each number in the set D where $D = \{0, 1, 2, 3\}$ is paired with one or more numbers in the set R, $R = \{1, 2, 3, 4\}$. The same pairing is shown in the adjoining table, and in the following roster of the set of ordered pairs of numbers:

$$\{(0, 1), (1, 2), (1, 4), (2, 3), (3, 2)\}$$

D	R
0	1
1	2
1	4
2	3
3	2

This pairing does *not* define a function with domain D and range R. In a function, each member of the domain is assigned *exactly one* partner in the range. The given pairing, however, assigns to the number 1, two range elements, namely, 2 *and* 4.

The pairing in the preceding example is a *relation*. A **relation** is any set of ordered pairs of elements. The set of first elements in the ordered pairs is called the **domain** of the relation, and the set of second elements is called the

range. The domain of the relation described on the previous page is

$$\{0, 1, 2, 3\},$$

and the range is

$$\{1, 2, 3, 4\}.$$

Recall (page 12) that in the roster of a set, no element is listed more than once. The listing $\{(0, 1), (1, 2), (1, 4), (2, 3), (3, 2)\}$ is called a *roster* of the relation.

A function is a special kind of relation. A function is a relation such that each element of the domain is paired with *one and only one* element of the range.

Figure 10–9 shows the graphs in a coordinate plane of all the ordered pairs that form the relation described above. We call this set of points the **graph** of the relation. Notice that domain elements are shown along the horizontal axis and range elements along the vertical axis.

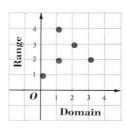

FIGURE 10–9

EXERCISES

Graph the relation whose ordered pairs are shown in the given table or roster. State the domain and the range of the relation. Is the relation a function?

1.
1	2
2	4
3	6
4	8
5	10

2.
1	3
2	2
3	9
4	12
5	15

3.
1	2
2	4
3	3
4	4
5	2

4.
1	5
2	4
3	3
4	4
5	5

5.
−2	−1
−1	−0.5
0	0
1	1.5
2	1
3	1.5

6.
−2	−0.7
−1	−0.3
0	0
1	0.3
2	0.7
3	1

7.
−2	4
−1	1
0	0
1	1
2	4
3	9

8.
−2	−8
−1	−1
0	0
1	1
2	8
3	27

9. {(1, 0), (2, 3), (3, 4), (1, 2)}
10. {(0, 1), (1, 2), (2, 1), (1, 3)}
11. {(−2, 2), (−1, 1), (0, 0), (1, 1), (2, 2)}
12. {(0, 0), (1, 1), (1, −1), (2, 2), (2, −2)}
13. {(−4, 1), (−1, 2), (4, 3), (4, −3), (−1, −2), (−4, −1), (−5, 0)}
14. {(0, 1), (1, 2), (2, 5), (3, 10), (−1, 2), (−2, 5), (−3, 10)}

Give the domain and range of each relation. Is the relation a function?

15.
16.
17.

18.
19.
20.

21.
22.
23.

24.
25.
26.

10–4 Open Sentences in Two Variables

Equations or inequalities that involve two variables, such as x and y, are called **open sentences in two variables**. To **solve an open sentence in two variables**, we must find all the ordered pairs of numbers (chosen from the replacement sets of the two variables) that make the sentence a true statement. Each such ordered pair is called a **solution** or **root** of the sentence and is said to **satisfy** the sentence. The set of all solutions of the open sentence is called the **solution set** over the given replacement sets of the variables.

Example 1. Find the solution set of $3x + 4y = 36$ if the replacement set of both x and y is the set of whole numbers.

Solution:

1. Transform the given sentence into an equivalent one having y as its left member.

$$3x + 4y = 36$$
$$y = \tfrac{1}{4}(36 - 3x)$$
$$y = 9 - \tfrac{3}{4}x$$

2. Replace x with each member of its replacement set in turn and determine the corresponding value of y. Since y must be a whole number, $\tfrac{3}{4}x$ must also be a whole number. Therefore, only multiples of 4 are acceptable replacements for x.

3. If the value of y determined in Step 2 belongs to the replacement set of y, then the pair of corresponding values is a root of the sentence. As shown in the table, values of x greater than 12 will yield negative numbers for y; therefore, 12 is the greatest acceptable value of x.

x	$y = 9 - \tfrac{3}{4}x$	y
0	$9 - \tfrac{3}{4} \cdot 0$	9
4	$9 - \tfrac{3}{4} \cdot 4$	6
8	$9 - \tfrac{3}{4} \cdot 8$	3
12	$9 - \tfrac{3}{4} \cdot 12$	0
16	$9 - \tfrac{3}{4} \cdot 16$	-3

∴ the solution set is $\{(0, 9), (4, 6), (8, 3), (12, 0)\}$.

The solution set of the equation in Example 1 is a function with domain $\{0, 4, 8, 12\}$ and range $\{9, 6, 3, 0\}$. In general, the solution set of any open sentence in two variables is a relation whose domain is the set of first elements and whose range is the set of second elements in the ordered pairs satisfying the sentence. In Example 2, below, the solution set of the given open sentence is a relation, but *not* a function.

Example 2. Find the solution set of $x - 2y \geq 6$ if $x \in \{-2, 0, 2\}$ and $y \in \{-1, -2, -3, -4\}$.

Solution:

$$x - 2y \geq 6$$
$$-2y \geq 6 - x$$
$$y \leq \tfrac{x}{2} - 3$$

x	$\tfrac{x}{2} - 3$	$y \leq \tfrac{x}{2} - 3$	y
-2	$\tfrac{-2}{2} - 3$	$y \leq -4$	-4
0	$\tfrac{0}{2} - 3$	$y \leq -3$	$-4, -3$
2	$\tfrac{2}{2} - 3$	$y \leq -2$	$-2, -3, -4$

∴ the solution set is

$$\{(-2, -4), (0, -4), (0, -3), (2, -2), (2, -3), (2, -4)\}.$$

EXERCISES

Find the solution set of each equation given that $\{-2, -1, 0, 1, 2\}$ is the replacement set of x and $\Re$, the replacement set of y. Graph the solution set.

1. $y = x$
2. $y = -x$
3. $y = x + 1$
4. $y = -x - 1$
5. $y = -2x$
6. $y = \frac{3}{2}x$
7. $y = x - 1$
8. $y = -x + 1$
9. $y = 2x + 2$
10. $y = 2x - 2$
11. $y = -x + 4$
12. $y = -x - 4$
13. $y = 2x^2$
14. $y = -x^2$
15. $y = x^2 + 3$
16. $y = x^2 - 5$
17. $y = x^2 - 3x$
18. $y = x^2 + 4x$

Find the solution set of each inequality given that $\{-1, 0, 1\}$ is the replacement set of x and $\{-2, -1, 0, 1, 2\}$ is the replacement set of y. Graph the solution set.

19. $y > x$
20. $y < x$
21. $y + 1 \leq 2x$
22. $y - 1 \leq 3x$
23. $x - y < x + y$
24. $1 - y > 2x + 1$

State the domain and the range of the function whose ordered pairs are given in the table. Then give a formula for the function.

Example.

x	1	2	3	4	5	6
y	-1	0	1	2	3	4

Solution: Domain: $\{1, 2, 3, 4, 5, 6\}$;
Range: $\{-1, 0, 1, 2, 3, 4\}$; $y = x - 2$.

25.
x	1	2	3	4	5
y	1	2	3	4	5

26.
x	1	2	3	4	5
y	-1	-2	-3	-4	-5

27.
x	1	2	3	4	5
y	4	5	6	7	8

28.
x	1	2	3	4	5
y	-2	-4	-6	-8	-10

Find the solution set over the given replacement sets for *x* and *y*.

29. $|x| + 3y = 2y + 4$; $x \in \{-3, 0, 4\}$, $y \in \{\text{positive integers}\}$
30. $|x| - 5y = 12 - y$; $x \in \{-1, 0, 4\}$, $y \in \{\text{negative integers}\}$
31. $y = x^2 - 4x + 3$; $y \in \{0, -1, 3\}$; $x \in \mathcal{R}$
32. $y = x^2 + 6x$; $y \in \{0, -9, 7\}$; $x \in \mathcal{R}$

For each problem write an equation in two variables. Restrict the replacement sets to the conditions of the problem and find the solution set.

33. Mr. Novak spent $1.50 for stamps, buying some 6¢ stamps and some 10¢ stamps. How many of each did he buy?
34. Bill Jorgensen sent $16 to a hatchery for chicks and ducklings. He specified that all the money be used and that he receive at least a dozen of each kind of fowl. The chicks cost 15¢ each and the ducklings 40¢ each. How many of each did he get?

10–5 The Graph of a Linear Equation in Two Variables

Every root of the equation

$$-2x + y = 1$$

is an ordered pair of numbers represented by (*x, y*). The graph of one such root, (0, 1), is the point *P* in Figure 10–10. Note that it is customary to take the horizontal axis as the *x*-axis and the vertical axis as the *y*-axis. To find other roots of this equation, substitute values for *x* and obtain corresponding values for *y*, as shown in the table.

$-2x + y = 1$		
x	y = 1 + 2x	y
−3	1 + 2(−3)	−5
−2	1 + 2(−2)	−3
−1	1 + 2(−1)	−1
0	1 + 2(0)	1
1	1 + 2(1)	3
2	1 + 2(2)	5
3	1 + 2(3)	7

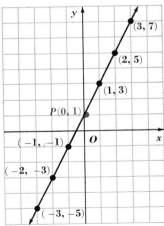

FIGURE 10–10

Figure 10-10 suggests that the points corresponding to these roots lie on a straight line, part of which is shown in the diagram. In fact, if the set $\Re$ (of real numbers) is the replacement set for both x and y, then each root of the equation does give the coordinates of a point on this line, and the coordinates of each point of the line satisfy the equation.

Because this line is the set of *all those points* and *only those points* whose coordinates satisfy the equation, the line is called the *graph of the equation* in the coordinate plane, and

$$-2x + y = 1$$

is called *an equation of the line*.

In general:

In the coordinate plane, the graph of any equation equivalent to one of the form

$$Ax + By = C, \quad x \in \Re, \quad y \in \Re$$

where A, B, and C are real numbers with A and B not both zero is a straight line. Any such equation is called a *linear equation in two variables*, x and y.

Notice that in a linear equation in standard form (page 185), each term is a constant or a monomial of degree 1. Thus, "$3x - 2y = 1$" is linear, but "$x + \frac{2}{y} = 5$," "$x^2 + y^2 = 3$," and "$xy = 7$" are not.

Although *only two points* are needed to graph a linear equation, it is good practice to plot a third point, as a check. Sometimes it is helpful to find roots of the form $(0, b)$ and $(a, 0)$. We obtain the latter by replacing y with "0" and finding the corresponding value of x.

In the following examples and throughout the rest of this book, assume, unless otherwise directed, that $\Re$ is the replacement set for each variable in an open sentence in two variables.

Example 1. Graph $x + 3y = 9$ in the coordinate plane.

Solution:

$3y = 9 - x$
$y = 3 - \frac{1}{3}x$

x	$3 - \frac{1}{3}x$	y
-3	$3 - \frac{1}{3}(-3)$	4
0	$3 - \frac{1}{3}(0)$	3
3	$3 - \frac{1}{3}(3)$	2
9	$3 - \frac{1}{3}(9)$	0

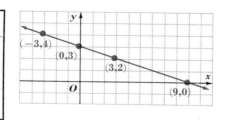

Example 2. Graph $y = 2$ in the coordinate plane.

Solution: Since the equation places no restrictions on x, every point having ordinate 2 corresponds to a root, regardless of its first coordinate.

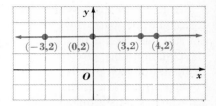

A function whose ordered pairs satisfy a linear equation is called a **linear function**. Thus,
$$f: x \to y = 2x - 4$$
is a linear function.

EXERCISES

Graph each equation in the coordinate plane.

1. $y = 4$
2. $y = -2$
3. $x = 3$
4. $x = -2$
5. $y = x + 1$
6. $y = x - 2$
7. $3x + y = 5$
8. $2x + y = 6$
9. $4x - y = 2$
10. $3x - y = 6$
11. $2x - 3y = 6$
12. $3x + 4y = 12$

Write a formula and then graph the function defined by each rule. In each case, let the domain be {the positive real numbers}.

13. The number of inches is 12 times the number of feet.
14. The number of feet is 3.3 times the number of meters.
15. The number of U.S. dollars is 2.4 times the number of British pounds.
16. The pressure of water (in pounds per square foot) on the bottom of a tank is 62.4 times the depth (in feet) of water.
17. The total number of dollars to be repaid on $100 borrowed at 7% per year is 100 more than 7 times the number of years of the loan.
18. Bradley's Rent-A-Car Service makes the following daily charge in dollars for a standard model automobile: 7 plus 0.07 times the number of miles driven.

Graph the function with the indicated domain.

19. $f: x \to y = \frac{1}{2}x - 7;\ x \in \mathcal{R}$
20. $g: x \to y = 4 - \frac{x}{2};\ x \in \mathcal{R}$
21. $P(l) = 2(l + 3);\ l \in$ {the nonnegative numbers}
22. $f(c) = \frac{9}{5}c + 32;\ c \in \mathcal{R}$
23. $C: r \to 2\pi r;\ 0 \leq r \leq 14$ (Use $\frac{22}{7}$ for π.)
24. $C: d \to \pi d;\ 0 \leq d \leq 7$ (Use $\frac{22}{7}$ for π.)

Find the coordinates of the point where the graph of each equation crosses (a) the x-axis and (b) the y-axis.

25. $7x + 5y = 35$
26. $4x - 6y = 24$
27. $5x = 50 - 10y$
28. $7x = 56 + 4y$
29. $2x = 9y$
30. $4y = -15x$

31. Show that the function $f: x \rightarrow y = (x - 4)(x + 3) - x^2 + 6$ is a linear function.
32. Show that the function $g: x \rightarrow y = (x - 1)(x + 1) - (x + 2)^2$ is a linear function.

Graph each pair of equations on the same set of axes. Name the coordinates of the point where the graphs intersect, and show by substitution that the coordinates satisfy both equations.

33. $x + y = 4$; $y = 2x + 1$
34. $2x + y = 4$; $y = x + 1$
35. $x - 2y = 6$; $2x - 3y = 5$
36. $x - y = 6$; $x + y = -2$

Graph each equation. Is it a linear equation?

37. $y = |x| + 1$
38. $y = |x + 1|$
39. $y = -|x| + 1$
40. $y = -|x| - 11$

Graph each function. Is it a linear function?

41. The function which assigns the number 2 to each nonnegative number, and the number -1 to each negative number.
42. The function which assigns the number 1 to each positive number, and the number 3 to each nonpositive number.
43. The function which assigns to each number twice its absolute value.
44. The function which assigns to each number half its absolute value.

10–6 Slope of a Line

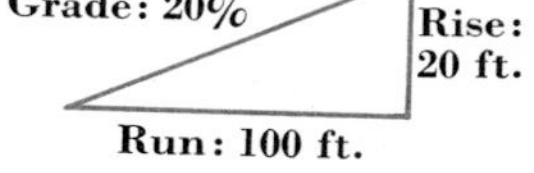

The steepness, or grade, of a hill is usually described in terms of the vertical *rise* compared with the horizontal *run*. For example, if a hill rises 20 feet for every 100 feet of horizontal distance, its grade is the ratio $\frac{20}{100}$, or 20% (Figure 10–11).

Grade: 20%
Rise: 20 ft.
Run: 100 ft.

FIGURE 10–11

Similarly, to describe the steepness, or *slope*, of a straight line, choose two points on it and compute the quotient

$$\text{slope} = \frac{\text{rise}}{\text{run}} = \frac{\text{vertical change}}{\text{horizontal change}}.$$

Because the vertical change in moving from one point to another is the difference of the ordinates, and the horizontal change is the *corresponding* difference of abscissas,

$$\text{slope} = \frac{\text{difference of ordinates}}{\text{difference of abscissas}}.$$

Thus, to find the slope of the line through $P(3, 2)$ and $Q(4, 5)$, notice that

ordinate of Q — ordinate of $P = 5 - 2$;
abscissa of Q — abscissa of $P = 4 - 3$.

Thus, using m to denote slope, we have

$$m = \frac{5-2}{4-3} = \frac{3}{1} = 3.$$

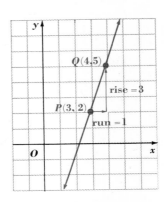

FIGURE 10–12

One way of checking the slope is to move on a line from left to right. On a line with slope 3, each horizontal change of 1 unit produces a positive change of 3 units in the vertical direction. For the line joining S and T in Figure 10–13,

$$m = \frac{1-3}{3-(-1)} = \frac{-2}{4} = -\frac{1}{2}.$$

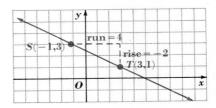

FIGURE 10–13

Check by counting units. From S to T are 4 units of horizontal change and -2 units of vertical change. For each positive change of one horizontal unit, therefore, there is a negative change of half a vertical unit, a rate of change equal to $-\frac{1}{2}$.

Whenever a line falls from left to right, its slope is a negative number. When it rises from left to right, its slope is a positive number. Can the slope of a line be 0? The slope of the line joining $K(-2, -2)$ and $M(3, -2)$ in Figure 10–14 is

$$\begin{aligned} m &= \frac{-2-(-2)}{3-(-2)} \\ &= \frac{-2+2}{3+2} \\ &= \frac{0}{5} = 0. \end{aligned}$$

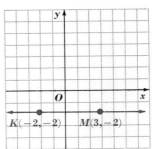

The slope of every horizontal line is 0.

FIGURE 10–14

If we use the formula on page 256 to try to compute the "slope" of the line pictured in Figure 10-15, we find

$$\frac{3-(-2)}{3-3} = \frac{5}{0}.$$

Since division by 0 is not defined, the "slope" of a vertical line is also not defined, and this line, like *every vertical line, has no slope.*

A basic property of a line is that its slope is constant. Thus, *any* two of its points may be used in computing its slope.

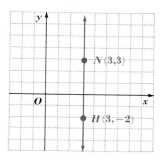

FIGURE 10-15

EXERCISES

Plot each pair of points, draw the line containing the points, and determine the slope of the line from the graph. Check by finding the slope algebraically.

1. (3, 1); (5, 4)
2. (1, 3); (4, 5)
3. (−2, 3); (0, 2)
4. (3, −1); (3, 4)
5. (4, 2); (−3, 2)
6. (2, 4); (−1, −1)

Through the given point, draw a line with the given slope.

Example. (2, −1); slope, $-\frac{3}{2}$

Solution: From (2, −1) measure 2 units to the right and 3 units down. Connect the point reached with (2, −1) to determine the line.

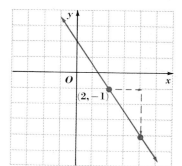

7. (3, 2); slope, 2
8. (0, −2); slope, 1
9. (−3, 1); slope, $\frac{3}{4}$
10. (2, −2); slope, $\frac{3}{5}$
11. (−3, −2); slope, $-\frac{1}{2}$
12. (1, 3); slope, 0

Determine a value for *b* so that the slope *m* of the line through each pair of points has the given value. Check your solution by graphing the line through the points.

13. (4, −1); (6, b); $m = 1$
14. (4, b); (9, 1); $m = -2$
15. (3, b); (2b, 7); $m = \frac{3}{5}$
16. (8, 3b); (b, 3); $m = \frac{1}{2}$
17. (2b, 3); (6, b); $m = -\frac{1}{2}$
18. (1, b); (−b, 7); $m = 5$

10–7 The Slope-Intercept Form of a Linear Equation

The graph of "$y = 2x$" is the straight line (Figure 10–16) containing the points whose coordinates are given in the table. Notice that when the abscissas of two points on the line differ by 1 their ordinates differ by 2, the slope of the line. Also notice that the line passes through the origin.

x	−1	0	1	2
y	−2	0	2	4

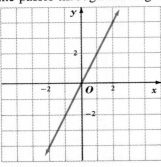

FIGURE 10–16   FIGURE 10–17

The slope of the line whose equation is "$y = -2x$" (Figure 10–17) is -2, because an increase of 1 in the abscissa produces a change of -2 in the ordinate. This line also contains the origin. In general:

For every real number *m*, the graph in the coordinate plane of the equation

$$y = mx$$

is the line that has slope *m* and passes through the origin.

In Figure 10–18, compare the graphs of "$y = 2x + 4$" and "$y = 2x$." They have equal slopes, but they cross the y-axis at different points. The ordinate of the point where a line crosses the y-axis is called the line's **y-intercept**. To determine the y-intercept, replace x with "0" in the equation of each line:

$y = 2x$ $y = 2x + 4$
$y = 2 \cdot 0$ $y = 2 \cdot 0 + 4$
$y = 0$ ← y-intercepts → $y = 4$

If we write "$y = 2x$" as "$y = 2x + 0$," we can see that the constant term in these equations is the y-intercept of each graph:

$y = 2x + 0$ and $y = 2x + 4$.

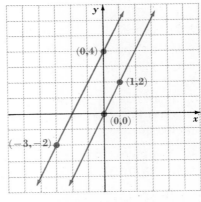

FIGURE 10–18

For all real numbers m and b, the graph in the coordinate plane of the equation

$$y = mx + b$$

is the line whose slope is m and y-intercept is b.

One way to describe a straight line is to write its equation in the form "$y = mx + b$," and then read the values of the slope m and the y-intercept b. This is called the **slope-intercept form**.

Equation	Transforming to $y = mx + b$	Describing the line	
		Slope	y-intercept
$x + 2y = 4$	$2y = -x + 4$, $y = -\frac{1}{2}x + 2$	$-\frac{1}{2}$	2
$6x - 3y = 8$	$3y = 6x - 8$, $y = 2x - \frac{8}{3}$	2	$-\frac{8}{3}$
$2y - 10 = 0$	$2y = 10$, $\quad y = 0x + 5$	0	5

Example. Draw the line with $m = \frac{2}{3}$, $b = 1$; then find its equation.

Solution: The y-intercept is 1; so label $(0, 1)$. Since the slope is $\frac{2}{3}$, move from $(0, 1)$ 3 units to the right and 2 units up to locate a second point on the line. Draw the line containing the two points.

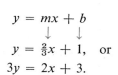

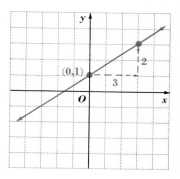

EXERCISES

Write a linear equation with integral coefficients whose graph has the given slope and y-intercept.

1. $m = 2$, $b = 5$
2. $m = -4$, $b = 1$
3. $m = -2$, $b = -3$
4. $m = 5$, $b = -4$
5. $m = \frac{1}{2}$, $b = 7$
6. $m = -\frac{1}{3}$, $b = 2$
7. $m = 0$, $b = 2$
8. $m = 0$, $b = -7$
9. $m = -\frac{2}{3}$, $b = -\frac{1}{6}$
10. $m = -\frac{4}{5}$, $b = \frac{3}{10}$

Use only the y-intercept and slope to graph each equation.

11. $y - x = 2$
12. $y - 2x = -2$
13. $2x - y = 3$
14. $3x + y = 1$
15. $2x + y = 0$
16. $-3x + y = 0$
17. $6x + 2y + 1 = 0$
18. $5y + 3 = 0$

10–8 Determining an Equation of a Line

The line in Figure 10–19 has slope -2 and passes through the point $(2, 1)$. The slope-intercept form of the equation of this line is

$$y = -2x + b.$$

Since the point $(2, 1)$ is on the line, its coordinates must satisfy the equation; that is,

$$1 = -2(2) + b$$
$$1 = -4 + b, \quad \text{or} \quad 5 = b.$$

Thus, an equation of the line is

$$y = -2x + 5.$$

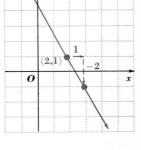

FIGURE 10–19

To determine an equation of a line containing two given points, find the slope of the line, and then find the y-intercept, as above. The following example illustrates the method.

Example. Find an equation of the line which passes through the points whose coordinates are $(5, 2)$ and $(2, -1)$.

Solution: 1. Slope $= m = \dfrac{-1 - 2}{2 - 5} = \dfrac{-3}{-3} = 1$

2. The slope-intercept form of the equation is $y = mx + b$:

$$y = 1x + b$$

Choose one point, say $(5, 2)$. Since it lies on the line:

$$2 = 1 \cdot 5 + b, \quad \text{or} \quad 2 = 5 + b$$
$$\therefore -3 = b.$$

3. To check, show that the coordinates of the other point $(2, -1)$ satisfy the equation:

$$y = x - 3$$
$$-1 \stackrel{?}{=} 2 - 3$$
$$-1 = -1$$

∴ an equation of the line is $y = x - 3$.

EXERCISES

Find an equation of the line through the given point having the given slope.

1. $(1, 2)$; 2
2. $(3, 2)$; 1
3. $(-1, 2)$; -2
4. $(3, -4)$; -3
5. $(2, 0)$; $\frac{2}{3}$
6. $(0, 4)$; $-\frac{1}{2}$
7. $(0, 0)$; $-\frac{3}{4}$
8. $(-2, -6)$; $\frac{4}{3}$

Find an equation of the line through the given points.

9. $(1, 4)$; $(4, 7)$
10. $(3, 2)$; $(5, 6)$
11. $(2, 1)$; $(2, 3)$
12. $(-1, 3)$; $(2, 3)$
13. $(-3, 2)$; $(5, -2)$
14. $(6, -5)$; $(1, 10)$
15. $(0, 0)$; $(-4, -2)$
16. $(-5, -3)$; $(0, 0)$

Determine the value of *a* so that the graph of the given equation contains the given point.

17. $ax + 2y = 3$; $(1, 2)$
18. $ax + 2y = 8$; $(0, 4)$
19. $5x + ay = 13$; $(2, -1)$
20. $-5x + ay = 4$; $(-2, 3)$
21. $6x - 5y + a = 0$; $(3, 4)$
22. $2x - ay + 2 = 0$; $(-5, -1)$

Find an equation of the line parallel to the given line containing the given point. (Parallel lines have the same slope.)

23. $x + y = 2$; $(1, 2)$
24. $x - y = 7$; $(2, -1)$
25. $2x - y = 5$; $(2, 2)$
26. $2x + 3y = 5$; $(-3, 1)$

Determine the coordinates of the point where the lines intersect.

27. $7x - 2y + 10 = 0$; the *x*-axis
28. $-4x + 3y - 8 = 0$; the *y*-axis
29. $3x - 6y = 27$; $x = 3$
30. $5x + 2y = 14$; $y = 2$

Draw a flow chart for each problem.

★31. Read each number in an array of 35 numbers. After each number is read, write the number, its square, and, if the number is not negative, its positive square root.

★32. Read in arrays A and B for 20 linear equations (see page 253). Compute $-A/B$ for each equation where $B \neq 0$, and write A, B, $-A/B$ for each equation.

10–9 Direct Variation and Proportion

The table shows the distance, d miles, an automobile would travel in t hours if it were able to maintain a constant speed. The ratio $\frac{d}{t}$ for every pair of numbers is the same, that is,

$$\frac{d}{t} = 60,$$

or

$$d = 60t.$$

t (in hours)	d (in miles)	$\frac{d}{t}$ (in miles per hour)
1	60	$\frac{60}{1} = 60$
2	120	$\frac{120}{2} = 60$
3	180	$\frac{180}{3} = 60$
4	240	$\frac{240}{4} = 60$
5	300	$\frac{300}{5} = 60$

Such a formula describes a function that is called a *linear direct variation*.

A **linear direct variation** (or simply, a **direct variation**) is a function in which the ratio between a number y of the range and the corresponding number x of the domain is the same for all pairs of the function other than $(0, 0)$; that is, $\frac{y}{x} = k$, for $x \neq 0$. Thus, in a direct variation

$$y = kx, \text{ where } k \text{ is a nonzero constant.}$$

We can say that y *varies directly as* x or y *is directly proportional to* x or y *varies with* x. k is called the **constant of proportionality** or the **constant of variation**.

Figure 10–20 shows the graph of $y = kx$ with $k = 2$. The line has slope 2 and passes through the origin.

In general, the graph of a linear direct variation with $\mathcal{R}$ as domain and range is a straight line passing through the origin and having a slope equal to the constant of proportionality.

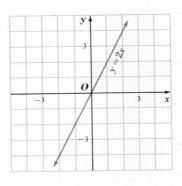

FIGURE 10–20

FUNCTIONS, RELATIONS, AND GRAPHS

Example 1. Given that r varies directly as s and that the value of r is 6 when the value of s is 9, find

    a. the constant of proportionality;

    b. the value of r when 24 is the value of s.

Solution: Let $r = ks$

    a. Replacing r with "6" and s with "9," we find

$$6 = k \cdot 9; \quad \therefore k = \tfrac{6}{9} = \tfrac{2}{3}.$$

    b. $r = \tfrac{2}{3}s$;

        If $s = 24$, then

$$r = \tfrac{2}{3} \cdot 24 = 16.$$

$\therefore$ when 24 is the value of s, 16 is the value of r.

If one ordered pair of a direct variation is (x_1, y_1) (read "x sub 1, y sub 1") and another of the same function is (x_2, y_2), neither $(0, 0)$, then.

$$y_1 = kx_1, \quad \text{and} \quad y_2 = kx_2,$$

or

$$\frac{y_1}{x_1} = k \quad \text{and} \quad \frac{y_2}{x_2} = k.$$

Therefore,

$$\frac{y_1}{x_1} = \frac{y_2}{x_2}.$$

Such an equality of ratios is called a **proportion** and can be read "y_1 is to x_1 as y_2 is to x_2." In this proportion, x_1 and y_2 are called the **means**, and y_1 and x_2 are called the **extremes**. Multiplying both members by $x_1 x_2$ gives

$$(x_1 x_2)\left(\frac{y_1}{x_1}\right) = (x_1 x_2)\left(\frac{y_2}{x_2}\right).$$

Then:

$$y_1 x_2 \left(\frac{x_1}{x_1}\right) = x_1 y_2 \left(\frac{x_2}{x_2}\right) \quad \text{Commutative and associative axioms of multiplication}$$

$$\therefore y_1 x_2 = x_1 y_2. \quad \text{Multiplicative axiom of 1}$$

In any proportion, the product of the means equals the product of the extremes.

Example 2. If the cost of some kinds of fruit varies directly with the weight and if 6 pounds cost $1.10,

    a. how much will 12 pounds cost?
    b. how much will 8 pounds cost?
    c. how much can be bought for $1.00?

Solution: **a.** Let c = cost in dollars,
w = weight in pounds.

$$\frac{c_1}{w_1} = \frac{c_2}{w_2} \; ; \; c_1 = 1.10, \; w_1 = 6, \; w_2 = 12$$

$$\frac{1.10}{6} = \frac{c_2}{12}$$

$$6c_2 = 13.20$$

$$c_2 = 2.20 \quad \therefore \text{ cost is } \$2.20.$$

b. $c_1 = 1.10, \; w_1 = 6, \; w_2 = 8$

$$\frac{1.10}{6} = \frac{c_2}{8}$$

$$6c_2 = 8.80$$

$$c_2 \doteq 1.466 \quad \therefore \text{ cost is } \$1.47.$$

c. $c_1 = 1.10, \; w_1 = 6, \; c_2 = 1.00$

$$\frac{1.10}{6} = \frac{1.00}{w_2}$$

$$1.10 w_2 = 6.00$$

$$w_2 \doteq 5.5 \quad \therefore \text{ about 5.5 pounds can be bought.}$$

Notice that in solving Example 2, it was not necessary to actually calculate the constant of proportionality since a proportion was used.

EXERCISES

1. y varies directly as x. y is 28 when x is 4. Find y when:
    a. x is 1   **b.** x is 20   **c.** x is -2

2. w is directly proportional to t. w is 6 when t is 30. Find w when:
    a. t is 15   **b.** t is -1 **c.** t is -10

In these direct variations, find the value of the indicated variable.

3. $x_2 = 10, \; y_1 = 12, \; x_1 = 5, \; y_2 = \underline{\;?\;}$
4. $x_1 = 4, \; x_2 = 16, \; y_2 = 36, \; y_1 = \underline{\;?\;}$
5. $n_2 = 35, \; n_1 = 14, \; m_1 = -10, \; m_2 = \underline{\;?\;}$
6. $y_1 = 4, \; x_1 = 1, \; y_2 = -16, \; x_2 = \underline{\;?\;}$

7. $s_1 = 6.5$, $r_1 = 3.9$, $r_2 = 3.6$, $s_2 = $ _?_
8. $u_1 = \frac{3}{10}$, $v_1 = 1\frac{1}{4}$, $u_2 = 1\frac{4}{5}$, $v_2 = $ _?_

Find all values of the variable for which each proportion is true.

9. $\dfrac{x-4}{x} = \dfrac{7}{9}$

10. $\dfrac{y+2}{y} = \dfrac{4}{3}$

11. $\dfrac{3w}{w+2} = \dfrac{5}{2}$

12. $\dfrac{6x}{x+7} = \dfrac{9}{5}$

13. $\dfrac{10y}{12y+7} = \dfrac{8}{11}$

14. $\dfrac{3w}{10w+2} = \dfrac{2}{7}$

15. $\dfrac{7z-2}{4z+13} = \dfrac{6}{5}$

16. $\dfrac{8x-5}{5x-4} = \dfrac{13}{8}$

17. $\dfrac{32}{x} = \dfrac{x}{2}$

18. $\dfrac{27}{y} = \dfrac{y}{3}$

19. $\dfrac{x-1}{x} = \dfrac{x+1}{x+3}$

20. $\dfrac{x+2}{x+1} = \dfrac{x}{x-2}$

21. If x varies directly as $y + 4$, and $x = 8$ when $y = 6$, find y when $x = 12$.
22. If y is proportional to $4x - 1$, and $y = 3$ when $x = 2$, find x when $y = 11$.
23. If $2y$ is directly proportional to $3x - 1$ and $y = 1$ when $x = 2$, find x when $y = 0.40$.
24. If $3x - 5$ and $y + 2$ are in a direct proportion, and $x = 3$ when $y = 1$, find x when $y = -5$.

Translate into formulas expressing direct variation.

Example. The circumference of a circle varies directly as its diameter. The constant of proportionality is π.

Solution: Let $C = $ circumference, and let $d = $ diameter. Then $\dfrac{C}{d} = \pi$, or $C = \pi d$, or $\dfrac{C_1}{d_1} = \dfrac{C_2}{d_2}$.

25. The time required to complete a job varies with the amount of work to be done.
26. The interest on a mortgage varies with the principal.
27. The distance traveled varies with the rate of speed.
28. The elongation of a coil spring varies with the weight suspended from it.
29. The pressure in a container of gas varies with the temperature.
30. The velocity of a freely falling body varies with the length of time it falls.

PROBLEMS

1. The ratio of an object's weight on Earth to its weight on Mars is 5:2. How much would a man who weighs 145 pounds here weigh on Mars?
2. The ratio of oil to vinegar in a certain brand of salad dressing is 7:3. How much oil must be blended with 51 quarts of vinegar for that recipe?
3. Fourteen gallons of heavy cream will yield 9 pounds of butter. How many gallons of cream are needed to make 2 pounds of butter?
4. Twelve grams of calcium chloride can absorb 5 cubic centimeters of water. How much calcium chloride is needed to absorb 138 cubic centimeters of water?
5. In a local election the winning candidate defeated his opponent by a margin of 8 to 5. If the loser's share of the votes was 7,425, how many votes were cast for the winner?
6. When an electric current is 35 amperes, the electromotive force is 315 volts. Find the force when the current is 50 amperes if the force varies directly as the current.
7. Find the resistance of 700 feet of wire having 0.00042 ohm resistance per foot if resistance varies directly as the length.
8. How heavy is a 500-foot cable if a 1-foot section weighs 0.87 pound?
9. What length represents 288 miles on a map scaled at $\frac{3}{16}$ inch = 15 miles?
10. What is the scale of a blueprint on which a 111-foot ceiling beam is drawn $9\frac{1}{4}$ inches long?
11. Rod A has 180 equal divisions, and rod B has 100, although both rods have the same length. A length equal to 66.6 divisions on rod A is equal to a length of how many divisions on rod B?
12. On a business trip Mr. Wilson decided to test the accuracy of his automobile's odometer. For the 220-mile trip from Oakdale to Alton the odometer registered only 216.7 miles. On the return trip Mr. Wilson had to detour for road repairs. If the odometer registered 453.1 miles for the round trip, how many actual miles was the detour?
13. One cubic centimeter of gold weighs 19.3 grams, while a cubic centimeter of silver weighs 10.5 grams. Which is heavier and by how much — a cube of gold 1.2 centimeters on an edge or a cube of silver 1.4 centimeters on an edge?
14. One cubic foot of mahogany wood weighs 32.2 pounds, while a cubic foot of white pine weighs 23.3 pounds. Which is heavier and by how much — a mahogany table top 6 feet by $3\frac{1}{2}$ feet by 3 inches, or a pine door $8\frac{1}{2}$ feet by 4 feet by $1\frac{1}{2}$ inches?
15. The cost of boring an automobile cylinder varies directly as the circumference of the cylinder. If it costs $14.20 to bore each cylinder with a 1.5-inch radius, how much more expensive is it to bore each cylinder with a radius of 1.9 inches?

16. The length of a shadow of a vertical object at a given time and location varies with the height of the object casting the shadow. A boy 5 feet 6 inches tall casts a shadow of 12 feet 9 inches. If the shadow of a nearby tree measures 153 feet at the same time, how tall is the tree?

17. If 1 inch represents 45 miles on a map, and Colorado is shown by a rectangle $6\frac{3}{4}$ inches by $8\frac{1}{8}$ inches, calculate its area to the nearest 100 square miles.

18. The Forestry Bureau plans to plant 1600 trees per square mile in a tract of land shown on a map as a rectangle $1\frac{3}{8}$ inches by $\frac{3}{4}$ inch. The scale of the map is 1 inch = 3 miles. How many trees will be needed for the area?

10–10 Quadratic Functions

Several points on the graph of the quadratic equation "$y = x^2$" have been plotted in Figure 10–21, and connected by a smooth curve.

$y = x^2$		
x	x^2	y
0	0^2	0
1	1^2	1
2	2^2	4
3	3^2	9
−1	$(-1)^2$	1
−2	$(-2)^2$	4
−3	$(-3)^2$	9

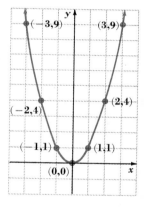

FIGURE 10–21

Figure 10–22 is the graph of another quadratic equation, "$y = -2x^2$."

$y = -2x^2$		
x	$-2x^2$	y
0	$-2(0)^2$	0
1	$-2(1)^2$	−2
$\frac{3}{2}$	$-2(\frac{3}{2})^2$	$-\frac{9}{2}$
2	$-2(2)^2$	−8
−1	$-2(-1)^2$	−2
$-\frac{3}{2}$	$-2(-\frac{3}{2})^2$	$-\frac{9}{2}$
−2	$-2(-2)^2$	−8

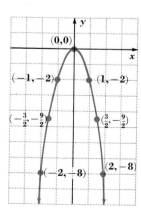

FIGURE 10–22

Quadratic equations are useful in solving a certain type of variation problem as in the following example.

Example. The value of a diamond varies as the square of its weight. If a 2-carat diamond is worth $1,600, what would be the value of a diamond weighing $1\frac{1}{2}$ carats?

Solution: Let d = value in dollars, and
c = weight in carats.

Method I

Use: $d = kc^2$
$1600 = k \cdot 2^2$; $1600 = 4k$
$400 = k$
$\therefore d = 400c^2$.
$d = 400(\frac{3}{2})^2 = 400 \cdot \frac{9}{4}$
$d = 900$

Method II

Use: $\dfrac{d_1}{c_1^2} = \dfrac{d_2}{c_2^2}$

$\dfrac{1600}{2^2} = \dfrac{d_2}{(\frac{3}{2})^2}$

$\dfrac{1600}{4} = \dfrac{d_2}{\frac{9}{4}}$

$900 = d_2$

$\therefore$ a $1\frac{1}{2}$-carat diamond would be worth $900.

This type of problem is called *quadratic direct variation*. In general, a *quadratic direct variation* is a function in which each number x in the domain and the corresponding number y in the range satisfy an equation of the form

$y = kx^2$, where k is a nonzero constant.

A quadratic direct variation is a special kind of quadratic function. A **quadratic function** is a function whose ordered pairs (x, y) satisfy a quadratic equation of the form

$$y = ax^2 + bx + c, \quad a \neq 0.$$

Consider the function

$$f: x \rightarrow y = x^2 - x - 6.$$

The graph of this function can be plotted by finding the coordinates of selected points as shown at the top of page 269.

To plot the graph of

$$f: x \rightarrow y = x^2 - x - 6,$$

assume that the domain is $\Re$ and that the range is the set of corresponding values of y.

$y = x^2 - x - 6$		
x	$x^2 - x - 6$	y
-3	9 + 3 - 6	6
-2	4 + 2 - 6	0
-1	1 + 1 - 6	-4
0	0 - 0 - 6	-6
1	1 - 1 - 6	-6
2	4 - 2 - 6	-4
3	9 - 3 - 6	0
4	16 - 4 - 6	6

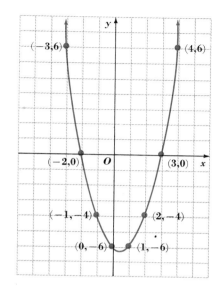

FIGURE 10-23

Notice that the graphs of "$y = x^2$" (Figure 10-21) and "$y = x^2 - x - 6$" (Figure 10-23) open upwards so that each has a lowest point, while the graph of "$y = -2x^2$" (Figure 10-22) opens downward and has a highest point.

Notice also, that in these examples the points, except the lowest or highest points, occur in pairs that have the same ordinate. Be sure to plot enough such points to enable us to draw a smooth curve and to show whether the curve opens upward or downward.

A curve such as those in Figures 10-21, 10-22, and 10-23 is called a **parabola**.

The graph of every quadratic equation of the form

$$y = ax^2 + bx + c, \quad a \neq 0, \quad x \in \mathcal{R}$$

is a parabola.

The path of a projectile moving in a vacuum, the cable supporting a suspension bridge, and certain cross sections of the reflector on a searchlight are examples of parabolas.

EXERCISES

Graph each equation in the coordinate plane.

1. $y = 2x^2$
2. $y = 4x^2$
3. $y = \frac{1}{2}x^2$
4. $y = \frac{1}{3}x^2$
5. $y = -2x^2$
6. $y = -3x^2$

7. $y = x^2 - 1$
8. $y = x^2 - 4$
9. $y = x^2 + 1$
10. $y = x^2 + 4$
11. $y = 2x^2 + 1$
12. $y = 1 - 2x^2$

Graph each function. Assume that the domain is $\mathcal{R}$.

13. $f: x \rightarrow y = x^2 - 2x$
14. $g: x \rightarrow y = x^2 + 2x$
15. $k: x \rightarrow y = 4x - x^2$
16. $r: x \rightarrow y = 4x + x^2$
17. $g(x) = x^2 - 2x - 3$
18. $t(x) = x^2 + 2x - 3$
19. $p(x) = x^2 + 2x + 1$
20. $q(x) = 3 - 2x - x^2$

PROBLEMS

1. The radii of two circles are 5 and 2, respectively. Find the ratio of their areas.
2. What is the ratio of the areas of two squares if one is 4 inches on a side and the other is 6 inches on a side?
3. If the distance needed to stop an automobile varies as the square of its speed, how much distance is needed to bring the car to a stop from 50 miles per hour, if it requires 22 feet to stop at 10 miles per hour?
4. The distance which a freely falling body falls varies directly as the square of the time it falls. If it falls 144 feet in 3 seconds, how far will it fall in 8 seconds?
5. Wind pressure on a flat surface varies as the square of the wind velocity. If 0.35 pounds per square foot results when the wind blows at 8 miles per hour, how much pressure is exerted by a gust of 40 miles per hour?
6. A diamond's price varies as the square of its weight. If one weighing $\frac{3}{8}$ carat is worth $360, find the cost of a similar diamond of $\frac{7}{8}$ carat.
7. A basketball has a radius 4 times larger than that of a tennis ball. If surface area varies as the square of the radius, find the surface area of the tennis ball if the surface area of the basketball is 144π inches.
8. The lift on an airplane wing is directly proportional to the square of the speed of the air moving over it. If the lift is 732 pounds per square foot when the air speed is 320 miles per hour, find the lift when the speed is increased by 80 miles per hour.
9. Find the ratio of the areas of two circular table tops if the circumference of the larger is 132 inches and that of the other is 88 inches.
10. Two spheres have a surface area of 1256 square inches and 113.04 square inches, respectively. If the radius of the larger sphere is 10 inches, find the radius of the smaller sphere. (See Problem 7 above.)
11. A pilot flying at an altitude of 14,400 feet released an object which fell to an altitude of 12,800 feet after 10 seconds. How many feet will the object have fallen 30 seconds after it is released? (See Problem 4 above.)

12. The pressure from a 20-mile-per-hour wind registered 0.028 grams per square centimeter on a very sensitive gauge. (See Problem 5 above.) Find the velocity of the wind which registered a 0.035-gram increase on the gauge.

10–11 Inverse Variation

Three rectangles whose lengths and widths are (15, 2), (10, 3), and (6, 5) have the same area:

$$15 \cdot 2 = 30, \quad 10 \cdot 3 = 30, \quad 6 \cdot 5 = 30,$$

or

$$lw = 30.$$

Such a formula describes a function that is called an *inverse variation*.

An **inverse variation** is a function in which the product of the coordinates of its ordered pairs is a nonzero constant. For any ordered pair (x, y) of the function,

$$xy = k \quad \text{or} \quad y = \frac{k}{x}, \quad \text{where } k \text{ is a nonzero constant.}$$

Because $y = k\left(\dfrac{1}{x}\right)$, we say that y varies *directly as the inverse of x*, or that y *varies inversely as x*, or y *is inversely proportional to x*.

We would not expect the graph of an inverse variation to be a straight line, because its equation,

$$xy = k,$$

is not linear; one term, xy, is of the second degree. The graph of $xy = 1$ is shown in Figure 10–24.

xy = 1

x	y
$\frac{1}{4}$	4
$\frac{1}{2}$	2
1	1
2	$\frac{1}{2}$
4	$\frac{1}{4}$

x	y
-4	$-\frac{1}{4}$
-2	$-\frac{1}{2}$
-1	-1
$-\frac{1}{2}$	-2
$-\frac{1}{4}$	-4

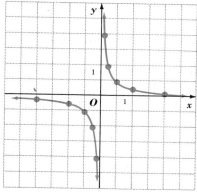

FIGURE 10–24

As x increases, y decreases so that the product is always 1. Neither x nor y can have the value 0. Notice that this graph consists of two separate branches neither of which intersects an axis.

For every nonzero value of k, the graph of "$xy = k$" has this shape and is called a **hyperbola**. The curve is in the first and third quadrants if k is positive and in the second and fourth quadrants if k is negative. If k were 0, what would be the limitation on the range and domain?

When negative answers are meaningless, as they often are in practical problems, the range and domain are limited to positive numbers. The graph of such an inverse variation has only one branch — the one in the first quadrant.

If (x_1, y_1) and (x_2, y_2) are ordered pairs of an inverse variation,

$$x_1 y_1 = k \quad \text{and} \quad x_2 y_2 = k.$$

Therefore,

$$x_1 y_1 = x_2 y_2.$$

Since neither y_1 nor x_2 is 0, we can divide both members by $x_2 y_1$, obtaining

$$\frac{x_1 y_1}{x_2 y_1} = \frac{x_2 y_2}{x_2 y_1} \quad \text{or} \quad \frac{x_1}{x_2} = \frac{y_2}{y_1}.$$

One instance of inverse variation is the law of the lever, a bar pivoted at a point called the fulcrum (Figure 10–25). If weights w_1 and w_2 are placed at distances d_1 and d_2 from the fulcrum, and the lever is in balance, then

$$d_1 w_1 = d_2 w_2$$

or

$$\frac{d_1}{d_2} = \frac{w_2}{w_1}.$$

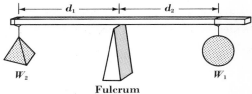

FIGURE 10–25

Example 1. If an eight-pound weight is 72 centimeters from the fulcrum of a lever, how far from the fulcrum is a nine-pound weight which balances it?

Solution: Let $w_1 = 8$, $d_1 = 72$, $w_2 = 9$.

$$d_1 w_1 = d_2 w_2$$
$$72(8) = d_2 \cdot 9$$
$$64 = d_2$$

$$\frac{8}{9} \stackrel{?}{=} \frac{64}{72}$$
$$\frac{8}{9} = \frac{8}{9}$$

∴ distance of nine-pound weight from fulcrum is 64 centimeters.

Several physical quantities vary *inversely as the square* of another quantity. We say that y varies inversely as x^2, or y is inversely proportional to x^2, if

$$y = \frac{k}{x^2} \quad \text{or} \quad x^2y = k, \quad k \neq 0.$$

The above equations are examples of inverse quadratic variation. If $x_1^2 y_1 = k$ and $x_2^2 y_2 = k$, then

$$\frac{x_1^2 y_1}{x_1^2 y_2} = \frac{x_2^2 y_2}{x_1^2 y_2} \quad \text{or} \quad \frac{y_1}{y_2} = \frac{x_2^2}{x_1^2}.$$

Example 2. The brightness of the illumination of an object varies inversely as the square of the distance from the source of light to the object. If the illumination of a book 9 feet from a lamp is 4 foot-candles, find the illumination of the book 3 feet closer to the lamp.

Solution: If I is the amount of illumination and d is the distance from the light source to the object,

$$\frac{I_1}{I_2} = \frac{d_2^2}{d_1^2}.$$

Let $I_1 = 4$, $d_1 = 9$, $d_2 = 6$.

$$I_1 d_1^2 = I_2 d_2^2$$
$$4(9)^2 = I_2(6)^2$$
$$9 = I_2$$

Check:
$$\frac{9}{4} \stackrel{?}{=} \frac{(9)^2}{(6)^2}$$
$$\frac{9}{4} = \frac{81}{36}$$

∴ in the second position the illumination is 9 foot-candles.

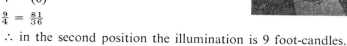

EXERCISES

Graph each of the following equations if the domain and range elements are limited to the set of positive numbers.

1. $xy = 4$
2. $2xy = 1$
3. $x = \dfrac{3}{y}$
4. $\dfrac{x}{4} = \dfrac{3}{y}$

In these inverse variations, find the value of the indicated variable.

5. $x_1 = 18$, $x_2 = 4$, $y_1 = 52$, $y_2 = $ __?__
6. $x_1 = \frac{4}{5}$, $y_1 = 15$, $y_2 = 24$, $x_2 = $ __?__
7. $p_2 = 3.90$, $q_2 = 0.35$, $q_1 = 0.65$, $p_1 = $ __?__
8. $q_2 = \frac{3}{14}$, $p_2 = \frac{8}{15}$, $p_1 = \frac{12}{5}$, $q_1 = $ __?__
9. If $w = rt$ and r is tripled while w remains constant, how does t change?
10. If $Fl = k$ and l is doubled while k remains constant, how does F change?
11. If $E = mc^2$, and c is tripled while E remains constant, how does m change?
12. If $Id^2 = k$, and d is halved while k remains constant, how does I change?
13. If x varies inversely as $t - 1$, and $x = 8$ when $t = 10$, find t when $x = 18$.
14. If y varies inversely as $3w + 2$, and $y = 24$ when $w = 1$, find w when $y = 15$.
15. How far from a lamp does a book receive 9 times as much illumination as a book 6 feet from the lamp? (See Example 2, page 273.)
16. If T varies inversely as s^2, and $T = 256$ when $s = 50$, find T when $s = 80$.
17. If P varies inversely as Q^2, what value of Q causes P to become one-sixteenth as much as it is when $Q = 10$?
18. If V varies inversely as the cube of s, and $V = 297$ when $s = 2$, what is V when $s = 3$?
★19. Plot the graph of $y = \dfrac{12}{x^2}$ for $-4 \leq x < 0$ and $0 < x \leq 4$.

PROBLEMS

1. At 60 miles per hour, how long does a journey take if it takes 8 hours at 45 miles per hour?
2. If 6 men do a job in 12 days, how long do 18 men take, working at the same rate?
3. How far from the seesaw support must John sit to balance Mary, who sits 6 feet from it, if he weighs 90 pounds and she weighs 60 pounds?
4. Tom, sitting 6 feet from the seesaw support, balances a friend who weighs 120 pounds and sits 7 feet from the support. How heavy is Tom?
5. At what rate does $9000 yield the same annual income as $15,000 at 3%?
6. What sum at 4% yields the same yearly income as $800 at 5%?

7. The altitude of a triangle is 24 inches, and the base is 4 inches. Find the altitude of a triangle of equal area whose base is 6 inches.
8. A rectangle has a base of 42 inches and a height of 12 inches. Find the base of another rectangle of equal area whose height is 14 inches.
9. The volume of a gas is 75 cubic feet under 7 pounds pressure. What is its volume at the same temperature when the pressure is 15 pounds?
10. If the current through a circuit is 24 amperes when the resistance is 15 ohms, what is the current when the resistance is reduced to 6 ohms?
11. A 12-inch pulley runs at 240 revolutions per minute (r.p.m.). How fast does the 8-inch pulley it drives revolve if the number of r.p.m. varies inversely as the diameter?
12. A gear with 28 teeth makes 45 r.p.m. and meshes with a gear having 20 teeth. What is the speed of the second gear if the number of r.p.m. varies inversely with the number of teeth?
13. Paul weighs 162 pounds and Philip weighs 135. How far from Paul, on a seesaw 12.1 feet long, is the support balancing them?
14. A meter stick is placed on a fulcrum at the 50-centimeter mark. If a 20-gram weight is suspended from the 15-centimeter mark and a 55-gram weight is suspended from the 90 centimeter mark, how much weight must be suspended at the 25 centimeter mark to bring the meter stick into balance?
15. The weight of a body at or above the earth's surface varies inversely as the square of the body's distance from the earth's center. What does a 445.5-pound projectile weigh 500 miles out from the earth's surface? (Use 4000 miles as the earth's radius.)
16. A three-eighths-inch wire has 48 ohms resistance. How much has the same length of half-inch wire if resistance varies inversely as the square of the diameter?

For the wave motion of sound, the following formula holds:

$$fl = v,$$

where f is the frequency (number of waves per second), l is the wavelength (in feet), and v is the speed of sound (about 1100 feet per second in air). Use this information in the following problems.

17. The frequency of a note an octave above a given note is twice that of the given note. How does the wavelength of the higher note compare with that of the lower note?
18. If the wavelength of a note is $\frac{3}{2}$ that of a given note, how do the frequencies compare?

19. An open organ pipe produces a sound wave that has a length that is twice the length of the pipe. Find the length of an open pipe that will produce the note A with the frequency 440.

20. A stopped organ pipe produces a sound wave that has a length that is four times the length of the pipe. What is the frequency of the sound produced by a stopped organ pipe $4\frac{1}{4}$ feet long?

★**21.** Read a number X and an array S of 100 different numbers. Print the number X and then test X against each number in the array S until a match, if any, is found. If X is equal to some number in the array, then the message "In the table" should be printed; otherwise, the output message should be "Not in the table." Draw a flow chart.

★**22.** Do Exercise 21 for each of the *ten* numbers in an *array X*. (*Hint:* Use a nested loop with J as the subscript for array X and I, the subscript for array S.)

10–12 Joint Variation and Combined Variation

The area A of a triangle depends upon its altitude a and its base b. If a and b are measured in the same units,

$$A = \frac{1}{2}ab \quad \text{or} \quad \frac{A}{ab} = \frac{1}{2}.$$

The area is directly proportional to the altitude and to the base, and we say that the area of a triangle *varies jointly* as its base and altitude.

Joint variation occurs when a variable z varies directly as the product of variables x and y. We say z *varies jointly as x and y*, and for a nonzero constant k, write

$$\frac{z}{xy} = k \quad \text{or} \quad z = kxy.$$

Therefore,

$$\frac{z_1}{x_1 y_1} = \frac{z_2}{x_2 y_2} \quad \text{or} \quad \frac{z_1}{z_2} = \frac{x_1 y_1}{x_2 y_2}.$$

Example 1. The volume of a right circular cylinder varies jointly as its height and the square of its radius. If a right circular cylinder of height 10 inches and radius 4 inches has a volume of 160π cubic inches, find the volume of one with a height of 8 inches and a radius of 3 inches.

FUNCTIONS, RELATIONS, AND GRAPHS 277

Solution: Let $h_1 = 10$, $r_1 = 4$, $h_2 = 8$, $r_2 = 3$, $V_1 = 160\pi$.

$$\frac{V_1}{h_1 r_1^2} = \frac{V_2}{h_2 r_2^2}$$

$$\frac{160\pi}{10(4)^2} = \frac{V_2}{8(3)^2}$$

$$72\pi = V_2$$

Check:
$$\frac{160\pi}{72\pi} \stackrel{?}{=} \frac{10(4)^2}{8(3)^2}$$

$$\frac{20}{9} = \frac{20}{9}$$

∴ the volume of the second cylinder is 72π cubic inches.

Another variation involving three variables is called *combined variation*. **Combined variation** is indicated when a variable z varies directly as one variable x and inversely as another y. For a nonzero constant k,

$$z = \frac{kx}{y} \quad \text{or} \quad zy = kx \quad \text{or} \quad \frac{zy}{x} = k.$$

Therefore,

$$\frac{z_1 y_1}{x_1} = \frac{z_2 y_2}{x_2} \quad \text{or} \quad \frac{z_1}{z_2} = \frac{x_1 y_2}{x_2 y_1}.$$

Example 2. If everyone available works at the same speed at a given task, then the number of persons needed to finish a given amount of work varies directly as the amount of work to be done and inversely as the amount of time desired. If 2 typists can type 210 pages of manuscript in 3 days, how many typists will be needed to type 700 pages in 2 days?

Solution: Let $t_1 = 2$, $p_1 = 210$, $d_1 = 3$, $p_2 = 700$, $d_2 = 2$.

$$\frac{t_1 d_1}{p_1} = \frac{t_2 d_2}{p_2}$$

$$\frac{(2)(3)}{210} = \frac{t_2(2)}{700}$$

$$2t_2 = \frac{4200}{210}$$

$$t_2 = 10$$

∴ 10 typists will be needed to type 700 pages in 2 days.

EXERCISES

Translate into formulas.

1. The rate of speed varies directly as the distance traveled and inversely as the time traveled.
2. The volume of a rectangular container varies jointly as the length, the width, and the depth.
3. The amount of time necessary to complete a job varies directly as the amount of work and inversely as the rate at which it is done.
4. The area of a trapezoid varies jointly as its altitude and the sum of its bases.
5. The temperature of a gas varies jointly as the volume and the pressure.
6. The area of a triangle varies jointly as its base and altitude.
7. The volume of a right circular cylinder varies jointly as its height and the square of the radius of its base.
8. The volume of a pyramid varies jointly as the altitude and the area of the base.
9. The force between two electrical charges varies jointly as the charges on the bodies and inversely as the square of the distance between them.
10. Centrifugal force varies directly as the square of the velocity of a moving body and inversely as the radius of its circular path.

PROBLEMS

1. In the formula $H = \dfrac{I^2 Rt}{4}$, R remains constant. If I is tripled, and t is made 4 times as large, how is H changed?
2. In the formula $F = \dfrac{mv^2}{r}$, m remains constant, v is quadrupled, and r is made $\frac{1}{3}$ as large. How does F change?
3. W varies jointly as x and y and inversely as the square of z. If $W = 189$, $x = 28$, $y = 16$, and $z = 8$, find (a) the constant k of variation, (b) the equation, and (c) W when $x = 24$, $y = 4$, and $z = 6$.
4. R varies directly as the cube of s and inversely as t and the square of u. If $R = 1$, $s = 4$, $t = 8$, and $u = 2$, find (a) the constant k of variation, (b) the equation of relation, and (c) t when $R = 12$, $s = 6$, and $u = 3$.
5. If 14 boys pick 294 crates of apples in 7 hours, how many boys pick 513 boxes in 3 hours?

6. A rod's weight varies jointly as its length and the area of its cross section. If a rod $5\frac{1}{3}$ feet long with a $\frac{3}{4}$-inch-square cross section is 5.60 pounds, what weight has a similar rod $3\frac{3}{4}$ feet long whose cross section is a half-inch square?

7. When a mass moves at 18 feet per second in a circle whose radius is 3 feet, the centrifugal force is 108 pounds. Find the force when that mass moves at 24 feet per second in a circle whose radius is 8 feet. (See Exercise 10, above.)

8. The cost of operating an appliance varies jointly as the number of watts drawn, the hours of operation, and the cost per kilowatt-hour. A thousand-watt waffle iron operates for 45 minutes for 3¢ at 4¢ per kilowatt-hour. What is the cost of cooking 30 waffles 4 minutes each, if the iron uses 875 watts?

9. The safe load on a horizontal beam supported at its ends varies directly as the square of the beam's depth and inversely as its length between supports. A beam 9.6 meters long and 4 centimeters deep bears 2170 grams. What load can one 3.5 meters long and 5 centimeters deep bear?

10. The heat developed in an electric wire varies jointly as the wire's resistance, the time the current flows, and the square of the current. In 8 minutes a current of 7 amperes develops 140 heat units in a wire of 0.1 ohm resistance. What resistance has a similar wire which develops 42,000 heat units with a current of 14 amperes in 2 minutes?

11. The wind pressure on a plane varies jointly as the surface area and the square of the wind's velocity. With a velocity of 12 miles per hour, the pressure on a 4-foot by $1\frac{1}{2}$-foot rectangle is 20 pounds. What is the velocity when the pressure on a surface 2 feet square is 30 pounds?

12. The heat lost through a windowpane varies jointly as the difference of the inside and outside temperatures and the window area, and inversely as the thickness of the pane. If 198 heat calories are lost through a pane 40 by 28 centimeters, $\frac{4}{5}$ centimeter thick, in one hour when the temperature difference is 44°C, how many are lost in one hour through a pane $\frac{1}{2}$ centimeter thick having $\frac{1}{4}$ the area, when the temperature difference is 40°C?

chapter summary

1. Bar and broken-line graphs and pictographs are employed for the visual presentation of statistics to display comparisons and trends in data.//
2. To set up a **rectangular coordinate system in a plane**, choose a vertical and a horizontal line, and scale them as number lines intersecting at zero. Each point in the coordinate plane corresponds to exactly one ordered pair of real numbers, and each ordered pair of real numbers corresponds to exactly one point in the plane.
3. The ordered pairing of the members of two sets is a **relation** that can be shown by table, graph, roster, or rule. The domain of definition (domain) and the range of values (range) of the relation must be specified in each case. A **function** is a relation which assigns only one element of the range to each element of the domain.
4. A plane coordinate system enables us to picture the **solution set of an open sentence in two variables** as the set of points whose coordinates satisfy the open sentence. To graph a linear equation in two variables, each having the set of real numbers as its replacement set, draw the straight line determined by plotting any two roots of the equation.
5. To measure the **slope** of a nonvertical straight line, choose two different points on the line, and compute the ratio of the difference between the ordinates of the points to the corresponding difference between the abscissas of the points. It is a property of a straight line that this ratio is the same for every pair of distinct points on the line. Vertical lines have no slope.
6. A line with slope m and y-intercept b is the graph of the equation $y = mx + b$. This **slope-intercept form** of a linear equation can be used to find an equation for a line
   a. with given slope and passing through a given point;
   b. passing through two different points.
7. **Direct variation** and **inverse variation** are special types of functions. If k is a nonzero constant, equations like

$$y = kx \quad \text{and} \quad y = \frac{k}{x}$$

are associated with, respectively, a direct and an inverse variation. In either case, the constant of proportionality, k, is found by substituting in the equation a pair of values for the variable. Direct and inverse quadratic variation are given by the equations

$$y = kx^2 \quad \text{and} \quad y = \frac{k}{x^2}.$$

CHAPTER REVIEW

10-1
1. Make a bar graph for the sales in millions of shares of stock: Monday 14, Tuesday 12, Wednesday 0, Thursday 9, Friday 11.
2. Make a broken-line graph from the table:

Time of Day	6:00 A.M.	10:00 A.M.	2:00 P.M.	4:00 P.M.
Temperature in Degrees Fahrenheit	45	55	75	70

10-2
3. Find all values of a and b for which $(a + 5, b - 2)$ is equal to $(2a - 3, 4 - b)$.
4. Give the coordinates of each point labeled in the adjoining figure.
5. Graph the following points in the coordinate plane.
   a. $(4, 0)$
   b. $(3, 6)$
   c. $(-3, 4)$
   d. $(-1, -1)$
   e. $(4, -2)$

a.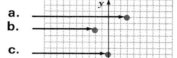
b.
c.
d.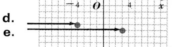
e.

10-3 Plot points to represent the ordered pairs listed in these tables. State the domain and the range of each relation. Is the relation a function?

6.
1	4
2	3
3	2
4	1

7.
0	5
1	4
2	4
3	2

8.
0	2
1	3
2	1
1	0

10-4
9. If the ordered pairs of numbers $(3, b)$ and $(a, -1)$ are equal, then $a = $ __?__ and $b = $ __?__.
10. The solution set of $2x - y = 8$ (does/does not) contain $(3, -2)$.

10-5
11. Graph the equation that is linear:

$$y = x^2 + 2, \quad 3x - xy = 2, \quad 2x - y = 1$$

12. Determine which of the points (6, 1), (3, 2), and (−3, 6) belong to the graph of $2x + 3y = 12$.

10–6 13. Determine the slope of the line containing (−2, −3) and (1, −1).

14. Draw the line through (2, −1) having slope −1.

15. Find the value of a for which (1, −3) is on the graph of $2x + ay = 8$.

10–7 16. Find the slope and y-intercept of the graph of $4x − 2y = 7$.

17. Write an equation for the line through (0, 3) with slope −2.

10–8 18. Write an equation for the line through (2, −2) and having slope $\frac{2}{3}$.

19. Write an equation for the line containing (3, −2) and (−2, 8).

10–9 20. Find the missing value in this direct variation:

$$x_1 = 35, \quad y_1 = 28, \quad x_2 = 5, \quad y_2 = \underline{\ ?\ }$$

21. In a certain college the women outnumber the men 5 to 4. If there are 2120 men in the school, how many women are there?

10–10 22. Graph $y = 3x^2 − 3$.

10–11 23. Find the missing value in this inverse variation:

$$x_1 = 14, \quad y_2 = 3, \quad x_2 = 21, \quad y_1 = \underline{\ ?\ }$$

24. If 3 farmers can harvest their crop in 24 days, how many days would be required to complete the harvest if 5 additional workers were hired?

10–12 25. If $I = prt$, and p is halved, r is made 3 times as large, and t is made $\frac{1}{4}$ as large, how is I changed?

26. If $P = k\dfrac{V^2}{d}$, and $P = 54$, $V = 12$, and $d = 4$, find:

   a. k, the constant of proportionality;
   b. d when $P = 18$ and $V = 6$.

11 systems of open sentences in two variables

11–1 The Graphic Method

In section 10–5 we saw that the graph of a linear equation in two variables is a straight line. When the graphs of two such equations in the same variables are drawn in the same coordinate system, the resulting lines may have in common:

A. No point — the lines are *parallel*. (*Parallel lines* are lines that lie in the same plane, but have no point in common.)

B. All their points — the lines *coincide*.

C. Just one point — the lines *intersect* and their common point is called their *point of intersection*.

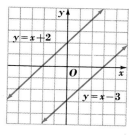
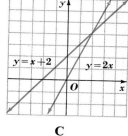

A
$y = x + 2$
$y = x - 3$

B
$y = x + 2$
$3y = 3x + 6$

C
$y = x + 2$
$y = 2x$

FIGURE 11–1

Because two equations impose two conditions on the variables at the same time, they are called a **system of simultaneous equations**. To solve such a system, we seek the ordered pairs of numbers that satisfy *both* equations of the system. Each such ordered pair is called a **solution of the system**; the set of all solutions is called the **solution set of the system**.

The graphs in Figure 11-1 show that:

$$\text{A.} \quad y = x + 2$$
$$y = x - 3$$

has no solution; the graphs do not intersect.

$$\text{B.} \quad y = x + 2$$
$$3y = 3x + 6$$

has an unlimited number of solutions; the graphs coincide.

$$\text{C.} \quad y = x + 2$$
$$y = 2x$$

has just one solution, (2, 4); the graphs intersect at one point.

To understand why the solution set of System **A** is ∅ (the empty set), notice that if "$y = x + 2$" and "$y = x - 3$" were *both* true statements for some ordered pair (x, y), then by substitution

$$x + 2 = x - 3,$$

and

$$2 = -3, \text{ a false statement.}$$

Simultaneous equations having no common root are called **inconsistent**. Because the equations of Systems **B** and **C** do have common roots, they are called **consistent equations**.

We can tell that the equations in System **B** have the same line as their graph by noticing that the slope-intercept form of each equation is

$$y = x + 2.$$

Thus, the solution set of System **B** is equal to the solution set of "$y = x + 2$."

We can check that the coordinates (2, 4) of the point of intersection of the graphs of the equation in System **C** satisfy both equations:

$$y = x + 2 \qquad\qquad y = 2x$$
$$4 \stackrel{?}{=} 2 + 2 \qquad\qquad 4 \stackrel{?}{=} 2 \cdot 2$$
$$4 = 4 \qquad\qquad\qquad 4 = 4$$

No other ordered pair satisfies both equations because no point other than the point (2, 4) lies on both graphs. Thus, the solution set of System **C** is {(2, 4)}.

A pair of linear equations can be solved by graphing the equations in the same coordinate system and determining the coordinates of all points common to the graphs.

Examine Figure 11–2. The ordered pair (2, 4) is the common root of any pair of linear equations whose graphs pass through that point. In particular, the pair of heavy lines in color in the figure pass through it. One of these lines is the horizontal line whose equation is "$y = 4$," and the other is the vertical line with equation "$x = 2$."

Because the system of equations

$$x = 2$$
$$y = 4$$

has the same solution set as the system

$$y = x + 2$$
$$y = 2x$$

these systems are said to be **equivalent systems**. The system $\begin{cases} x = 2 \\ y = 4 \end{cases}$ is also equivalent to the systems

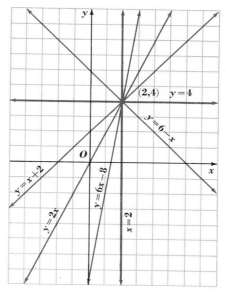

FIGURE 11–2

| $y = 2x$ | $y = 2x$ | $y = 6 - x$ |
| $y = 6x - 8$ | $y = 6 - x$ | $y = 6x - 8$ | and so on.

EXERCISES

Graph the system and from your graph determine the solution set of the system.

1. $y = x$
   $y = 2 - x$
2. $y = 2x$
   $y = 6 - x$
3. $x + y = 3$
   $x - y = 1$
4. $x + y = 4$
   $x - y = 2$
5. $x + y = 0$
   $x + 2y = 2$
6. $2x + y = 3$
   $x + 2y = 0$
7. $y = 3 - x$
   $x + y = 5$
8. $y = x - 2$
   $2x - 2y = 4$
9. $y = 2 - x$
   $x = 2 + y$

10. $x = y - 1$
    $y = x - 1$

11. $y = 2x + 1$
    $x + y = -2$

12. $y = \frac{2}{3}x + 1$
    $y = -\frac{2}{3}x + 5$

Solve graphically and estimate the coordinates of the point of intersection to the nearest tenth.

13. $x + 2y = 7$
    $x - y = 5$

14. $2x + y = 8$
    $x - 2y = 1$

15. $2x + 3y = -2$
    $2x - y = -9$

16. $3x + 2y = 4$
    $x - 2y = 6$

17. Where on the graph of $2x + 3y = 5$ is the abscissa equal to the ordinate?

18. Where on the graph of $x - 2y = 9$ is the ordinate twice the abscissa?

★19. Find the area of the triangle whose vertices are determined by the graphs of $y = x + 4$, $y = 10 - x$, $y = 2$.

★20. Find the area of the triangle whose vertices are determined by the graphs of $2x + y = -2$, $x - 2y = -6$, $x = 2$.

11–2 The Addition or Subtraction Method

When the solution of a system of equations is an ordered pair of integers, it can usually be found by the graphic method of the preceding section as in the diagram below. However, there are algebraic methods that enable us to compute the solution for any pair of simultaneous equations.

Example 1. Solve: $x + 3y = 19$
                   $x - y = -1$

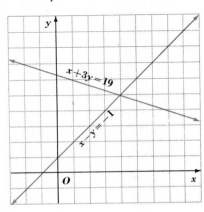

SYSTEMS OF OPEN SENTENCES IN TWO VARIABLES **287**

Solution: We can use the addition property of equality (page 76) to obtain the equivalent system made up of the equations of the horizontal and vertical lines through the point of intersection.

1. To obtain an equation that does not involve x, subtract (add the opposite of) each member of the second given equation from the corresponding member of the first equation:

$$\begin{aligned} x + 3y &= 19 \\ x - y &= -1 \\ \hline 4y &= 20 \\ y &= 5 \end{aligned}$$

(horizontal line through the intersection, shown below)

Now, substitute "5" for y in either of the original equations:

$x + 3y = 19$
$x + 3(5) = 19$
$x + 15 = 19$
$x = 4$ (vertical line through the intersection)

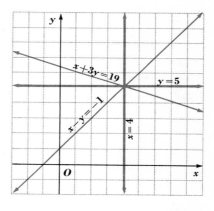

Check: Substitute "4" for x and "5" for y in both original equations.
∴ the solution set is $\{(4, 5)\}$.

Example 2. Solve: $3a + 4b = 7$
$a - 4b = 5$.

Solution:
1. Add the corresponding members of the given equations.

$$\begin{aligned} 3a + 4b &= 7 \\ a - 4b &= 5 \\ \hline 4a &= 12 \\ a &= 3 \end{aligned}$$

2. Substitute "3" for a in one of the given equations.

$3a + 4b = 7$
$3 \cdot 3 + 4b = 7$
$4b = -2$
$b = -\frac{1}{2}$

Check:
$3a + 4b = 7$ $a - 4b = 5$
$3(3) + 4(-\frac{1}{2}) \stackrel{?}{=} 7$ $(3) - 4(-\frac{1}{2}) \stackrel{?}{=} 5$
$9 - 2 \stackrel{?}{=} 7$ $3 + 2 \stackrel{?}{=} 5$
$7 = 7$ $5 = 5$

∴ the solution set is $\{(3, -\frac{1}{2})\}$.

EXERCISES

Solve by addition or subtraction.

1. $x + y = 7$
   $x - y = 9$
2. $r - s = -5$
   $r + s = 25$
3. $2t + u = 7$
   $2t - u = 13$
4. $m + 3n = -26$
   $m - 3n = 22$
5. $x - 3y = 2$
   $x + 4y = 16$
6. $7w - z = 18$
   $-5w + z = -14$
7. $2A - 3B = 20$
   $A - 3B = 13$
8. $5x + 3y = 8$
   $-7x - 3y = -10$
9. $6r + 5s = -8$
   $2r - 5s = -16$
10. $3m - 7n = 16$
    $5m - 7n = 36$
11. $7w + 11z = -25$
    $w - 11z = 9$
12. $5x + 3y = 10$
    $2x - 3y = 4$
13. $3C + 2D = -6$
    $C + 2D = -6$
14. $3k - 7g = 15$
    $3k + 2g = 15$
15. $46 = 4x + 3z$
    $14 = 2x - 3z$
16. $32 = 5s - 3t$
    $-8 = 5s + 7t$
17. $8g + 7h = 26$
    $8g - 10h = 60$
18. $12x - 9y = 126$
    $12x + 13y = 170$

Clear the equations of fractions before adding or subtracting.

19. $\frac{1}{8}(x + y) = 1$
    $x - y = 4$
20. $3y - 2x = 4$
    $\frac{1}{6}(3y - 4x) = 1$
21. $\frac{x}{3} - y = 0$
    $\frac{x}{5} + \frac{2y}{5} = 1$
22. $\frac{2u}{5} - \frac{v}{2} = 1$
    $\frac{2u}{5} + v = -2$
23. $0.3(x + y) = 22.2$
    $0.4(x - y) = 6.4$
24. $0.5(x - y) = 2$
    $0.75(x + y) = 9$
25. $\frac{5a}{6} + \frac{b}{4} = 7$
    $\frac{2a}{3} - \frac{b}{8} = 3$
26. $\frac{2r}{5} + \frac{6s}{20} = \frac{4}{5}$
    $\frac{2r}{3} + \frac{5s}{12} = \frac{7}{6}$

11–3 Problems with Two Variables

Problems concerning two numbers can be solved by using one or two variables. A solution using two variables to form two open sentences is often the more direct.

Example. Cathy spent 40 minutes longer on her algebra assignment than she did on her biology lab report. If she spent 1 hour and 50 minutes on both subjects, how long did she spend on each?

Solution:

1. Choose two variables to represent the desired numbers.

    Let x = number of minutes spent on algebra and y = number of minutes spent on biology.

2. Form two open sentences using the facts of the problem.

    $x + y = 110$ (minutes spent on both subjects)

    $x - y = 40$ (difference in time spent on each subject)

3. Solve the equations.

    $$x + y = 110$$
    $$x - y = 40$$
    $$2x = 150$$
    $$x = 75$$

    $75 + y = 110$
    $y = 35$

4. Check your results in the words of the problem.

    75 minutes is 40 minutes more than 35 minutes.
    Total time is $75 + 35$, or 110, minutes, or 1 hour and 50 minutes.

∴ Cathy spent 1 hour and 15 minutes on algebra and 35 minutes on biology.

PROBLEMS

Use two variables and a system of two equations to solve each problem.

1. Half the perimeter of a rectangular lot, which is 50 feet longer than it is wide, is 350 feet. What are the dimensions of the lot?
2. Mr. Smith's house has 3 rooms more than Mr. Tripp's. The two houses together contain 17 rooms. How many rooms are in each house?
3. The difference between three times one number and a smaller one is 23. The sum of the smaller and twice the larger is 27. Name the numbers.
4. If Jane were 20 years older, she would be twice as old as Jim, and their combined ages would be 54. How old are Jane and Jim?
5. Fred and Bill bowl together and have a combined score of 425. Twice the difference between Fred's score and Bill's is 50. Find their scores.

6. In a game of cards, Judy scored 2 points more than twice the number of points Ann scored. If a total of 26 points were scored, how many points did each score?

7. Large boxes of a certain kind of tea sell for 82¢ and small boxes for 56¢. Ted buys several boxes for a total of $3.58. If he spent $1.34 more for the large boxes than for the small boxes, how many boxes of each size did he buy?

8. A shoe store is selling all shoes for $12 and all slippers for $8. If Linda spent $44 more on shoes than on slippers, and spent a total of $76, how many pairs of each did she buy?

9. Three hamburgers and four hot dogs cost a total of $2.15. If the three hamburgers cost $.55 more than the four hot dogs, what is the cost of each hamburger and each hot dog?

10. A store received $823 one month from the sale of 5 tape recorders and 7 radios. If the receipts from the tape recorders exceeded the receipts from the radios by $137, what is the cost of a tape recorder?

11. Anne has 15 coins, all nickels and dimes, with a total value of $1.20. Find the number of each kind of coin.

12. A shipment of 18 cars, some weighing 3,000 pounds each and the others 5,000 pounds each, has a total weight of 30 tons. Find the number of each kind of car.

11–4 Multiplication in the Addition or Subtraction Method

Sometimes adding or subtracting the members will not eliminate either variable because the coefficients of a pair of corresponding terms do not have the same absolute value. We then can use the multiplication property of equality (page 83) to make the coefficient of a variable in one equation have the same absolute value as the corresponding coefficient in the other.

Example 1. Solve: $2r + 3s = 12$
$r - 4s = -5$

Solution:

1. Multiply both members of the second equation by 2.

2. Subtract the second equation from the first and solve for s.

3. Find the value of r by substituting "2" for s in one of the given equations.

$2r + 3s = 12$
$2r - 8s = -10$
$11s = 22$
$s = 2$

$r - 4(2) = -5$
$r - 8 = -5$
$r = 3$

Check:

4. Substitute in original equations. (This is left to you.)

Example 2. Solve: $5x - 2y = -5$
$3x - 7y = -32$

Solution:
$$7(5x - 2y) = 7(-5) \longrightarrow 35x - 14y = -35$$
$$-2(3x - 7y) = -2(-32) \longrightarrow \underline{-6x + 14y = 64}$$
$$29x = 29$$
$$x = 1$$

$$5(1) - 2y = -5$$
$$-2y = -10$$
$$y = 5$$

Check: $5(1) - 2(5) \stackrel{?}{=} -5 \quad 3(1) - 7(5) \stackrel{?}{=} -32$
$-5 = -5 \quad\quad\quad -32 = -32$

∴ the solution set is $\{(1, 5)\}$.

EXERCISES

Solve each system of equations algebraically.

1. $d + f = 3$
   $3d - 5f = 17$

2. $4a - 3b = -1$
   $a - b = -1$

3. $\dfrac{m}{6} + \dfrac{n}{4} = \dfrac{3}{2}$
   $\dfrac{2m}{3} - \dfrac{n}{2} = 0$

4. $\dfrac{u}{5} + \dfrac{2v}{5} = 2$
   $\dfrac{u}{2} - v = 1$

5. $2a + 3b = -1$
   $3a + 5b = -2$

6. $2w - 3z = -1$
   $3w + 4z = 24$

7. $2m + 3n = 0$
   $5m - 2n = -19$

8. $5a - 2b = 0$
   $2a - 3b = -11$

9. $\dfrac{5x}{4} + y = \dfrac{11}{2}$
   $x + \dfrac{y}{3} = 3$

10. $2r - \dfrac{5s}{2} = 13$
    $\dfrac{r}{3} + \dfrac{s}{5} = \dfrac{14}{15}$

11. $3c - 2d = 13$
    $7c + 3d = 15$

12. $7p + 5q = 2$
    $8p - 9q = 17$

13. $\dfrac{1}{x} + \dfrac{1}{y} = 7$
    $\dfrac{2}{x} + \dfrac{3}{y} = 16$

14. $\dfrac{1}{a} + \dfrac{2}{b} = 11$
    $\dfrac{1}{a} - \dfrac{2}{b} = -1$

Hint: To solve Exercises 13–16, make the equations linear by letting $a = \dfrac{1}{x}$ and $b = \dfrac{1}{y}$. Solve for a and b; then $x = \dfrac{1}{a}$ and $y = \dfrac{1}{b}$.

15. $\dfrac{5}{c} - \dfrac{6}{d} = -3$
    $\dfrac{10}{c} + \dfrac{9}{d} = 1$

16. $\dfrac{1}{r} - \dfrac{1}{s} = 4$
    $\dfrac{2}{r} - \dfrac{1}{2s} = 11$

17. $\dfrac{a-1}{3} + \dfrac{b-1}{3} = 2$
    $\dfrac{a-1}{2} + \dfrac{b-1}{6} = \dfrac{5}{3}$

18. $\dfrac{2s+1}{7} + \dfrac{3t+2}{5} = \dfrac{1}{5}$
    $\dfrac{3s-2}{2} + \dfrac{t+4}{4} = 4$

PROBLEMS

Solve, using a system of two equations in two variables.

1. Mr. Waite's rent and utilities bills each month total $160. If his utilities bill increases $8 per month, it will be $\frac{1}{5}$ of his rent. What is his rent?

2. A certain recipe requires a total of 5 cups of sugar and flour together. If the recipe had called for $\frac{1}{4}$ cup more sugar, there would be twice as much flour as sugar. How much sugar does the recipe call for?

3. The ages of Jim and his father total 57 years. If Jim's age were doubled, Jim's father would be 12 years older than Jim. How old is Jim?

4. The federal tax on a $10,000 salary was $500 less than 10 times the state tax. If the taxes total $2800, find the state's share.

5. A square house, 24 feet on a side, is located on a lot which is 50 feet longer than it is wide. The perimeter of the lot is 20 feet more than 5 times the perimeter of the house. Find the length of the lot.

6. A rectangular rug which is 3 feet longer than it is wide is in a room of perimeter 54 feet. The perimeter of the rug is 12 feet less than the perimeter of the room. Find the dimensions of the rug.

7. Judy buys 2 bags of potato chips and 3 boxes of pretzels for $2.35. She then buys another bag of potato chips and 2 more boxes of pretzels for $1.37. Find the cost of potato chips and pretzels.

8. A group of 4 couples are going out for refreshments. If 4 people have ice cream sodas and 4 people have sundaes, the bill will total $3.60. However, if only 2 people have sundaes and 6 people have sodas, the bill will be $3.20. What is the cost of each soda and each sundae?

9. Mr. Britten's income from two stocks each year totals $280. Stock A pays dividends at the rate of 5% and stock B at the rate of 6%. If he has invested a total of $5000, how much capital is invested in each stock?

10. Mr. Towne takes loans from two banks. He borrows $300 more from the bank which charges 7% interest than from the bank which charges 8% interest. If his interest payments for one year are $126, how much does he borrow at each rate?

11-5 The Substitution Method

We can solve either equation of a system of two equations for one variable in terms of the other, and use the substitution principle to obtain a third equation involving only one variable. This method is sometimes easier to use than the addition and subtraction method.

Example. Solve: $x - 2y = 4$
$3x + 4y = 2$

Solution:
1. Solve for x in the first equation.
2. Substitute this expression for x in the other equation.
3. Solve for y.

Solving for x and checking are left for you. The solution set is $\{(2, -1)\}$.

$$x - 2y = 4$$
$$x = 4 + 2y$$
$$3x + 4y = 2$$
$$3(4 + 2y) + 4y = 2$$
$$12 + 6y + 4y = 2$$
$$10y = -10$$
$$y = -1$$

EXERCISES

Solve each system of equations by substitution.

1. $3x + y = 5$
$y = 2x$

2. $z + 5r = 2$
$z = -3r$

3. $-a + b = 1$
$a + b = -5$

4. $m - 3n = -4$
$2m + 6n = 5$

5. $x + y = 2$
$3x + 2y = 5$

6. $5a - 3b = -1$
$a + b = 3$

7. $x + 3y = 2$
$2x + 3y = 7$

8. $3t - 2s = 5$
$t + 2s = 15$

9. $3c - 2d = 3$
$2c - d = 2$

10. $3r - 5s = 8$
$r + 2s = 1$

11. $3a - 4b = 5$
$a + 7b = 10$

12. $x - 2y = 0$
$4x - 3y = 15$

13. $\frac{1}{5}(x + y) = 2$
$\frac{1}{2}(x - y) = 1$

14. $\frac{s}{3} - t = 2$
$s - t = 20$

15. $x + y = 7$
$x - \frac{y}{2} = 4$

16. $x - y = 16$
$\frac{1}{2}(x + y) = 37$

17. $3a - 2b = 11$
$a - \frac{b}{2} = 4$

18. $\frac{1}{3}(x + y) = 2$
$\frac{1}{5}(x - y) = 2$

PROBLEMS

Solve, using a system of two equations in two variables and the substitution method.

1. Half the sum of two numbers is $\frac{15}{2}$. Half their difference is $-\frac{3}{2}$. Find the two numbers.

2. Steve and Harry live 250 miles apart. They decide to meet at a town which is 30 miles further from Steve's house than it is from Harry's. How far will each travel?

3. Janet and Lynn live 8 blocks apart in opposite directions from their office. If Lynn lives 1 block less than twice as far from the office as Janet does, how far does each girl live from the office?

4. A mother took her 3 children on an airplane flight. Her ticket cost $2 more than twice each of the children's. If the total cost of the tickets was $53.25, find the mother's fare and the fare for each of the children.

5. On a jury there are 3 fewer men than twice the number of women. If there were 2 more women on the jury, there would be an equal number of men and women. How many men are on the jury?

6. Mrs. Birch invested $1200, some at 6%, the rest at $4\frac{1}{2}$% per year. The return from the $4\frac{1}{2}$% investment exceeded that from the 6% investment by $12. How much was invested at each rate?

7. The average of two numbers is $\frac{11}{24}$. One-third of their difference is $\frac{1}{12}$. Find the two numbers.

8. One sum invested at 6% and another at 5% yield a total of $37.50. If the investments were interchanged, their income would increase by $2.00. Find the sums.

9. Mr. Bowen receives a total of $135 a year interest from a regular savings account, paying 5% per year, and from a special notice account, paying $5\frac{1}{2}$% per year. If he had interchanged the amounts deposited in each type of account, his income would have decreased by $7.50. Find the amount he deposited in each type of account.

10. A certain laundromat has washing machines available for $.25 per load and $.35 per load. Using both types of machines, several loads can be laundered for $2.65. If only the $.35-per-load type of machines is used, the laundry could be done in 2 fewer loads and cost $.20 less. Find the number of loads of laundry done if only the $.35-per-load machines are used.

★11. Draw a flow chart of a program to determine and print all positive two-digit numbers, if any, which are equal to twice the product of the two digits.

11-6 Digit Problems

The digits 0, 1, 2, 3, 4, 5, 6, 7, 8, 9 not only represent different values themselves, but represent different values in different positions within a numeral:

$$76 = 7 \cdot 10 + 6 \cdot 1; \quad 67 = 6 \cdot 10 + 7 \cdot 1$$

All two-digit decimal numerals have the same form in general, $10t + u$, where t denotes the value of the tens digit and u denotes the value of the ones (units) digit. Thus,

$$t \in \{1, 2, 3, 4, 5, 6, 7, 8, 9\} \quad \text{and} \quad u \in \{0, 1, 2, 3, 4, 5, 6, 7, 8, 9\}.$$

To represent a number with the same digits in reverse order, write $10u + t$. In either case, the sum of the values of the digits themselves is represented by $t + u$.

Frequently, to avoid clumsy wording, we refer to "the sum of the digits" rather than "the sum of the values of the digits." Similarly, we may say "reverse the digits of the number" instead of "reverse the digits of the numeral."

Example. The sum of the digits in a two-digit numeral is 8. When the digits are interchanged, the number designated is 18 more than the original number. Find the original number.

Solution: 1. Let $t =$ the value of the tens digit in the original numeral and $u =$ the value of the units digit in the original numeral.

Then $10t + u =$ the original number
and $10u + t =$ the new number.

2. The sum of the values of the digits is 8: $t + u = 8$
The new number is 18 more than the original number.

$$10u + t = 18 + 10t + u$$
$$(10u + t) - (10t + u) = 18$$
$$9u - 9t = 18$$
$$u - t = 2$$

3. $u + t = 8$
$\underline{u - t = 2}$
$2u = 10$
$u = 5$

$5 - t = 2$
$t = 3$

Check: 4. Is the sum of the digits 8? $3 + 5 = 8$
Is 53 eighteen more than 35? $53 - 35 = 18$

∴ the original number was 35.

PROBLEMS

1. The sum of the digits of a two-digit numeral is 11. If 45 is added to the number, the result is the number with its digits reversed. Find the original number.

2. The sum of the digits of a two-digit numeral is 9. If the order of the digits is reversed, the result names a number exceeding the original by 9. Find the original number.

3. The sum of the digits of a two-digit numeral is 12. The value of the number is 13 times the tens digit. Find the number.

4. The sum of the digits of a two-digit numeral is 8. The number with its digits interchanged is 11 times the original units digit. Find the original number.

5. The units digit of a two-digit numeral is three times the tens digit. The sum of the digits is 12. Find the number.

6. The tens digit of a two-digit numeral exceeds three times the units digit by 2. The sum of the digits is 10. Find the number.

7. The sum of the digits of a two-digit numeral is 7. The number with the digits interchanged is 5 times the tens digit of the original number. Find the original number.

8. The sum of the digits of a two-digit numeral is 10. The number with the digits in reverse order is 16 times the original tens digit. Find the original number.

9. A clerk mistakenly reversed the two digits in the price of a paperback book, overcharging the customer 18¢. If the sum of the digits is 16, determine the correct price of the paperback book.

10. Find a number less than 100 whose tens digit exceeds twice its units digit by 1 and whose digits in reverse order give a number which is 7 more than 3 times their sum.

11. Find a three-digit number whose hundreds digit is twice its tens digit and three times its units digit, and whose digits total 11.

12. A three-digit number is 198 more than itself reversed. The hundreds digit is three more than the tens digit, and the sum of the digits is 16. Find the original number.

13. If a two-digit number is divided by its tens digit, the quotient is 12 and the remainder is 1. If the number with its digits interchanged is divided by its original units digit, the quotient is 10 and the remainder is 4. Find the original number.

14. Show that the difference between a three-digit number and the number with the order of the digits reversed is always divisible by 99.

Problems 15–18 refer to two-place decimal fractions between 0 and 1.

15. The sum of the digits of a two-place fraction is 9. When its digits are reversed, the new fraction exceeds the original by 0.27. Find the original fraction.

16. When the digits of a two-place fraction are reversed, the new fraction is $\frac{4}{7}$ the original fraction. If the sum of the digits is 12, find the original fraction.

17. The sum of the digits of a two-place fraction is 7. The fraction with its digits reversed is 0.02 more than twice the original fraction. Find the original fraction.

18. The tenths digit of a two-place fraction exceeds twice the hundredths digit by 1. If the digits are reversed, the original is 0.02 more than twice the new fraction. Find the original fraction.

11–7 Motion Problems

We can solve some motion problems conveniently by using a system of two equations in two variables.

Example. Jack took 45 minutes to go 6 miles down Taylor's River in a motorboat, and it took him $1\frac{1}{2}$ hours to return. What was the speed (in miles per hour) of Jack's boat in still water, and what was the speed of the current in the river?

Solution: 1. Let $m=$ the speed, in m.p.h., of Jack's boat
and $r=$ the speed, in m.p.h., of the river's current.

2.

	r	$\cdot$	t	$=$	d	
Downstream	$m + r$		$\frac{3}{4}$		6	$\frac{3}{4}(m + r) = 6$
Upstream	$m - r$		$\frac{3}{2}$		6	$\frac{3}{2}(m - r) = 6$

3. $\frac{3}{4}(m + r) = 6 \quad \rightarrow \quad m + r = 8$
   $\frac{3}{2}(m - r) = 6 \quad \rightarrow \quad m - r = 4$
   $\phantom{\frac{3}{2}(m - r) = 6 \quad \rightarrow \quad } 2m = 12$
   $\phantom{\frac{3}{2}(m - r) = 6 \quad \rightarrow \quad \ \ } m = 6$

Solving for r and checking the roots are left to you.

The speed of the boat is 6 m.p.h.
The speed of the current is 2 m.p.h.

PROBLEMS

1. A motorboat goes 15 miles downstream in 45 minutes. The return trip against the current takes $1\frac{1}{2}$ hours. Find the boat's speed in still water.

2. A cyclist rode 5 miles in 15 minutes with the wind, and returned in 20 minutes against the same wind. Find his speed without a wind.

3. A speed boat traveled 60 miles with the current in $1\frac{1}{2}$ hours. The return trip took 2 hours against the current. What was the speed of the current?

4. It required $1\frac{1}{3}$ hours for a 400-mile plane trip and $1\frac{3}{5}$ hours for the return against the same wind. What would have been the speed of the plane without wind?

5. Nick took 36 minutes to row 3 miles. When he returned, he took 2 hours. What was the river's current?

6. A fish swims 12 miles downstream in $2\frac{2}{5}$ hours and returns in 6 hours. At the same rate, how fast does he go in still water?

7. A man rows 3 miles upstream and 3 miles back in $2\frac{1}{4}$ hours. He rows 1 mile against the current in the time he rows 2 miles with it. At what rate does he row in still water? What is his average rate of travel?

8. A swimmer takes $1\frac{1}{4}$ hours on a 1-mile round trip. On the return trip against the current he did $\frac{1}{8}$ mile in the time that he did $\frac{1}{2}$ mile on the trip downstream. Find his average rate. What was the rate of the current?

9. A round trip flight of 1037.5 miles takes $7\frac{1}{2}$ hours. The part of the flight with the wind takes $1\frac{1}{2}$ hours less than the other part of the trip. Find the speed of the plane in still air and the speed of the wind.

10. A motorboat races a distance up the river in the time it races $1\frac{1}{4}$ that distance downstream. If the speed of the boat is s and that of the current is c, find the relationship between s and c.

★11. Draw a flow chart to compute and print the solution of the system $\begin{cases} ax + by = c \\ dx + ey = f \end{cases}$. Assume that $ae - bd \neq 0$.

  (*Hint:* The system is equivalent to $x = \dfrac{ce - bf}{ae - bd}$ and $y = \dfrac{af - cd}{ae - bd}$.)

11–8 Age Problems

The solution of age problems can be simplified by using two variables and by organizing the facts in chart form.

Example. Five years ago, Cliff was $\frac{2}{3}$ as old as Judith. Ten years from now, he will be $\frac{5}{6}$ as old as Judith. How old is each now?

Solution:

Time	Cliff	Judith
5 years ago	$x - 5$	$y - 5$
This year	x	y
10 years hence	$x + 10$	$y + 10$

$x - 5 = \frac{2}{3}(y - 5)$

$x + 10 = \frac{5}{6}(y + 10)$

Solve these equations and check. Cliff is 15 years old; Judith is 20 years old.

PROBLEMS

1. A man is 13 times as old as his son. In ten years he will be 3 times as old as his son will be then. How old is the son now?

2. Karen's mother is twice as old as Karen is. Ten years ago she was 3 times as old as Karen was. Find Karen's present age.

3. Two years ago, Rick's age was 4 years less than twice Christopher's. Four years from now, Rick will be 14 years more than half Christopher's age. How old is Rick?

4. Five years ago, Barbara was $\frac{4}{5}$ as old as Fred. Ten years from now, she will be $\frac{7}{8}$ as old as Fred. How old is each now?

5. Ann is $\frac{3}{4}$ as old as Judy. Four years ago Ann was $\frac{7}{10}$ as old as Judy. How old is each?

6. A man said, "My son is four times as old as my daughter. My wife is 3 times as old as the combined ages of both, and I am as old as my wife and daughter together. My father, who is as old as all of us together, is 72." How old is the speaker's son?

7. Nick said, "If I were $\frac{1}{2}$ as old as I am, and Bill were $\frac{3}{4}$ as old as he is, we would be 3 years older together than I am alone. But if I were $\frac{2}{3}$ as old as I am, and Bill were $\frac{1}{4}$ as old as he is, together we would be 2 years younger than I am alone." How old is Nick?

8. Laura is 3 times as old as Maria was when Laura was as old as Maria is now. In 2 years, Laura will be twice as old as Maria was 2 years ago. Find their present ages.

9. Mary is 3 times as old as Linda was when Mary was as old as Linda is now. Find the relationship between Mary's present age (*m*) and Linda's (*l*).

10. Meg is twice as old as Peg will be when Meg is 5 times as old as Peg is now. Find the relationship between Meg's present age (m) and Peg's (p).

11. Mary is twice as old as Jane was at the time when Mary was as old as Jane is now. The sum of the present ages of Mary and Jane is 28 years. How old is each person now?

12. A man is three times as old as his son was at the time when the father was twice as old as his son will be 2 years from now. Find the present age of each person if the sum of their ages is 55 years.

11–9 Problems about Fractions

Among the problems that can be solved by using two variables are those about fractions, like this one:

Example. The value of a fraction is $\frac{3}{4}$. When 7 is added to its numerator, the resulting fraction is equal to the reciprocal of the original fraction. Find the original fraction.

Solution:

1. Let $\frac{n}{d}$ = the original fraction.

2. $\frac{n}{d} = \frac{3}{4}$ and $\frac{n+7}{d} = \frac{4}{3}$.

3. Multiply by d:

$$d\left(\frac{n}{d}\right) = d\left(\frac{3}{4}\right)$$

$$n = \frac{3}{4}d$$

Multiply by LCD, $3d$:

$$3d\left(\frac{n+7}{d}\right) = 3d\left(\frac{4}{3}\right)$$

$$3(n+7) = d(4)$$

$$3n + 21 = 4d$$

$$3\left(\frac{3}{4}d\right) + 21 = 4d$$

$$9d + 84 = 16d$$

$$-7d = -84$$

$$d = 12$$

$$n = \frac{3}{4}(12)$$

$$n = 9$$

4. Check is left to you.

∴ the original fraction is $\frac{9}{12}$.

PROBLEMS

Using two variables, find the original fraction.

1. The denominator is 4 more than the numerator. If each is increased by 1, the value of the resulting fraction is $\frac{1}{2}$.

2. The denominator is 2 more than the numerator. If 1 is subtracted from each, the value of the resulting fraction is $\frac{1}{2}$.

3. The denominator exceeds the numerator by 1. If 1 is subtracted from the numerator, and the denominator is unchanged, the resulting fraction has value $\frac{3}{4}$.

4. The denominator exceeds the numerator by 3. If 1 is added to the denominator, a fraction is obtained whose value is $\frac{2}{3}$.

5. A fraction has value $\frac{2}{3}$. When 15 is added to its numerator, the resulting fraction equals the reciprocal of the value of the original fraction.

6. A fraction's value is $\frac{4}{5}$. When its numerator is increased by 9, the new fraction equals the reciprocal of the value of the original fraction.

7. The two digits in the numerator of a fraction whose value is $\frac{2}{9}$ are reversed in its denominator. The reciprocal of the fraction is the value of the fraction obtained when 27 is added to the original numerator and 71 is subtracted from the original denominator.

8. The numerator equals the sum of the two digits in the denominator. The value of the fraction is $\frac{1}{4}$. When both numerator and denominator are increased by 3, the resulting fraction has the value $\frac{1}{3}$.

9. The two digits in the numerator of a fraction are reversed in its denominator. If 1 is subtracted from both the numerator and the denominator, the value of the resulting fraction is $\frac{1}{2}$. The fraction whose numerator is the difference and whose denominator is the sum of the units and tens digits equals $\frac{2}{5}$.

10. The numerator is a three-digit number whose hundreds digit is 4. The denominator is the numerator with the digits reversed. If 31 is subtracted from the numerator, the value of the fraction is $\frac{1}{2}$. If 167 is subtracted from the denominator, the resulting fraction equals $\frac{2}{3}$.

List all possible members of each solution set.

11. The numerator is a two-digit number and the denominator is that number with the digits reversed. The value of the fraction is $\frac{7}{4}$.

12. The numerator of a fraction whose value is $\frac{334}{667}$ is a four-digit number whose hundreds and tens digits are 0. The denominator contains the same digits in reverse order.

11–10 Graph of an Inequality in Two Variables

In Figure 11–3 the graph (line *l*) of "$y = 2$" separates the coordinate plane into two regions. If we start at any point on line *l*, say (1, 2) and move vertically upward, the *y*-coordinate *increases* as we move. If we move vertically downward from this point, the value of *y decreases*. In either case, the value of *x* remains 1.

The equation "$y = 2$" is the boundary of two **half-planes**. In Figure 11–3 the half-plane above the line consists of all points for which

$$y > 2,$$

and is the **graph** of that inequality. The half-plane below the line is the graph of

$$y < 2.$$

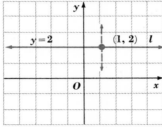

FIGURE 11–3

The half-plane above the line together with the **boundary line** forms the graph of

$$y \geq 2,$$

while the boundary line together with the half-plane below it is the graph of

$$y \leq 2.$$

A half-plane without its boundary is called an **open half-plane**, while the union of an open half-plane and its boundary is called a **closed half-plane**.

These graphs are indicated by shading. If the graph is a closed half-plane, the boundary line is drawn as a solid line. If the graph is an open half-plane, a dashed line is used for the boundary line. See the graphs in Figure 11–4.

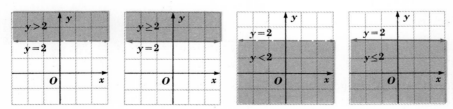

FIGURE 11–4

The inequalities "$y > 2$" and "$x > 3$" are graphed on the same coordinate plane in Figure 11-5. The points in the upper right-hand section of the plane represent those points whose coordinates satisfy *both* inequalities. That is, these are the points for which $y > 2$ *and* $x > 3$, and so this region is the graph of the solution set of the **system of inequalities**

$$y > 2$$
$$x > 3.$$

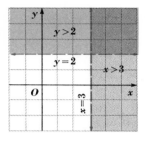

FIGURE 11-5

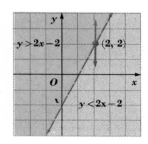

FIGURE 11-6

Figure 11-6 shows the graph of "$y = 2x - 2$" separating the plane into two half-planes. For each x, all the points on the line satisfy the equation "$y = 2x - 2$." All the points in the half-plane above the line satisfy the inequality "$y > 2x - 2$," and all the points in the half-plane below the line satisfy the inequality "$y < 2x - 2$."

$y = 2x - 2$

x	3	2	1	0	-1	-2	-3
y	4	2	0	-2	-4	-6	-8

Example. Graph the inequality "$2x - y < 1$."

Solution: 1. Transform this into an equivalent inequality (page 108) having y as one member:

$$2x - y < 1$$
$$-y < 1 - 2x$$
$$y > 2x - 1$$

2. Graph "$y = 2x - 1$" and show it as a *dashed line*.

3. Shade the half-plane *above* the line.

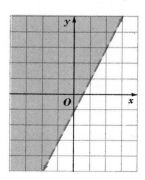

EXERCISES

Graph each inequality in the coordinate plane.

1. $y \leq 4$
2. $y \geq -2$
3. $x > 0$
4. $y < 1$
5. $y \leq x$
6. $y \geq 3x$
7. $y \geq -x$
8. $y \leq -\dfrac{x}{3}$
9. $y \leq x + 2$
10. $y > 3 - x$
11. $x + 2y < 4$
12. $2x + y \geq 1$

In the coordinate plane, indicate the region consisting of all points whose coordinates satisfy both inequalities.

13. $y \geq 0$ and $x \leq 0$
14. $y < 3$ and $x > -2$
15. $x + y > 0$ and $x > 1$
16. $x + y \leq 3$ and $y \geq 1$
17. $y \leq 2x + 1$ and $x > 0$
18. $y > 2 - x$ and $y \geq 0$
19. $|x| \leq 1$
20. $|y| \geq 2$
21. $y \geq |x|$
22. $y + |x| \leq 0$
23. $-1 < x \leq 2$
24. $-3 \leq y < 1$

11-11 Graphs of Systems of Linear Inequalities

Not only can graphs be used to solve systems of equations; they can also be employed in determining the solution sets of systems of simultaneous inequalities such as:

$$x + y > 4$$
$$2x - y > 2$$

First, draw the graphs of "$x + y = 4$" and "$2x - y = 2$" (Figure 11-7) on the same axes. Use broken lines here, since the graphs of the given inequalities are both open half-planes.

Since "$x + y > 4$" is equivalent to "$y > 4 - x$," the graph of the solution set of "$x + y > 4$" consists of all points in the open half-plane *above* the graph of "$x + y = 4$." Since "$2x - y > 2$" is equivalent to "$y < 2x - 2$," the graph of the solution set of "$2x - y > 2$" consists of all points in the open half-plane *below* the graph of "$2x - y = 2$." The graph of the *intersection* (common points) of the solution sets is the double-shaded region which contains all, and only those, points with coordinates satisfying both inequalities. Some points in the graph of the common solution set are (4, 2), (5, 0), and (7, −1).

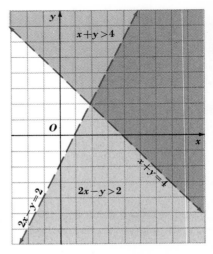

FIGURE 11-7

EXERCISES

Graph each pair of inequalities, indicating their solution set with cross-hatching.

1. $y \geq 0$
   $x \geq 0$

2. $y \leq 1$
   $x \geq -1$

3. $y < x$
   $x > 1$

4. $y > 2x$
   $x < 3$

5. $y \geq x$
   $y \leq x + 2$

6. $y \leq x - 1$
   $y \geq 1 - x$

7. $y > 2x - 4$
   $y < 3x + 6$

8. $y < 3x + 3$
   $y > 3 - 3x$

9. $y \leq 4$
   $y \geq 0$

10. $x + 2 > 0$
    $x - 3 < 0$

11. $2x + y \geq 1$
    $x - 2y \geq 2$

12. $2x + y \geq -1$
    $2x + y \leq 3$

13. $2x + y \leq 3$
    $4x + 2y \leq 6$

14. $x + 2y \geq 3$
    $4y \geq 6 - 2x$

Graph each inequality in the given system. Show the solution set of the system as points in a three-way shaded region.

Example 1. $y \leq x$
$y \geq -x$
$x \leq 5$

Solution:

a. The graph of the solution set of "$y \leq x$" consists of points on the graph of "$y = x$" and in the diagonally shaded region below it.

b. The graph of the solution set of "$y \geq -x$" consists of points on the graph of "$y = -x$" and in the diagonally shaded region above it.

c. The graph of the solution set of "$x \leq 5$" consists of points on the graph of "$x = 5$" and in the horizontally shaded region to the left of it.

d. The intersection of these three sets is the three-way shaded region, triangle AOB, including points on its sides as well as in its interior.

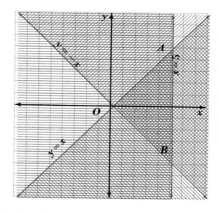

15. $x + y \leq 2$
    $x \geq 0$
    $y \geq 0$

16. $y \leq 2 - x$
    $y \leq x + 2$
    $y \geq -1$

17. $y \geq x$
    $y \leq -x$
    $x \geq -2$

18. $y \geq -1$
    $y \leq 1$
    $x + y \leq 2$

19. $2y \geq x + 2$
    $2y + x \leq 8$
    $x > 0$

20. $2x - y \leq 2$
    $x \geq 0$
    $y > 1$

In Exercises 21–24, (a) solve each pair graphically, and (b) check by solving algebraically.

Example 2. $y = 2x$
$x + y \geq 3$

Solution:

a. The heavy ray including (1, 2) is the graph of the solution set.

b. Substitute $2x$ for y in $x + y \geq 3$:
$$x + 2x \geq 3$$
Solve for x: $3x \geq 3$
$$x \geq 1$$
Since $y = 2x$, $x = \tfrac{1}{2}y$.

Substitute $\tfrac{1}{2}y$ for x in $x \geq 1$: $\tfrac{1}{2}y \geq 1$, or $y \geq 2$

∴ the given system is equivalent to the system: $\begin{cases} y = 2x \\ x \geq 1 \\ y \geq 2 \end{cases}$

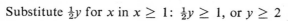

21. $y = x$
    $x + y \leq 2$

22. $y \geq 2 - x$
    $y = x + 1$

23. $y = 2 - x$
    $x < 0$

24. $2x - y = 4$
    $y \geq 1$

chapter summary

1. The **solution set** of a system of open sentences in two variables consists of **ordered pairs** of numbers.

2. If the graphs of the two equations of a system of linear equations in two variables **intersect** in exactly one point, then the solution set is **one ordered pair of numbers** and the equations in the system are said to be **consistent**;

   if the graphs are **two parallel lines**, then the solution set is the **empty set** and the equations in the system are said to be **inconsistent**;

   if the graphs are the same line, then the solution set is an **infinite set** and the equations in the system are said to be **consistent**.

3. When the coefficients of one variable have the same absolute value, use the **addition** or **subtraction property of equality** to eliminate that variable; then solve for the other variable. When the coefficients of both variables have different absolute values in the two equations, use the **multiplication property of equality** before adding or subtracting.

4. A system of simultaneous linear equations in the same variables can be solved by applying the **substitution principle**.

5. To solve a system of linear inequalities, graph each inequality in the same coordinate system; the intersection of their graphs contains all points which satisfy all the inequalities.
6. **Digit** problems, **motion** problems, **age** problems, and problems about **fractions** can usually best be solved by using two variables to form two equations.

CHAPTER REVIEW

11–1 1. Solve "$x - y = 4$ and $x + y = 6$" graphically.

11–2 Solve each system by addition or subtraction.

   2. $x + y = 12$
      $x - y = -4$

   3. $2z + 3t = 7$
      $z - 3t = 8$

   4. $5a + 2b = 9$
      $5a - b = 3$

11–3 5. Two packages, weighed together, total 42 pounds. If one package weighs 4 pounds more than the other, what is the weight of each?

11–4 Solve by using multiplication with addition or subtraction.

   6. $r + 2s = -5$
      $3r - s = -1$

   7. $2p + 5q = 9$
      $3p - 2q = 4$

11–5 Solve by substitution.

   8. $m - 2n = 0$
      $2m + 3n = 14$

   9. $3a + b = 4$
      $2a - 3b = 10$

   10. $x + 3y = 1$
       $3x + y = 11$

11–6 11. The sum of the digits of a two-digit numeral is 6. If the order of the digits is reversed, the result names a number 6 less than twice the original number. Find the original number.

11–7 12. Two trains left Endicott and Central City, which are 300 miles apart, at the same time, moving toward each other on parallel tracks. When they met, the train from Endicott had traveled 40 miles farther than the train from Central City. How far from Central City did they meet?

11–8 13. Maria is 16 years older than Carl. In 4 years, she will be twice as old as Carl. How old is each person now?

11–9 14. If 6 is added to the numerator of a fraction, the resulting fraction equals 2. The sum of the numerator and denominator of the original fraction is 9. Find the original fraction.

11–10 15. Graph: $y - 3x \leq 2$ 16. Graph: $x + 2y > 5$

11–11 17. Solve the system $\begin{cases} y \leq x + 2 \\ x - y < 1 \end{cases}$ graphically.

12 rational and irrational numbers

12–1 The Nature of Rational Numbers

The set $\mathcal{R}$ of real numbers, together with the operations of addition and multiplication, is a *number system* — the **real number system**. A **number system** is defined by specifying a set of numbers and rules for adding and multiplying them.

For example, the set $N = \{1, 2, 3, 4, \ldots\}$, together with the operations of addition and multiplication, forms the **system of natural numbers** (or **positive integers**). This system is closed under addition and multiplication, but not under subtraction or division.

The system of positive numbers named by fractions is closed under addition, multiplication, and division. When we extend our idea of number to include zero, the negative integers, and the negative numbers named by fractions, we have a system of numbers that is closed under subtraction, too. This is the **system of rational numbers**, and we learned how to operate with these numbers in Chapter 8. Because the system of rational numbers is closed under addition, subtraction, multiplication, and division (except by zero), these operations are called the **rational operations**.

Recall that a rational number can be expressed in an unlimited number of ways:

$$0 = \frac{0}{8} = \frac{0}{-3} \cdots \qquad 7 = \frac{7}{1} = \frac{-7}{-1} \cdots \qquad -\frac{4}{5} = \frac{-4}{5} = \frac{8}{-10} \cdots$$

$$\frac{3}{14} = \frac{-3}{-14} = \frac{9}{42} \cdots \qquad 2.3 = \frac{23}{10} = \frac{69}{30} \cdots \qquad 26\% = \frac{26}{100} = \frac{13}{50} \cdots$$

Furthermore, we always can tell which of two rational numbers is the greater by writing them with the same positive denominator and comparing their numerators. For example, to compare $-\frac{3}{4}$ and $-\frac{2}{3}$, we write $-\frac{3}{4} = \frac{-9}{12}$ and $-\frac{2}{3} = \frac{-8}{12}$. Then

$$-\tfrac{3}{4} < -\tfrac{2}{3} \quad \text{because} \quad -9 < -8.$$

Similarly,

$$\tfrac{7}{3} > \tfrac{13}{6} \quad \text{because} \quad \tfrac{7}{3} = \tfrac{14}{6} \text{ and } 14 > 13.$$

This test can be developed in another form. Let a and b be integers and c and d be positive integers. Then if

$$\frac{a}{c} > \frac{b}{d},$$

we have

$$\frac{a}{c}(cd) > \frac{b}{d}(cd)$$

and so

$$ad > bc.$$

On the other hand, if

$$ad > bc,$$

then since $cd > 0$,

$$\frac{ad}{cd} > \frac{bc}{cd}$$

and

$$\frac{a}{c} > \frac{b}{d}.$$

For all integers a and b and all positive integers c and d:

$$\frac{a}{c} > \frac{b}{d} \quad \text{if and only if} \quad ad > bc.$$

Thus,

$$\tfrac{5}{6} > \tfrac{3}{4} \quad \text{because} \quad 5(4) > 3(6);$$
$$-\tfrac{1}{2} > -\tfrac{3}{2} \quad \text{because} \quad -1(2) > -3(2).$$

For each integer, there is a next larger one. For example, -4 follows -5, 0 follows -1, and 4 follows 3. This is not true for the set of rational numbers. There is no "next larger" rational number after $\frac{2}{3}$, for instance. Instead, the set of rational numbers has a property which the set of integers does not have, namely:

The Property of Density

Between every pair of different rational numbers there is another rational number.

Example. Find a rational number between $\frac{5}{4}$ and $\frac{4}{3}$.

Solution: 1. Find the difference of the numbers: $\frac{4}{3} - \frac{5}{4} = \frac{16}{12} - \frac{15}{12} = \frac{1}{12}$

2. Add half this difference to the smaller: $\frac{5}{4} + \frac{1}{2}(\frac{1}{12}) = \frac{31}{24}$

3. Is $\frac{5}{4} < \frac{31}{24} < \frac{4}{3}$?

$5(24) \stackrel{?}{<} 4(31)$ $\qquad$ $3(31) \stackrel{?}{<} 4(24)$

$120 < 124$ $\qquad\qquad$ $93 < 96$

∴ a rational number between (exactly halfway between) $\frac{5}{4}$ and $\frac{4}{3}$ is $\frac{31}{24}$.

Another way of finding the rational number that was given as the answer to the preceding Example is to take the average of the given numbers:

$$\frac{1}{2}\left(\frac{5}{4} + \frac{4}{3}\right) = \frac{1}{2}\left(\frac{15 + 16}{12}\right) = \frac{31}{24}$$

However, the method used in the solution of the Example suggests a way of finding other rational numbers between $\frac{5}{4}$ and $\frac{4}{3}$. For example, we can add $\frac{1}{3}$ the difference to $\frac{5}{4}$:

$$\frac{5}{4} + \frac{1}{3}(\frac{1}{12}) = \frac{45}{36} + \frac{1}{36} = \frac{46}{36} = \frac{23}{18}$$

Check that $\frac{5}{4} < \frac{23}{18} < \frac{4}{3}$.

The number of rational numbers between $\frac{5}{4}$ and $\frac{4}{3}$ is unlimited. The property of density implies that between every pair of rational numbers there is an infinite set of rational numbers.

EXERCISES

Replace the __?__ with =, <, or > to make a true statement.

1. $\frac{7}{8}$ __?__ $\frac{11}{14}$
2. $\frac{3}{5}$ __?__ $\frac{8}{13}$
3. $\frac{3}{14}$ __?__ $\frac{5}{24}$
4. $-\frac{7}{16}$ __?__ $-\frac{11}{25}$
5. $\frac{23}{47}$ __?__ $\frac{41}{83}$
6. $\frac{215}{103}$ __?__ $\frac{311}{148}$
7. $\frac{312}{12}$ __?__ $28\frac{1}{2}$
8. $-18\frac{1}{5}$ __?__ $-\frac{127}{7}$

Arrange the members of each set in increasing order.

9. $\{\frac{2}{3}, -\frac{3}{4}, \frac{5}{9}\}$
10. $\{\frac{4}{7}, \frac{3}{5}, -\frac{3}{5}\}$
11. $\{-2.1, -\frac{4}{3}, -2.0\}$
12. $\{-1.6, -\frac{33}{20}, -\frac{37}{22}\}$
13. $\{\frac{4}{15}, \frac{5}{16}, \frac{7}{24}, \frac{1}{4}\}$
14. $\{\frac{7}{8}, \frac{9}{7}, \frac{19}{18}, \frac{13}{12}\}$

Find the number halfway between the given numbers.

15. $\frac{2}{3}, \frac{3}{5}$
16. $\frac{4}{7}, \frac{5}{9}$
17. $-\frac{2}{25}, -\frac{3}{100}$
18. $-\frac{7}{1000}, -\frac{7}{100}$
19. $4\frac{1}{6}, 5\frac{1}{8}$
20. $2\frac{3}{5}, 3\frac{5}{8}$

21. Find the number one-fourth of the way from $\frac{5}{6}$ to $1\frac{1}{4}$.
22. Find the number one-third of the way from $-\frac{3}{2}$ to $-\frac{3}{4}$.
23. Show that the number halfway between x and y is $\frac{x+y}{2}$.
24. What number is one-fourth of the way from p to q:
    a. if $p < q$?
    b. if $p > q$?

Explain why each statement is true.

25. 5 is the smallest integer greater than 4.
26. There is no smallest rational number greater than 0.

In each case, tell whether the given statement is true with blanks filled as indicated.

27. If a and b are different __(x)__, then there are as many __(y)__ between a and b as we please.
    a. (x) integers; (y) integers
    b. (x) integers; (y) rational numbers
    c. (x) rational numbers; (y) rational numbers
    d. (x) rational numbers; (y) integers

28. If a and b are different __(x)__, then the number halfway between a and b is always a(n) __(y)__.
    a. (x) integers; (y) integer
    b. (x) integers; (y) rational number
    c. (x) rational numbers; (y) rational number
    d. (x) rational numbers; (y) integer

In Exercises 29–32, *a*, *b*, *c*, and *d* are nonzero integers. Explain why each expression represents a rational number.

29. $\dfrac{a}{b} \cdot \dfrac{c}{d}$ **30.** $\dfrac{a}{c} + \dfrac{b}{d}$ **31.** $\dfrac{a}{b} - \dfrac{c}{d}$ **32.** $\dfrac{a}{b} \div \dfrac{c}{d}$

12–2 Decimal Forms for Rational Numbers

To change a common fraction to a decimal, carry out the indicated division.

$$\tfrac{5}{16} = 5 \div 16 \qquad \tfrac{1}{6} = 1 \div 6 \qquad \tfrac{3}{11} = 3 \div 11$$

```
       0.3125              0.1666              0.2727
  16)5.0000            6)1.0000           11)3.0000
     4 8                   6                    2 2
     ———                   ——                   ———
       20                   40                    80
       16                   36                    77
       ——                   ——                   ——
        40                    40                    30
        32                    36                    22
        ——                    ——                    ——
         80                     40                    80
         80                     36                    77
         ——                     ——                    ——
          0                      4                     3
```

A decimal with a finite number of places, like 0.3125 above, is called **terminating**, **ending**, or **finite**. Such a decimal represents a rational number; for example,

$$0.3125 = \tfrac{3125}{10000} \qquad \text{or} \qquad 0.3125 = \tfrac{5}{16}.$$

In the division of 1 by 6, however, we never have a remainder of 0, but the remainder 4 repeats step after step, and 6 repeats in the quotient. A decimal which continues indefinitely is called **nonterminating** or **unending**. A nonterminating decimal like the one for $\tfrac{1}{6}$ is called **repeating** or **periodic**, because the same digit (or block of digits) repeats unendingly. We may write

$$\tfrac{1}{6} = 0.1666\ldots \qquad \text{or} \qquad \tfrac{1}{6} = 0.1\overline{6},$$

where the dots and the bar indicate "continue unendingly."

When 3 is divided by 11, the successive remainders are 8, 3, 8, 3, ... and the quotient is a repeating decimal:

$$\tfrac{3}{11} = 0.272727\ldots \qquad \text{or} \qquad \tfrac{3}{11} = 0.\overline{27}.$$

When we divide an integer by 11, the remainder at each step belongs to {0, 1, 2, 3, 4, 5, 6, 7, 8, 9, 10}. Within no more than ten steps after only zeros are left in the dividend, either the remainder is 0 and the division terminates, or a sequence of other remainders repeats unendingly. This sort of reasoning can be applied to division by any positive integer and thus leads to the following result.

For every integer r and every positive integer s, the decimal numeral of the rational number $\frac{r}{s}$ either terminates or eventually repeats in a block of fewer than s digits.

On the other hand, the following statement is also true.

All terminating decimals and all repeating decimals represent rational numbers which can be written in the form $\frac{r}{s}$ where r is an integer and s is a positive integer.

The preceding conversion of 0.3125 to $\frac{5}{16}$ shows how a terminating decimal can be written as a common fraction. The following examples show how to convert a repeating decimal into a common fraction.

Example 1. Write $0.5\overline{16}$ as a common fraction.

Solution: Let $N =$ the number.

$$100N = 51.61\overline{616}$$
Subtract: $\quad N = 0.51\overline{616}$
$$99N = 51.1000\overline{0}$$
$$N = \frac{51.1}{99} = \frac{511}{990}$$

Example 2. Write $0.\overline{234}$ as a common fraction.

Solution: Let $N =$ the number.

$$1000N = 234.\overline{234}$$
Subtract: $\quad N = 0.\overline{234}$
$$999N = 234.00\overline{0}$$
$$N = \tfrac{234}{999} = \tfrac{26}{111}$$

In general, if the number of digits in the block of repeating digits is p, multiply the given number N by 10^p, producing a number with the same repeating block as the given number. Then subtracting the given number from this product yields a terminating decimal.

It often is convenient to break off a lengthy decimal, leaving an approximation of the number represented. We may write, for example,

$$\tfrac{1}{12} \doteq 0.08333 \quad \text{or} \quad \tfrac{1}{12} \doteq 0.083 \quad \text{or} \quad \tfrac{1}{12} \doteq 0.08.$$

To round a decimal, add 1 to the value of the last digit kept if the first digit dropped is 5 or more; otherwise, leave the digits unchanged.

Thus,

$$\tfrac{2}{3} = 0.666\ldots \quad \text{or} \quad \tfrac{2}{3} \doteq 0.67 \quad \text{or} \quad \tfrac{2}{3} \doteq 0.7;$$
$$\tfrac{5}{9} = 0.555\ldots \quad \text{or} \quad \tfrac{5}{9} \doteq 0.56 \quad \text{or} \quad \tfrac{5}{9} \doteq 0.6;$$
$$\tfrac{161}{110} = 1.4\overline{63} \quad \text{or} \quad \tfrac{161}{110} \doteq 1.5 \quad \text{or} \quad \tfrac{161}{110} \doteq 1.$$

EXERCISES

Write as terminating or repeating decimals.

1. $\tfrac{7}{50}$
2. $\tfrac{6}{25}$
3. $\tfrac{3}{32}$
4. $\tfrac{11}{8}$
5. $-\tfrac{9}{11}$
6. $-\tfrac{4}{3}$
7. $\tfrac{-3}{7}$
8. $\tfrac{-19}{80}$
9. $\tfrac{41}{20}$
10. $\tfrac{48}{70}$
11. $-\tfrac{3}{16}$
12. $-\tfrac{10}{21}$

Write as common fractions.

13. 0.33
14. 0.6
15. $0.11\overline{8}$
16. $-3.\overline{148}$
17. $0.1\overline{18}$
18. $-1.30\overline{4}$
19. $0.21212121\ldots$
20. $0.202202202\ldots$

Find the difference of the given numbers, and name a number between them.

21. 0.18 and $0.\overline{18}$
22. 0.66 and $0.\overline{66}$
23. 0.8 and $0.\overline{8}$
24. 0.14 and $0.\overline{14}$
25. 0.126 and $\tfrac{1}{8}$
26. 0.101 and $\tfrac{1}{9}$

Compare the decimal forms of the members of each set.

27. $\{\tfrac{1}{7}, \tfrac{2}{7}, \tfrac{3}{7}, \ldots, \tfrac{6}{7}\}$
28. $\{\tfrac{1}{9}, \tfrac{2}{9}, \tfrac{3}{9}, \ldots, \tfrac{8}{9}\}$
29. $\{\tfrac{1}{11}, \tfrac{2}{11}, \tfrac{3}{11}, \ldots, \tfrac{10}{11}\}$
30. $\{\tfrac{1}{15}, \tfrac{2}{15}, \tfrac{3}{15}, \ldots, \tfrac{14}{15}\}$

12–3 Roots of Numbers

Recall that the power of a number is the product of factors each equal to that number:

$$5^2 = 5(5), \quad 5^3 = 5(5)(5), \quad \text{and} \quad 5^n = 5(5) \cdots (5), \; n \text{ factors}$$

This operation is called *raising to a power*.

Just as addition and multiplication have inverse operations, so has raising to a power. Its inverse operation is called *extracting a root*. For any positive integer n, a number x is an nth **root** of the number a if it satisfies

$$x^n = a.$$

For example, since $3^4 = 81$, 3 is a *fourth root* of 81.

To indicate the nth root of a, we use the expression $\sqrt[n]{a}$, which is called a **radical** (in Latin *radix* means "root"). The symbol $\sqrt{}$ indicates that a root is to be extracted; n is the **root index**, signifying the root to be taken; the bar, usually incorporated in the radical symbol, covers the **radicand**, the expression for the number whose root is to be extracted. With no root index, $\sqrt{}$ indicates square root:

$$\sqrt{81} = 9, \quad \sqrt[3]{125} = 5, \quad \sqrt[4]{81} = 3$$

When we square a positive or a negative number, we get a positive result. That is, $5^2 = 25$ and $(-5)^2 = 25$. Thus, every positive number has two square roots, one positive and the other negative. Zero, however, has only one square root, zero. We use the expression $\sqrt{25}$ to indicate the positive root 5 (the *principal square root*), $-\sqrt{25}$ to indicate the negative root -5, and $\pm\sqrt{25}$ (read "positive and negative square root of 25") to represent both roots. Thus,

$$\sqrt{\tfrac{4}{9}} = \tfrac{2}{3}, \quad -\sqrt{\tfrac{4}{9}} = -\tfrac{2}{3}, \quad \pm\sqrt{\tfrac{4}{9}} = \pm\tfrac{2}{3}.$$

Since the square of every real number is either positive or zero, negative numbers do not have square roots in the set of real numbers.

One method of finding the square root of a large number is to determine its factors, and then to express it as a product of powers, and take the square roots of the powers.

Example 1. Evaluate $\sqrt{3969}$.

Solution:
$$3969 = 9(441) = 9(9)(49) = 9^2(7^2)$$
$$\sqrt{3969} = \sqrt{9^2(7)^2} = 9(7) = 63$$

Check:
$$63(63) \stackrel{?}{=} 3969$$
$$3969 = 3969 \qquad \therefore \sqrt{3969} = 63.$$

This method of solution just shown is based on the following:

Product Property of Square Roots

For any real numbers *a* and *b*:

if $a \geq 0$ and $b \geq 0$, then $\sqrt{ab} = \sqrt{a} \cdot \sqrt{b}$.

To prove this property, we show that $\sqrt{a} \cdot \sqrt{b}$ is a nonnegative number and that its square is ab. Notice that only principal roots are used.

To show that $\sqrt{a} \cdot \sqrt{b} \geq 0$:	To show that $(\sqrt{a} \cdot \sqrt{b})^2 = ab$:
If $a \geq 0$, and $b \geq 0$,	$(\sqrt{a} \cdot \sqrt{b})^2 = (\sqrt{a} \cdot \sqrt{b})(\sqrt{a} \cdot \sqrt{b})$
then $\sqrt{a} \geq 0$, $\sqrt{b} \geq 0$,	$= (\sqrt{a} \cdot \sqrt{a})(\sqrt{b} \cdot \sqrt{b})$
and $\sqrt{a} \cdot \sqrt{b} \geq 0$.	$= a \cdot b$
	$\therefore (\sqrt{a} \cdot \sqrt{b})^2 = ab$

Similarly, we can prove (Exercise 33, page 318) the following.

Quotient Property of Square Roots

For any real numbers *a* and *b*:

if $a \geq 0$ and $b > 0$, then $\sqrt{\dfrac{a}{b}} = \dfrac{\sqrt{a}}{\sqrt{b}}$.

Example 2. Evaluate $\sqrt{\dfrac{64}{2025}}$.

Solution: $64 = 8^2$; $2025 = (25)(81) = 5^2(9^2)$

$$\sqrt{\frac{64}{2025}} = \sqrt{\frac{8^2}{5^2(9^2)}} = \frac{8}{5 \cdot 9} = \frac{8}{45}$$

Check: $\frac{8}{45} \cdot \frac{8}{45} = \frac{64}{2025}$ $\qquad \therefore \sqrt{\dfrac{64}{2025}} = \dfrac{8}{45}$.

Example 3. What is the principal square root of $49x^2$, if $x \in \mathcal{R}$?

Solution: $\sqrt{49x^2} = \sqrt{7^2 x^2} = 7|x|$.

Check: For every real number x, $7|x| \geq 0$.
Also $(7|x|)^2 = 49|x|^2 = 49x^2$.
$\therefore \sqrt{49x^2} = 7|x|$.

Notice that in Example 3, "$7x$" would *not* be an acceptable answer unless we knew that $x \geq 0$. Remember that the principal square root of a number is *never* a negative number.

EXERCISES

Evaluate each expression.

1. $\sqrt{225}$
2. $\sqrt{324}$
3. $\sqrt{676}$
4. $\sqrt{1024}$
5. $-\sqrt{1296}$
6. $-\sqrt{1600}$
7. $-\sqrt{\frac{256}{49}}$
8. $-\sqrt{\frac{576}{25}}$
9. $\pm\sqrt{\frac{1}{256}}$
10. $\pm\sqrt{\frac{1}{784}}$
11. $\pm\sqrt{\frac{25}{484}}$
12. $\pm\sqrt{\frac{36}{729}}$

Simplify each expression.

13. $-\sqrt{49a^2}$
14. $-\sqrt{64r^2s^4}$
15. $\sqrt{81a^8b^2}$
16. $\sqrt{\frac{n^2}{144}}$
17. $\pm\sqrt{\frac{k^4}{16}}$
18. $\sqrt{6^2 + 8^2}$

Solve. Use {the rational numbers} as the replacement set of the variable.

Example 1. $t^2 = 25$

Solution: $t^2 = 25$
$t = \pm\sqrt{25} = \pm 5$

Check: $(5)^2 \stackrel{?}{=} 25 \qquad (-5)^2 \stackrel{?}{=} 25$
$25 = 25 \qquad 25 = 25$

∴ the solution set is $\{5, -5\}$.

19. $n^2 = 16$
20. $r^2 = 36$
21. $9x^2 - 4 = 0$
22. $25t^2 - 49 = 0$
23. $5a^2 - 80 = 0$
24. $7z^2 - 175 = 0$

Example 2. If $r^2 + s^2 = 12$ and $rs = 2$, find the positive value of $r + s$.

Solution:
$$2rs = 2 \cdot 2$$
Add: $\quad r^2 + s^2 = 12$
$$r^2 + 2rs + s^2 = 12 + 4$$
$$(r + s)^2 = 16$$
∴ $r + s = \pm\sqrt{16} = \pm 4$.

∴ the positive value of $r + s$ is 4.

25. If $r^2 + s^2 = 15$ and $rs = 5$, find the positive value of $r + s$.
26. If $x^2 + y^2 = 20$ and $xy = 8$, find the negative value of $x + y$.
27. If $p - q = 5$ and $pq = 3$, find the value of $p^2 + q^2$.

28. If $(a - b)^2 = 5$ and $ab = 1$, find the positive value of $a + b$.
29. If $r + s = 8$ and $r^2 + s^2 = 40$, find the positive value of $r - s$.
30. If $c - d = 3$ and $c^2 + d^2 = 29$, find the positive value of $c + d$.
★31. Solve the equation $\sqrt{a^2 - a} = 6$.
★32. Solve the equation $n - \sqrt{n^2} = -18$.
★33. Prove the quotient property of square roots (page 316).
★34. Prove that if $a > b > 0$, then $a^2 > b^2$.
  Hint: Show that $a^2 > ab$, and $ab > b^2$.
★35. Use the result of Exercise 34 to explain why a positive number cannot have two different positive square roots.
★36. Find the fallacy in this "proof" that every real number is 1.
  Let r be any real number. Then:

$$r - 1 = -(1 - r)$$
$$\therefore (r - 1)^2 = (1 - r)^2.$$
$$\therefore r - 1 = 1 - r.$$
$$\therefore 2r = 2.$$
$$\therefore r = 1.$$

12–4 Properties of Irrational Numbers

Rational numbers, like 25, 36, and $\frac{49}{16}$, which are squares of rational numbers are called **perfect squares**. However, not every positive rational number is a perfect square.

Do those integers which are not squares of integers have rational square roots? Consider some positive integer n. Assume that its square root is named by a fraction $\frac{a}{b}$ in lowest terms; that is, $\sqrt{n} = \frac{a}{b}$, where a, b, and n are positive integers, and a and b have no common integral factors.

If $\sqrt{n} = \frac{a}{b}$, then $n = \frac{a^2}{b^2}$. Since a^2 has the same prime factors as a, and b^2 has the same prime factors as b, if a and b have no factors in common, neither do a^2 and b^2, and $\frac{a^2}{b^2}$ is in lowest terms. *If a fraction in lowest terms is equal to an integer, the denominator of the fraction must be 1.* Thus, since n is an integer, and $\frac{a^2}{b^2}$ is in lowest terms, $b^2 = 1$, and $b = 1$, which means that $\frac{a}{b} = \frac{a}{1} = a$. Therefore, if the square root of a positive integer is a rational number, $\frac{a}{b}$, the root is in fact an integer, $\frac{a}{1}$. Thus, only integers which are squares of integers can have rational square roots.

Since integers such as 3, 5, and 7 are not the squares of integers, we must seek numbers like $\sqrt{3}$, $\sqrt{5}$, and $\sqrt{7}$ outside the set of rational numbers. Such numbers are found in another major subset of the real numbers called the set of *irrational numbers*. **Irrational numbers** are real numbers which cannot be expressed in the form $\frac{r}{s}$, where r and s are integers. Thus, the set of real numbers is the union of the set of rational numbers and the set of irrational numbers.

Since every rational number is a real number, the real numbers share with the set of rational numbers the property of density (page 310). In addition, the set of real numbers has the following property:

Property of Completeness

Every decimal represents a real number, and every real number has a decimal representation.

Terminating and repeating decimals represent rational numbers; therefore, the decimals for irrational numbers must neither terminate nor repeat. One method of finding successive digits in the decimal approximations of irrational numbers which are square roots is based on the following.

Property of Pairs of Divisors of Any Real Number

If we divide a positive number by a positive number which is smaller than the square root of that number, the quotient will be larger than the square root.

Consider 100 and its square root 10:

$$100 \div 10 = 10;$$

but if the divisor is less than 10, then the quotient is greater than 10:

$$100 \div 2 = 50, \quad 100 \div 4 = 25, \quad 100 \div 5 = 20.$$

Of course, if the divisor is greater than 10, then the quotient is less than 10:

$$100 \div 50 = 2, \quad 100 \div 25 = 4, \quad 100 \div 20 = 5.$$

Example 1. Find the decimal approximation of $\sqrt{15}$ to 4 digits.

Solution:

1. As a first approximation select the integer whose square is nearest 15.

   First $a = 4$

2. Divide 15 by a. Carry the quotient to twice as many digits as are in the divisor.

   $15 \div 4 \doteq 3.7$

3. From the property of pairs of divisors, we know that $\sqrt{15}$ is between a and $\dfrac{15}{a}$. Take their average to find a better approximation to $\sqrt{15}$.

   $\tfrac{1}{2}(4 + 3.7) \doteq 3.8$

4. Use this average as the new a. Repeat Steps 2, 3, 4 until the approximation is as close as is desired.

   Second $a = 3.8$
   $15 \div 3.8 \doteq 3.947$
   $\tfrac{1}{2}(3.8 + 3.947) \doteq 3.873$

The approximation is accurate to at least as many digits as match in a and $15 \div a$.

Third $a = 3.873$
$15 \div 3.873 \doteq 3.8729666$
Since $3.87296 < \sqrt{15} < 3.873$,
$\sqrt{15} \doteq 3.873$ to 4 digits.

The product and quotient properties of square roots, too, may be useful in finding an approximation for roots of numbers less than 1 and numbers greater than 100. Notice how even powers of 10 are used in Example 2.

Example 2. Evaluate: **a.** $\sqrt{13924}$ **b.** $\sqrt{0.013924}$

Solution:

a. $13924 = 10^4(1.3924)$; $\sqrt{13924} = 100\sqrt{1.3924}$

b. $0.013924 = \dfrac{1}{10^2}(1.3924)$; $\sqrt{0.013924} = \tfrac{1}{10}\sqrt{1.3924}$

Both solutions require us to find $\sqrt{1.3924}$. To find this:

First approx. $= 1.2$, since $(1.2)^2 = 1.44$
$1.3924 \div 1.2 \doteq 1.160$

Second approx. $= \dfrac{2.360}{2} = 1.180$

$1.3924 \div 1.180 = 1.180$

$\therefore \sqrt{1.3924} = 1.18$, the true root, since the last remainder was 0. Substituting this value in the given expressions gives:

a. $\sqrt{13924} = 100(1.18) = 118$
b. $\sqrt{0.013924} = \tfrac{1}{10}(1.18) = 0.118$

The next example shows how the values in the table of square roots on page 359 can be used to solve some square-root problems.

Example 3. Approximate the square roots of: **a.** 0.39 **b.** 390

Solution: Use the table of square roots on page 359.

a. $0.39 = \dfrac{39}{100}$

$\sqrt{\dfrac{39}{100}} \doteq \dfrac{6.245}{10}$

$\therefore \sqrt{0.39} \doteq 0.6245$

b. $390 = 39(10)$

$\sqrt{(39)(10)} \doteq (6.245)(3.162)$

$\therefore \sqrt{390} \doteq 19.75$

EXERCISES

Find the indicated square roots.

1. $\sqrt{2.89}$
2. $\sqrt{5.29}$
3. $\sqrt{12.96}$
4. $\sqrt{33.64}$
5. $\sqrt{1849}$
6. $\sqrt{5184}$
7. $-\sqrt{309.76}$
8. $-\sqrt{416.16}$

Find each square root to the nearest hundredth.

9. $\sqrt{3}$
10. $\sqrt{5}$
11. $\sqrt{40.7}$
12. $\sqrt{11.9}$
13. $-\sqrt{273}$
14. $-\sqrt{408}$
15. $\sqrt{190}$
16. $\sqrt{230}$

Find both roots to the nearest tenth.

17. $r^2 = 840$
18. $t^2 = 382$
19. $k^2 - 5.7 = 0$
20. $n^2 - 11.2 = 0$
21. $600 = 5x^2$
22. $900 = 4u^2$
23. Find $\sqrt{41}$ by taking $a = 6$ and by taking $a = 7$.
24. Find $\sqrt{72}$ by taking $a = 8$ and by taking $a = 9$.

Solve to the nearest tenth.

25. $0.7r^2 = 5.81$
26. $(y + 1)^2 + (y - 1)^2 = 50$
27. $12x^2 = 38$
28. $11z^2 - 9 = 0$

PROBLEMS

Find each answer to the nearest tenth, unless otherwise directed.

1. Find the side of a square whose area is 68 square inches.
2. The area of a square is 51 square feet. How long is one side?

3. The area of a circle is $A \doteq 3.14r^2$. Find the radius, correct to hundredths, of a circle whose area is 74.34 square centimeters.
4. The area of a circle is $A \doteq 0.7854d^2$. Find the diameter, correct to hundredths, of a circle whose area is 13.636 square centimeters.
5. Find the side of a square whose area is $\frac{43}{25}$ square meters.
6. The area of a square is $\frac{46}{9}$ square meters. Find its side.
7. A rectangle whose area is 174 square meters has a length four times its width. Find the length and width of this rectangle.
8. The width of a rectangle is $\frac{1}{3}$ its length. Its area is 1783 square centimeters. Find the dimensions of the rectangle.

12–5 Geometric Interpretation of Square Roots

Is it possible to locate irrational square roots on the number line without using approximations? Pythagoras proved the existence of distances which could not be measured by rational numbers in the *Pythagorean theorem*.

Pythagorean Theorem

In any right triangle, the square of the length of the hypotenuse equals the sum of the squares of the lengths of the other two sides.

The **hypotenuse** of a right triangle is the longest side and is opposite the right angle.

Figure 12–1 illustrates the Pythagorean theorem, $c^2 = a^2 + b^2$, where c is the length of the hypotenuse, and a and b are the lengths of the other two sides.

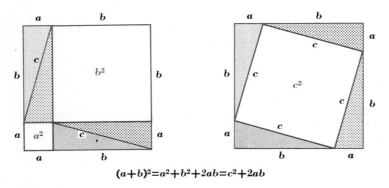

$(a+b)^2 = a^2 + b^2 + 2ab = c^2 + 2ab$

FIGURE 12–1

The first square shows that $(a + b)^2 = a^2 + b^2 + 2ab$ while the second square shows that $(a + b)^2 = c^2 + 2ab$. Thus

$$a^2 + b^2 + 2ab = c^2 + 2ab$$

or

$$a^2 + b^2 = c^2.$$

To find a length equal to $\sqrt{2}$, draw a square whose sides are 1 unit long (Figure 12–2). The diagonal $\overline{OP}$ separates it into two right triangles in which $a = 1$ and $b = 1$.

$$c^2 = a^2 + b^2$$
$$c^2 = 1^2 + 1^2$$
$$c^2 = 1 + 1$$
$$c^2 = 2$$
$$c = \sqrt{2}$$

FIGURE 12–2

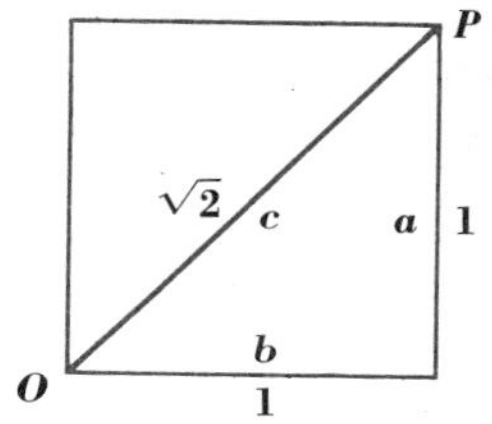

Figure 12–3 combines this square with the coordinate axes. The semicircle has the origin as its center and $\sqrt{2}$ as its radius.

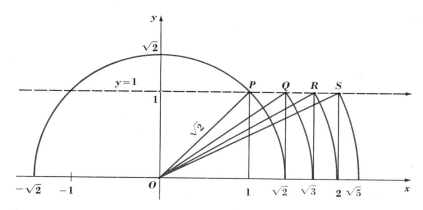

FIGURE 12–3

Points Q, R, and S are on the line $y = 1$ at distances of $\sqrt{3}$, $\sqrt{4}$, and $\sqrt{5}$ from the origin. Each can be found by using the previously constructed square root and drawing perpendiculars to form new right triangles. If OQ, OR, and OS represent lengths, then:

$(OQ)^2 = (\sqrt{2})^2 + 1^2$ $(OR)^2 = (\sqrt{3})^2 + 1^2$ $(OS)^2 = (\sqrt{4})^2 + 1^2$
$(OQ)^2 = 2 + 1$ $(OR)^2 = 3 + 1$ $(OS)^2 = 4 + 1$
$(OQ)^2 = 3$ $(OR)^2 = 4$ $(OS)^2 = 5$
$(OQ) = \sqrt{3}$ $(OR) = \sqrt{4} = 2$ $(OS) = \sqrt{5}$

The next Example applies the *converse* of the Pythagorean theorem. The **converse** of a theorem is obtained by interchanging the hypothesis and conclusion.

Converse of the Pythagorean Theorem

If the sum of the squares of the lengths of the two shorter sides of a triangle is equal to the square of the length of the longest side, then the triangle is a right triangle, with the right angle opposite the longest side.

Example. Is a triangle whose sides measure 3, 4, and 5 units a right triangle?

Solution: $c^2 = a^2 + b^2$
$5^2 \stackrel{?}{=} 3^2 + 4^2$
$25 = 25.$ ∴ a 3-4-5 triangle is a right triangle.

EXERCISES

Determine whether or not each triangle described is a right triangle.

1. The three sides measure 5, 12, and 13 inches.
2. The three sides measure 12, 16, and 20 feet.
3. The three sides measure 10, 24, and 26 inches.
4. The three sides measure 9, 12, and 14 centimeters.

In each right triangle, find the missing dimension to the nearest hundredth of a unit.

5. $a = 6$ feet; $b = 8$ feet
6. $a = 12$ meters; $b = 9$ meters
7. $a = 12$ yards; $b = 3\frac{1}{2}$ yards
8. $a = 5$ feet; $b = 2\frac{3}{4}$ feet
9. $b = 5$ miles; $c = 13$ miles
10. $a = 12$ yards; $c = 22$ yards

PROBLEMS

Make a sketch for each problem. If the number asked for is irrational, approximate it to the nearest hundredth.

1. If the bottom of a 17-foot ramp is 15 feet from a loading platform, how high is the platform?
2. A rope from the top of a mast on a sailboat attached to a point 7 feet from the mast is 25 feet long. How high is the mast?

3. A rectangular flower garden is 30 feet wide and 50 feet long. How long is a straight path running from one corner to the corner diagonally opposite?
4. Two sides of a plastic draftsman's triangle are 9 inches long. How long is the third side?
5. The length of one side of a right triangle, expressed in inches, is one inch less than twice the length of the other. The hypotenuse is one inch more than twice the length of the latter side. Find the length of each side.
6. The length of the longer side of a right triangle, expressed in inches, is 3 more than the height of the shorter side, and the length of the hypotenuse is 3 more than the length of the longer side. Find the length of each side.
7. The lengths of the sides of a right triangle have a ratio of 8:15. The hypotenuse is 34 centimeters in length. Find the length of each side.
8. A group of hikers walks 6 miles due west from their base camp, then north for 2 miles, then due east. They spend the night at a point 4 miles northwest of base camp. How far east did they walk?
9. An arched doorway is a rectangle 4 feet wide by 8 feet high, surmounted by a semicircular arch. How high a cabinet, $2\frac{1}{2}$ feet deep, can be passed through the doorway upright?
10. Can a 40-inch long fencing foil be stored in a box 35 inches long by 15 inches wide by 6 inches deep?

12–6 Multiplication, Division, and Simplification of Square-Root Radicals

The product and quotient properties of square roots (page 316) together with the commutative and associative axioms enable us to multiply, divide, and simplify square-root radicals quickly.

$$\sqrt{5} \cdot \sqrt{6} = \sqrt{5 \cdot 6} = \sqrt{30} \qquad \sqrt{2} \cdot \sqrt{8} = \sqrt{2 \cdot 8} = \sqrt{16} = 4$$
$$(3 \cdot \sqrt{7}) \cdot (5 \cdot \sqrt{2}) = (3 \cdot 5)(\sqrt{7} \cdot \sqrt{2}) = 15\sqrt{14}$$
$$\frac{\sqrt{15}}{\sqrt{3}} = \sqrt{\frac{15}{3}} = \sqrt{5} \qquad \sqrt{\frac{5}{12}} = \sqrt{\frac{5 \cdot 3}{12 \cdot 3}} = \sqrt{\frac{15}{36}} = \frac{\sqrt{15}}{6}$$

An expression having a square-root radical is in **simplest form** when
1. no integral radicand has a square factor other than 1,
2. no fractions are under a radical sign, and
3. no radicals are in a denominator.

The following are examples of simplifying square-root radicals:

$$\sqrt{12} = \sqrt{4} \cdot \sqrt{3} = 2\sqrt{3}; \quad 2\sqrt{45} = 2\sqrt{9 \cdot 5} = 6\sqrt{5}$$

$$\sqrt{\frac{2}{3}} = \frac{\sqrt{2}}{\sqrt{3}} = \frac{\sqrt{2} \cdot \sqrt{3}}{\sqrt{3} \cdot \sqrt{3}} = \frac{\sqrt{6}}{3}; \quad \frac{1}{\sqrt{5}} = \frac{1 \cdot \sqrt{5}}{\sqrt{5} \cdot \sqrt{5}} = \frac{\sqrt{5}}{5}$$

$$\frac{3\sqrt{7}}{2\sqrt{32}} = \frac{3\sqrt{7} \cdot \sqrt{2}}{2\sqrt{32} \cdot \sqrt{2}} = \frac{3\sqrt{14}}{2\sqrt{64}} = \frac{3\sqrt{14}}{2 \cdot 8} = \frac{3\sqrt{14}}{16}$$

The process of changing the form of a fraction with an irrational denominator such as $\frac{3\sqrt{7}}{2\sqrt{32}}$ to an equal fraction with a rational denominator such as $\frac{3\sqrt{14}}{16}$ is called **rationalizing the denominator**. Rationalizing the denominator of a radical expression helps in approximating its value, since division by a whole number is easier than division by a decimal which is an approximation for the square root of an irrational number.

EXERCISES

Express in simplest form. (Assume that all radicands are nonnegative real numbers.)

1. $2\sqrt{5} \cdot 3\sqrt{5}$
2. $4\sqrt{7} \cdot 2\sqrt{7}$
3. $\sqrt{2} \cdot \sqrt{5} \cdot \sqrt{10}$
4. $\sqrt{15} \cdot \sqrt{3} \cdot \sqrt{5}$
5. $2\sqrt{3} \cdot \sqrt{5} \cdot \sqrt{7}$
6. $3\sqrt{2} \cdot \sqrt{7} \cdot \sqrt{3}$
7. $\sqrt{\frac{2}{5}} \cdot \sqrt{\frac{5}{2}}$
8. $\sqrt{\frac{4}{3}} \cdot \sqrt{\frac{12}{4}}$
9. $\sqrt{\frac{3}{5}} \cdot \sqrt{\frac{25}{12}}$
10. $\sqrt{\frac{5}{9}} \cdot 2\sqrt{\frac{9}{20}}$
11. $\sqrt{5\frac{1}{7}} \cdot \sqrt{1\frac{3}{4}}$
12. $\frac{1}{2}\sqrt{\frac{4}{15}} \cdot \frac{1}{3}\sqrt{\frac{3}{20}}$
13. $\frac{12\sqrt{6}}{3\sqrt{2}}$
14. $\frac{4\sqrt{24}}{3\sqrt{4}}$
15. $\sqrt{150}$
16. $5\sqrt{24}$
17. $3\sqrt{72}$
18. $5\sqrt{75}$
19. $\frac{\sqrt{80}}{\sqrt{5}}$
20. $\frac{\sqrt{7}}{\sqrt{63}}$
21. $3\sqrt{\frac{32}{9}}$
22. $5\sqrt{\frac{45}{4}}$
23. $\frac{12\sqrt{3}}{4\sqrt{27}}$
24. $\frac{\sqrt{x^7}}{\sqrt{x}}$
25. $15\sqrt{\frac{8}{3}}$
26. $6\sqrt{\frac{5}{18}}$
27. $8\sqrt{\frac{3}{50}}$
28. $\frac{\sqrt{5x}}{\sqrt{x}}$
29. $\frac{\sqrt{50}}{3\sqrt{10}}$
30. $\frac{\sqrt{72}}{2\sqrt{6}}$
31. $\sqrt{54a^2}$
32. $\sqrt{63x^5}$

33. $(-2\sqrt{x^3y})(4\sqrt{xy})$
34. $(-5\sqrt{np^2})(-3\sqrt{n})$
35. $\sqrt{a}(\sqrt{a}+2)$
36. $\sqrt{b}(3-\sqrt{b})$
37. $(4\sqrt{3})(\sqrt{6})(-\sqrt{8})$
38. $(-5\sqrt{28})(-\sqrt{14})(2\sqrt{2})$
39. $(\sqrt{14x})(2\sqrt{x})(\sqrt{7})$
40. $(\sqrt{2x})(\sqrt{3x})(\sqrt{6})$
41. $(3x\sqrt{2x})^2$
42. $(4y\sqrt{3y})^2$
43. $3\sqrt{2}(2\sqrt{6}+\sqrt{3})$
44. $5\sqrt{3}(-2\sqrt{6}+\sqrt{15})$
45. $-3\sqrt{2\tfrac{1}{3}}$
46. $4\sqrt{1\tfrac{1}{5}}$
47. $-3\sqrt{1.2}$
48. $2.38\sqrt{0.26}$
49. $\dfrac{2\sqrt{3}+3\sqrt{6}}{\sqrt{3}}$
50. $\dfrac{4\sqrt{28}}{5\sqrt{7}-3\sqrt{7}}$
51. $\dfrac{2\sqrt{5}-\sqrt{10}}{\sqrt{5}}$
52. $\dfrac{\sqrt{7}+2\sqrt{14}}{\sqrt{7}}$
53. $\dfrac{5\sqrt{x^3}+\sqrt{x}}{\sqrt{x}}$
54. $\dfrac{\sqrt{y}-5\sqrt{y^3}}{\sqrt{y}}$

PROBLEMS

Find answers to the nearest tenth unless otherwise directed.

1. One positive number is five times another, and the difference of their squares is 18. Find the numbers.
2. Find two positive numbers in the ratio of 3 to 2 whose squares differ by 10.
3. The length and width of a rectangle are in the ratio 4 to 3. If the area of the rectangle is 36 square inches, what are its dimensions?
4. Find the length of a diagonal of a square with area 81 square inches.
5. Use Heron's formula, $A = \sqrt{s(s-a)(s-b)(s-c)}$, where a, b, and c are lengths of sides and s is half the perimeter, to find the area of a triangle with sides of length 3 inches, 6 inches, and 7 inches.
6. If a circle is inscribed in a right triangle as shown in the figure at the right, $c = a + b - 2r$, and so the radius r of the circle is determined by $r = \tfrac{1}{2}(a + b - c)$, where a and b are the lengths of the shorter sides of the triangle and c is the length of the hypotenuse. Find r, if a is 8 feet and b is 10 feet.
7. Find the length s of each side of a square if a diameter of the square is 5.3 feet long.

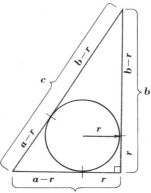

8. A square is inscribed in a circle as shown at the right. Find the radius of the circle if the area of the square is 36 square inches.

9. Will a square whose area is 36 square inches fit inside a circle whose area is 54 square inches? Support your answer with calculations.

10. Show that an equilateral triangle with sides of length 9 inches is smaller in area than a square inscribed in a circle with a diameter of 9 inches.

11. An altitude of an equilateral triangle separates the triangle into two congruent triangles as pictured at the right. Find the length h of the altitude if the length s of a side is 1 foot.

12. Express the length h of an altitude of an equilateral triangle in terms of the length s of a side of the triangle.

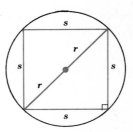

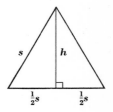

12–7 Addition and Subtraction of Square-Root Radicals

Because $3\sqrt{7}$ and $2\sqrt{7}$ have the common factor $\sqrt{7}$, the expression for their sum can be simplified by using the distributive axiom:

$$3\sqrt{7} + 2\sqrt{7} = (3 + 2)\sqrt{7} = 5\sqrt{7}$$

The sum or difference of square-root radicals having the same radicand is the sum or difference of the coefficients of the radicals, multiplied by the common radical. On the other hand, the addition or subtraction of radicals having unlike radicands can only be indicated.

$$3\sqrt{5} - 5\sqrt{11} + 2\sqrt{5} + \sqrt{11} = 5\sqrt{5} - 4\sqrt{11}$$

By reducing each radical to simplest form, we sometimes can combine terms in a sum of radicals.

Example. Simplify: $4\sqrt{8} - 2\sqrt{50} + 3\sqrt{128}$

Solution:
$$\begin{aligned}
4\sqrt{8} - 2\sqrt{50} + 3\sqrt{128} &= 4\sqrt{4 \cdot 2} - 2\sqrt{25 \cdot 2} + 3\sqrt{64 \cdot 2} \\
&= 4(2\sqrt{2}) - 2(5\sqrt{2}) + 3(8\sqrt{2}) \\
&= 8\sqrt{2} - 10\sqrt{2} + 24\sqrt{2} \\
&= 22\sqrt{2}
\end{aligned}$$

To simplify sums or differences of square-root radicals:
1. **Express each radical in simplest form.**
2. **Use the distributive axiom to add or subtract radicals with like radicands.**
3. **Indicate the sum or difference of radicals with unlike radicands.**

EXERCISES

Simplify each expression.

1. $2\sqrt{3} + 3\sqrt{3} - \sqrt{3}$
2. $5\sqrt{6} - 8\sqrt{6} - 2\sqrt{6}$
3. $\sqrt{5} - 2\sqrt{7} + 3\sqrt{7} - \sqrt{5}$
4. $2\sqrt{3} + \sqrt{11} - 3\sqrt{11} + \sqrt{3}$
5. $2\sqrt{8} + \sqrt{2}$
6. $\sqrt{3} - \sqrt{27}$
7. $\sqrt{12} - 2\sqrt{3}$
8. $\sqrt{32} + 2\sqrt{2}$
9. $\sqrt{3} + \sqrt{\frac{1}{3}}$
10. $\sqrt{5} - \sqrt{\frac{1}{5}}$
11. $2\sqrt{54} + \sqrt{24}$
12. $5\sqrt{48} - 8\sqrt{27}$
13. $\sqrt{10} + \sqrt{\frac{2}{5}}$
14. $2\sqrt{21} - \sqrt{\frac{3}{7}}$
15. $\sqrt{200} - \frac{2}{3}\sqrt{162}$
16. $\sqrt{192} + \frac{1}{5}\sqrt{50}$
17. $2\sqrt{2} - \sqrt{32} + 2\sqrt{\frac{1}{2}}$
18. $2\sqrt{54} + \sqrt{96} - 9\sqrt{\frac{2}{3}}$
19. $15\sqrt{\frac{2}{5}} + 6\sqrt{\frac{5}{2}} - \sqrt{160}$
20. $8\sqrt{\frac{3}{2}} - 15\sqrt{\frac{2}{3}} + \sqrt{96}$
21. $6\sqrt{\frac{5}{4}} - 15\sqrt{\frac{1}{5}} + 5\sqrt{45}$
22. $2\sqrt{\frac{3}{8}} + \sqrt{\frac{8}{3}} - 4\sqrt{24}$

In Exercises 23–26, assume that all radicands are nonnegative real numbers, and simplify.

23. $3\sqrt{18y} - \frac{3y}{4}\sqrt{\frac{32}{y}}$
24. $x^3\sqrt{\frac{3}{x^3}} + 4x\sqrt{27x}$
25. $\sqrt{\frac{x^2}{16} + \frac{x^2}{9}}$
26. $\sqrt{\frac{3n^2}{5} - \frac{n^2}{20}}$

Solve each equation.

27. $x\sqrt{3} - 2\sqrt{12} = 2\sqrt{27} - x\sqrt{27}$
28. $\sqrt{288} - y\sqrt{18} = y\sqrt{72} + \sqrt{162}$
29. $3\sqrt{a} + 15 = 35 - \sqrt{a}$
30. $3(2 + \sqrt{b}) - \sqrt{b} = 14$

12–8 Multiplication of Binomials Containing Square-Root Radicals

Sometimes in dealing with radicals, we may wish to find a product like $(2 + \sqrt{5})(2 - \sqrt{5})$. Notice that this product resembles $(a + b)(a - b) = a^2 - b^2$. Two binomials of the form $x + \sqrt{y}$ and $x - \sqrt{y}$ are called **conjugates** of each other. They differ only in the sign before the radical. In case x and y are rational numbers, their product is also a rational number, as this example shows.

Example 1. $(2 + \sqrt{5})(2 - \sqrt{5})$

Solution: $(2 + \sqrt{5})(2 - \sqrt{5}) = 2^2 - (\sqrt{5})^2$
$= 4 - 5 = -1$

Consider the product $(3 + \sqrt{7})(3 + \sqrt{7})$ which is of the form

$$(a + b)(a + b) = a^2 + 2ab + b^2.$$

Example 2. $(3 + \sqrt{7})(3 + \sqrt{7})$

Solution: $(3 + \sqrt{7})(3 + \sqrt{7}) = 3^2 + 2(3\sqrt{7}) + (\sqrt{7})^2$
$= 9 + 6\sqrt{7} + 7 = 16 + 6\sqrt{7}$

Noting conjugates or their negatives can help us *rationalize a binomial denominator*.

Example 3. Rationalize the denominator: $\dfrac{1}{3\sqrt{2} - 2}$

Solution: Here the conjugate of $2 + 3\sqrt{2}$ is $2 - 3\sqrt{2}$, which is $-(3\sqrt{2} - 2)$. Therefore, multiply the numerator and denominator by $3\sqrt{2} + 2$:

$$\frac{1}{3\sqrt{2} - 2} = \frac{1(3\sqrt{2} + 2)}{(3\sqrt{2} - 2)(3\sqrt{2} + 2)} = \frac{3\sqrt{2} + 2}{(3\sqrt{2})^2 - (2)^2}$$
$$= \frac{3\sqrt{2} + 2}{18 - 4} = \frac{3\sqrt{2} + 2}{14}$$

EXERCISES

Express in simplest form.

1. $(3 + \sqrt{2})(3 - \sqrt{2})$
2. $(5 - \sqrt{7})(5 + \sqrt{7})$
3. $(\sqrt{5} - \sqrt{2})(\sqrt{5} + \sqrt{2})$
4. $(\sqrt{3} + \sqrt{6})(\sqrt{3} - \sqrt{6})$

5. $(3\sqrt{5} + 1)(\sqrt{5} - 2)$
6. $(2\sqrt{3} - 1)(3\sqrt{3} + 2)$
7. $(5 + \sqrt{5})^2$
8. $(3 - \sqrt{6})^2$
9. $(2\sqrt{3} - 3)^2$
10. $(4\sqrt{5} + 3)^2$
11. $3\sqrt{5}(2\sqrt{10} + \sqrt{5})$
12. $4\sqrt{6}(2\sqrt{3} + 5\sqrt{2})$
13. $(3\sqrt{5} + \sqrt{2})(2\sqrt{5} - \sqrt{2})$
14. $(4\sqrt{3} + 5)(3\sqrt{3} - 4)$
15. $(4\sqrt{2} + 3\sqrt{6})(2\sqrt{2} - \sqrt{6})$
16. $(2\sqrt{10} + \sqrt{5})(3\sqrt{10} - 2\sqrt{5})$

Rationalize the denominator of each fraction.

17. $\dfrac{1}{1 + \sqrt{2}}$
18. $\dfrac{2}{\sqrt{5} - 1}$
19. $\dfrac{\sqrt{3}}{\sqrt{3} - 2}$
20. $\dfrac{\sqrt{5}}{3 + \sqrt{5}}$
21. $\dfrac{2 + \sqrt{3}}{1 - \sqrt{3}}$
22. $\dfrac{3 + \sqrt{2}}{2 - \sqrt{2}}$
23. $\dfrac{5}{2\sqrt{7} + 3}$
24. $\dfrac{4}{3\sqrt{5} - 2}$

If $f(x) = x^2 - 4x + 1$, find:

25. $f(\sqrt{2})$
26. $f(\sqrt{2} - 1)$
27. $f(2 - \sqrt{3})$
28. $f(2 + \sqrt{3})$

29. Show that $1 + 2\sqrt{3}$ and $1 - 2\sqrt{3}$ are roots of the equation "$x^2 - 2x - 11 = 0$."

30. Write an expression in simplest form for the area of a square whose perimeter is $8\sqrt{5} + 4$ centimeters.

Simplify. Assume that the value of each variable is nonnegative.

31. $(\sqrt{a} - b)(\sqrt{a} + b)$
32. $(x + 2\sqrt{3})^2$
33. $(2a\sqrt{b} - c)(3a\sqrt{b} + 4c)$
34. $2\sqrt{\dfrac{x}{y}} - 3\sqrt{\dfrac{y}{x}} + \sqrt{xy}$

12–9 Radical Equations

An equation having a variable in a radicand is a **radical equation**. The simplest kind of radical equation is one like $\sqrt{y} = 5$, which is solved by squaring each of its members:

$$\sqrt{y} = 5, \quad (\sqrt{y})^2 = 5^2, \quad y = 25$$
$$\sqrt{25} \stackrel{?}{=} 5$$
$$5 = 5$$

∴ the solution set is $\{25\}$.

To solve radical equations which have several terms in each member but only one radical term, we first isolate the radical term in one member. Then we can square each member and solve the resulting equation.

Example 1. Solve $2 = 3\sqrt{x} - x$.

Solution:

1. Isolate the radical term in one member of the equation.

$$2 = 3\sqrt{x} - x$$
$$x + 2 = 3\sqrt{x}$$

2. Square both members.

$$(x + 2)^2 = (3\sqrt{x})^2$$
$$x^2 + 4x + 4 = 9x$$

3. Solve the resulting equation.

$$x^2 - 5x + 4 = 0$$
$$(x - 4)(x - 1) = 0$$
$$x - 4 = 0 \quad | \quad x - 1 = 0$$
$$x = 4 \quad | \quad x = 1$$

Check:

4. Substitute. Be sure to take the principal root of the number in the radicand.

$$2 \stackrel{?}{=} 3\sqrt{4} - 4 \quad | \quad 2 \stackrel{?}{=} 3\sqrt{1} - 1$$
$$2 \stackrel{?}{=} 3 \cdot 2 - 4 \quad | \quad 2 \stackrel{?}{=} 3 \cdot 1 - 1$$
$$2 = 6 - 4 \quad | \quad 2 = 3 - 1$$

∴ the solution set is $\{4, 1\}$.

The "squared" equation in Step 2 may not be equivalent to the given equation. Notice that:

If $a = b$, then $a^2 = b^2$; but if $a^2 = b^2$, it need not be true that $a = b$.

For example, $5^2 = (-5)^2$, but $5 \neq -5$.

Thus, when we square the members of an equation, the most that we can say is that the solution set of the original equation is a *subset* of the solution set of the new equation. Therefore, we *must* check each root of the new equation in the original equation to see which, if any, satisfy the equation.

Example 2. Solve $10 = 3\sqrt{x} + x$.

Solution:

$$10 = 3\sqrt{x} + x$$
$$10 - x = 3\sqrt{x}$$
$$(10 - x)^2 = (3\sqrt{x})^2$$
$$100 - 20x + x^2 = 9x$$
$$x^2 - 29x + 100 = 0$$
$$(x - 4)(x - 25) = 0$$
$$x - 4 = 0 \quad | \quad x - 25 = 0$$
$$x = 4 \quad | \quad x = 25$$

Check:
$$10 \stackrel{?}{=} 3\sqrt{4} + 4$$
$$10 \stackrel{?}{=} 3 \cdot 2 + 4$$
$$10 = 10$$
$$10 \stackrel{?}{=} 3\sqrt{25} + 4$$
$$10 \stackrel{?}{=} 3 \cdot 5 + 4$$
$$10 \neq 19$$

25 is not a root.   ∴ the solution set is {4}.

EXERCISES

Solve each equation.

1. $\sqrt{3y} = 6$
2. $\sqrt{2z} = 8$
3. $\sqrt{7a} = \frac{1}{2}$
4. $\sqrt{6t} = \frac{2}{3}$
5. $\sqrt{n} - 2 = -1$
6. $\sqrt{r} + 3 = 5$
7. $\sqrt{x} - \frac{2}{3} = 3$
8. $\sqrt{b} + \frac{3}{5} = 2$
9. $\sqrt{\frac{z}{3}} = 1$
10. $\sqrt{\frac{k}{5}} = 2$
11. $\sqrt{n + 2} = 4$
12. $\sqrt{m - 3} = 3$
13. $3\sqrt{2x} = 6$
14. $5\sqrt{3x} = 15$
15. $\sqrt{2t + 3} = 1$
16. $\sqrt{6n + 5} = 2$
17. $\sqrt{3x + 2} - 1 = 1$
18. $\sqrt{5y - 2} + 3 = 6$
19. $\sqrt{\frac{4a}{3}} - 2 = 6$
20. $\sqrt{\frac{7t}{2}} + 3 = 12$
21. $\sqrt{\frac{2n + 6}{5}} = 4$
22. $\sqrt{\frac{3t - 1}{6}} = 3$
23. $\sqrt{n} = 3\sqrt{2}$
24. $2\sqrt{t} = 4\sqrt{3}$
25. $3\sqrt{2t^2 - 28} = 6$
26. $3\sqrt{5x^2 - 11} = 9$
27. Solve for m: $v = \sqrt{\frac{2E}{m}}$
28. Solve for v: $r = \sqrt{\frac{7v}{22h}}$
29. $\sqrt{y^2 + 2} = 2 - y$
30. $\sqrt{x^2 - 2} = x + 10$
31. $\sqrt{a} = -\frac{a}{4}$
32. $\sqrt{b} = -\frac{b}{3}$
33. $\sqrt{y^2 + 11} - 1 = y$
34. $\sqrt{z^2 - 13} + 1 = z$
35. $\sqrt{y + 4} = y + 2$
36. $\sqrt{x + 2} = x - 4$
37. $\sqrt{7t - 3} = 2t - 3$
38. $\sqrt{3b + 10} = b + 4$
39. $\sqrt{n - 3} + 5 = 0$
40. $\sqrt{r + 2} + 3 = 0$

PROBLEMS

1. Twice the square root of a number is 22. Find the number.
2. One-fifth the square root of a number is 2. Find the number.
3. The square root of 5 less than one third a number is 10. What is the number?
4. When 7 is added to three times a certain number, the square root of the result is 5. Find the number.
5. The diameter of a circle in terms of its area is $d \doteq \sqrt{\dfrac{14A}{11}}$. Solve for A in terms of d, and find the area of a circle of diameter 7 inches.
6. The distance in miles to the horizon from a submarine's periscope that is h feet above a calm sea is $d = \sqrt{\dfrac{3h}{2}}$. Solve this for h in terms of d, and determine how far above the surface the periscope is if the lookout sees 3 miles to the horizon.
7. Solve this form of the Pythagorean theorem for the positive value of b: $c = \sqrt{a^2 + b^2}$. Find b when $c = 10$ and $a = 6$.
8. The current I (measured in amperes) which flows through an electrical appliance is expressed by $I = \sqrt{\dfrac{P}{R}}$, where P is power consumed (in watts) and R is the resistance of the appliance (in ohms). If an electric iron has a resistance of 20 ohms and draws 8 amperes of current, how much power does it consume?
9. A 1-carat diamond costs $700. The weight in carats of a diamond of equal quality may be found by $w = \sqrt{\dfrac{C}{700}}$, where C is the cost in dollars. Using this formula, find the cost of a 1.7 carat diamond.
10. The number of times a pendulum swings in one second is given by the formula $f \doteq \dfrac{1}{2\pi}\sqrt{\dfrac{32}{l}}$, where l is its length in feet. If a pendulum completes half a swing in each second, how long is it? (Use $\pi \doteq 3.14$.)
11. The velocity of sound in air, in meters per second, at a temperature of t degrees, is $V = 333\sqrt{14.0037t}$. At what temperature will sound travel at 11,544 meters per second?
12. $s = 16t^2$ gives the approximate distance (in feet) traveled by a falling object in t seconds. How long does it take a rock to fall 900 feet?
13. The thickness (in inches) of a beam of fixed width and length in terms of the weight (in pounds) which the beam can support is $t = \dfrac{1}{5}\sqrt{\dfrac{w}{6}}$. How much weight can a 2-inch-thick beam support?

14. The acceleration in meters/sec² of an object moving in a circular path is expressed by $a = \dfrac{v^2}{r}$, where v is velocity in meters/sec and r is the radius of the path in meters. How fast does an object tied to a 12-meter string need to move in order to have the same acceleration as an object tied to a 10-meter string and moving at 10 meters/sec?

15. Let f be the function with domain {3 and the real numbers greater than 3} such that $f(x) = \sqrt{x - 3} - 5$. For what value of x is $f(x) = 2$?

16. Two positive numbers x and y are related as follows: $x = \sqrt{y^2 - 5y - 1}$. Find y when x is 7.

chapter summary

1. A **rational number** can be expressed as a fraction in an unlimited number of ways, and as either a **terminating decimal** or a **repeating decimal**. Between every pair of rational numbers is another rational number (**Property of Density**).

2. To round a decimal, retain the digits unchanged if the first digit dropped is less than 5; increase the value of the last digit by one if the first digit dropped is 5 or more.

3. If $a \geq 0$ and $b > 0$, then $\sqrt{ab} = \sqrt{a} \cdot \sqrt{b}$ and $\sqrt{\dfrac{a}{b}} = \dfrac{\sqrt{a}}{\sqrt{b}}$.

4. In any right triangle, if c is the length of hypotenuse, and a and b are the lengths of the other sides, then $a^2 + b^2 = c^2$. This is the **Pythagorean theorem**.

5. Roots of rational numbers are not all rational numbers. **Irrational numbers** are represented by **unending, nonrepeating decimals**. The set of real numbers is the union of the set of rational numbers and the set of irrational numbers.

6. If a positive number is divided by a positive number which is smaller than the square root of the number, the quotient is larger than the square root. (**Property of Pairs of Divisors**)

7. Sums and differences of square roots having the same radicand can be simplified by applying the distributive axiom.

8. Squaring both members of an equation produces a new equation whose solution set includes but may not be identical with the solution set of the given equation.

CHAPTER REVIEW

12–1 Arrange in order, from greatest to least.

1. $\frac{3}{20}, \frac{4}{23}, \frac{7}{25}$
2. $\frac{77}{32}, \frac{56}{25}, \frac{64}{27}$

Find a number halfway between the given pair.

3. $\frac{23}{24}, 1$
4. $-\frac{11}{7}, 3$

12–2 Write as a terminating or repeating decimal.

5. $\frac{3}{40}$
6. $\frac{8}{9}$
7. $-\frac{5}{11}$
8. $3\frac{2}{7}$

Write as a common fraction.

9. $0.1818\ldots$
10. $0.\overline{5}$
11. $3.\overline{13}$
12. $5.2\overline{34}$

12–3 Evaluate each expression using the product and quotient properties.

13. $\sqrt{1600}$
14. $-\sqrt{\frac{1}{1296}}$

Find each solution set over $\mathcal{R}$.

15. $x^2 = 2500$
16. $5z^2 - 405 = 0$

12–4 Compute the given square root to the nearest hundredth.

17. $\sqrt{3.18}$
18. $-\sqrt{0.285}$

12–5 Find the length of the third side of right triangle ABC if its hypotenuse has length AB. $\angle C$ is the right angle.

19. $BC = 16$, $AB = 20$
20. $AC = 20$, $BC = 15$

12–6 Express each product or quotient in simplest form.

21. $2\sqrt{11} \cdot 3\sqrt{11}$
22. $2\sqrt{5} \cdot \sqrt{10} \cdot 3\sqrt{8}$
23. $\frac{5\sqrt{6}}{2\sqrt{75}}$
24. $\frac{6\sqrt{x^3 y}}{\sqrt{9xy}}$

12–7 Simplify each expression.

25. $2\sqrt{7} + \sqrt{7} - 5\sqrt{7}$
26. $7\sqrt{8} - 2\sqrt{\frac{1}{2}}$
27. $4\sqrt{\frac{1}{12}} + \sqrt{\frac{1}{3}}$
28. $\frac{1}{3}\sqrt{\frac{2}{27}} - \frac{1}{6}\sqrt{\frac{1}{3}}$

12–8 Express in simplest form.

29. $(2 + \sqrt{3})(1 - \sqrt{3})$
30. $(4 - \sqrt{2})^2$
31. $\frac{\sqrt{5} + 1}{1 - \sqrt{5}}$

12–9 Find the solution set.

32. $\sqrt{2x} - 4 = 0$
33. $\sqrt{t} = t - 6$

13 quadratic equations and inequalities

13–1 The Square-Root Property

A quadratic equation can be put into the standard form

$$ax^2 + bx + c = 0,$$

where a, b, and c are real numbers and $a \neq 0$. At present our chief tool in solving such equations is factoring (page 182). If the equation is of the form $ax^2 + c = 0$, it is a *pure quadratic* and may be solved by using the following:

Property of Square Roots of Equal Numbers

If r and s are any real numbers, $r^2 = s^2$ if and only if $r = s$ or $r = -s$.

Example 1. Solve $t^2 - 18 = 0$.

Solution: $t^2 - 18 = 0$
$t^2 = 18$
$t = \sqrt{18}$ or $t = -\sqrt{18}$
$t = 3\sqrt{2}$ $t = -3\sqrt{2}$

Check: $(3\sqrt{2})^2 - 18 \stackrel{?}{=} 0$ | $(-3\sqrt{2})^2 - 18 \stackrel{?}{=} 0$
$18 - 18 = 0$ | $18 - 18 = 0$

∴ the solution set is $\{3\sqrt{2}, -3\sqrt{2}\}$.

Example 2. Solve $36r^2 + 9 = 0$.

Solution:
$$36r^2 + 9 = 0$$
$$36r^2 = -9$$
$$r^2 = -\tfrac{9}{36}$$

Since negative numbers have no square roots in the set of real numbers, $36r^2 + 9 = 0$ is not solvable in the real number system.

This method may also be used to solve quadratic equations having a trinomial square as one member and a nonnegative constant as the other.

Example 3. Solve $x^2 - 4x + 4 = 16$.

Solution:
$$x^2 - 4x + 4 = 16$$
$$(x - 2)^2 = 16$$
$$\therefore x - 2 = \pm\sqrt{16}.$$
$$x - 2 = 4 \quad \text{or} \quad x - 2 = -4$$
$$x = 6 \qquad\qquad x = -2$$

Do the elements of $\{-2, 6\}$ check as the roots of the given equation?

EXERCISES

Solve each equation.

1. $t^2 = 121$
2. $r^2 = 144$
3. $4x^2 = 81$
4. $49y^2 = 64$
5. $12 - x^2 = 0$
6. $32 - a^2 = 0$
7. $50k^2 - 2 = 0$
8. $72n^2 - 2 = 0$
9. $6t^2 - \tfrac{1}{24} = 0$
10. $3n^2 - \tfrac{1}{27} = 0$
11. $\tfrac{1}{4}x^2 - 81 = 0$
12. $\tfrac{4}{9}z^2 - 1 = 0$
13. $(t + 2)^2 = 9$
14. $(s - 3)^2 = 16$
15. $(x - 1)^2 = 25$
16. $(y + 1)^2 = 49$
17. $9(z + 3)^2 = 81$
18. $4(k - 5)^2 = 64$
19. $(x - \tfrac{2}{5})^2 = \tfrac{9}{25}$
20. $(t + \tfrac{1}{3})^2 = \tfrac{4}{9}$

21. $n^2 + 2n + 1 = 9$
22. $k^2 - 4k + 4 = 36$
23. $t^2 - 10t + 25 = 4$
24. $r^2 + 14r + 49 = 100$
25. $(y - 4)^2 = 3$
26. $(t - 2)^2 = 5$
27. $(s + 5)^2 = 7$
28. $(n + 3)^2 = 6$
29. $(x + \frac{1}{2})^2 = \frac{1}{2}$
30. $(x - \frac{2}{3})^2 = \frac{1}{3}$
31. $n^3 - n = 0$
32. $4k - k^3 = 0$
33. $\frac{1}{3}p^3 - 3p = 0$
34. $7y - \frac{1}{7}y^3 = 0$
35. $x^3 = 5x$
36. $s^3 = 3s$
37. $r^2 - 2\sqrt{5}\,r + 5 = 0$
38. $z^2 + 2\sqrt{2}\,z + 2 = 0$
39. $a^2 + 2\sqrt{6}\,a + 6 = 0$
40. $b^2 - 2\sqrt{7}\,b + 7 = 0$
41. Explain why $x^2 + 8 = 0$ is not solvable over $\Re$.
42. Explain why $x^4 + 1 = 0$ is not solvable over $\Re$.

13–2 Sum and Product of the Roots of a Quadratic Equation

If r and s are the roots of a quadratic equation in x, then the equation is equivalent to the disjunction:

$$x - r = 0 \quad \text{or} \quad x - s = 0$$

Therefore, by the zero-product property (page 184) the equation is also equivalent to

$$(x - r)(x - s) = 0$$

or

$$x^2 - (r + s)x + rs = 0.$$

This latter equation has the form "$x^2 + bx + c = 0$" where $-(r + s) = b$ and $rs = c$. Thus, we have:

The Property of the Sum and Product of the Roots of a Quadratic Equation

If the roots of a quadratic equation of the form

$$x^2 + bx + c = 0$$

are r and s, then

$$r + s = -b \quad \text{and} \quad rs = c.$$

Example 1. Is $\{2, 7\}$ the solution set of $y^2 - 9y + 14 = 0$?

Solution: $y^2 - 9y + 14 = 0; \quad b = -9, \quad c = 14$

$r + s = 2 + 7 = 9 = -(-9) = -b$

$rs = 2 \cdot 7 = 14 = c$

$\therefore \{2, 7\}$ is the solution set of $y^2 - 9y + 14 = 0$.

Example 2. Is $\{5 + \sqrt{2}, 5 - \sqrt{2}\}$ the solution set of

$$n^2 - 10n + 23 = 0?$$

Solution:

Is the sum of the roots 10?	Is the product of the roots 23?
$5 + \sqrt{2} + 5 - \sqrt{2} \stackrel{?}{=} 10$	$(5 + \sqrt{2})(5 - \sqrt{2}) \stackrel{?}{=} 23$
$10 = 10$	$25 - 2 = 23$

$\therefore \{5 + \sqrt{2}, 5 - \sqrt{2}\}$ is the solution set of

$$n^2 - 10n + 23 = 0.$$

The expressions $5 + \sqrt{2}$ and $5 - \sqrt{2}$ in Example 2 may be written as the single expression $5 \pm \sqrt{2}$ (read "5 plus or minus $\sqrt{2}$").

EXERCISES

Determine whether or not the given set is the solution set of the given equation.

1. $x^2 - 8x - 20 = 0; \{10, -2\}$
2. $t^2 + 12t - 28 = 0; \{4, -7\}$
3. $2s^2 - 9s - 5 = 0; \{-\frac{1}{2}, 5\}$
4. $3n^2 - n - 10 = 0; \{-\frac{5}{3}, 2\}$
5. $4z^2 + 8z + 3 = 0; \{-\frac{3}{2}, -\frac{1}{2}\}$
6. $7t^2 - 33t - 10 = 0; \{-\frac{2}{7}, 5\}$
7. $6k^2 - 11k + 3 = 0; \{\frac{1}{3}, \frac{3}{2}\}$
8. $4t^2 + 4t + 1 = 0; \{-\frac{1}{2}, \frac{1}{2}\}$
9. $x^2 - 2x - 1 = 0; \{1 \pm \sqrt{2}\}$
10. $y^2 + 6y - 1 = 10; \{-3 \pm \sqrt{10}\}$
11. $5n^2 - 8n + 1 = 0; \left\{\frac{4 \pm \sqrt{10}}{5}\right\}$
12. $3t^2 - 2t - 9 = 0; \left\{\frac{1 \pm \sqrt{7}}{2}\right\}$

Find a quadratic equation having the given solution set.

Example. $\{\sqrt{3}, -\sqrt{3}\}$

Solution: $(x - \sqrt{3})[x - (-\sqrt{3})] = 0$

$(x - \sqrt{3})(x + \sqrt{3}) = 0$

$x^2 - 3 = 0$

13. $\{1, 3\}$
15. $\{-3, 2\}$
17. $\{5, 0\}$
19. $\{\sqrt{11}\}$
21. $\{\sqrt{6}, -\sqrt{6}\}$
23. $\{2 - \sqrt{3}, 2 + \sqrt{3}\}$
14. $\{7, 5\}$
16. $\{-6, -3\}$
18. $\{0, -7\}$
20. $\{3\sqrt{2}\}$
22. $\{2\sqrt{5}, -2\sqrt{5}\}$
24. $\{5 - \sqrt{2}, 5 + \sqrt{2}\}$

13–3 Solution by Completing a Trinomial Square

If a quadratic equation can be transformed into one having a trinomial square as a member, its solution set can easily be found.

Example 1. Solve $x^2 + 4x - 12 = 0$.

Solution:
1. Write an equivalent equation with the constant term as right member.
2. Add to both members the number making the left member a trinomial square.
3. Use the property of square roots.
4. Form two linear equations.
5. Solve.

$$x^2 + 4x - 12 = 0$$
$$x^2 + 4x = 12$$
$$x^2 + 4x + 4 = 12 + 4$$
$$(x + 2)^2 = 16$$
$$x + 2 = \pm 4$$

$x + 2 = 4$ or $x + 2 = -4$
$x = 2$ $x = -6$

Check:
$2^2 + 4(2) - 12 \stackrel{?}{=} 0$ | $(-6)^2 + 4(-6) - 12 = 0$
$0 = 0$ $0 = 0$

$\therefore$ the solution set is $\{2, -6\}$.

The only unfamiliar step in this solution is the second, called **completing a trinomial square**. To apply the method of this example to any quadratic equation in the form $x^2 + bx = k$, we must be able to determine what to add to $x^2 + bx$ to produce a trinomial square.

Analyze the following trinomials, which are squares of binomials.

$(x + 7)^2 = x^2 + 2(7)x + 7^2$ $(x - n)^2 = x^2 + 2(-n)x + (-n)^2$

$(x - 5)^2 = x^2 + 2(-5)x + (-5)^2$ $\left(x + \dfrac{b}{2}\right)^2 = x^2 + 2\left(\dfrac{b}{2}\right)x + \left(\dfrac{b}{2}\right)^2$

Notice that in each case, the *constant term is the square of half the coefficient of the linear term*.

Example 2. What value of c makes $t^2 - \frac{3}{4}t + c$ a trinomial square?

Solution: Half the coefficient of the linear term is
$\frac{1}{2}(-\frac{3}{4}) = -\frac{3}{8}$.

$\therefore c = (-\frac{3}{8})^2 = \frac{9}{64}$.

Check: Is $t^2 - \frac{3}{4}t + \frac{9}{64}$ a trinomial square?

$t^2 - \frac{3}{4}t + \frac{9}{64} = (t - \frac{3}{8})(t - \frac{3}{8})$
$= (t - \frac{3}{8})^2$

$\therefore c = \frac{9}{64}$.

To solve an equation whose quadratic term has a coefficient other than 1, first use the multiplication property of equality (page 83) and multiply each term by the reciprocal of the coefficient of the quadratic term.

Example 3. Solve $2x^2 - 3x - 3 = 0$.

Solution:
$$2x^2 - 3x - 3 = 0$$
$$x^2 - \tfrac{3}{2}x - \tfrac{3}{2} = 0$$
$$x^2 - \tfrac{3}{2}x = \tfrac{3}{2}$$
$$x^2 - \tfrac{3}{2}x + \tfrac{9}{16} = \tfrac{3}{2} + \tfrac{9}{16}$$
$$(x - \tfrac{3}{4})^2 = \tfrac{33}{16}$$
$$x - \tfrac{3}{4} = \pm\sqrt{\tfrac{33}{16}}$$

$x - \frac{3}{4} = \sqrt{\frac{33}{16}}$ | $x - \frac{3}{4} = -\sqrt{\frac{33}{16}}$
$x = \frac{3}{4} + \frac{\sqrt{33}}{4}$ | $x = \frac{3}{4} - \frac{\sqrt{33}}{4}$
$x = \frac{3 + \sqrt{33}}{4}$ | $x = \frac{3 - \sqrt{33}}{4}$

Is the sum of the roots $-(-\frac{3}{2})$? | Is the product of the roots $-\frac{3}{2}$?

$\frac{3 + \sqrt{33}}{4} + \frac{3 - \sqrt{33}}{4} \stackrel{?}{=} -\left(\frac{-3}{2}\right)$ | $\left(\frac{3 + \sqrt{33}}{4}\right)\left(\frac{3 - \sqrt{33}}{4}\right) \stackrel{?}{=} -\frac{3}{2}$

$\frac{3}{2} = \frac{3}{2}$ | $\frac{9 - 33}{16} \stackrel{?}{=} -\frac{3}{2}$

$-\frac{3}{2} = -\frac{3}{2}$

$\therefore$ the solution set is $\left\{\frac{3 + \sqrt{33}}{4}, \frac{3 - \sqrt{33}}{4}\right\}$.

For computational purposes, decimal approximations of such roots (section 12-4) are often needed. To approximate them to the nearest tenth, use a two-decimal-place approximation of $\sqrt{33}$ from the table on page 359, and perform the indicated operations:

$$\frac{3 + \sqrt{33}}{4} \doteq \frac{3 + 5.75}{4} = \frac{8.75}{4} \doteq 2.19 \doteq 2.2$$

$$\frac{3 - \sqrt{33}}{4} \doteq \frac{3 - 5.75}{4} = \frac{-2.75}{4} \doteq -0.69 \doteq -0.7$$

∴ to the nearest tenth, the roots are 2.2 and −0.7.

EXERCISES

Solve by completing the square. Give irrational roots in radical form and also approximate them to the nearest tenth.

1. $n^2 + 2n = 24$
2. $t^2 - 4t = 21$
3. $y^2 - 8y = 4$
4. $z^2 - 4z = 2$
5. $x^2 + 2x - 1 = 0$
6. $a^2 + 2a - 7 = 0$
7. $t^2 - 5t = 0$
8. $r^2 + 3r = 0$
9. $s^2 - s - 3 = 0$
10. $u^2 - 5u + 3 = 0$
11. $2z^2 - 6z - 5 = 0$
12. $4p^2 + 4p - 3 = 0$
13. $2x^2 = x + 2$
14. $2y^2 = 3y + 1$
15. $3n^2 + n = 1$
16. $4p^2 - 2p = 1$
17. $x^2 - \frac{4}{3}x = 1$
18. $a^2 - \frac{5}{2}a = 1$
19. $3y^2 + y - \frac{1}{2} = 0$
20. $\frac{1}{3}n^2 + \frac{3}{2}n = 3$
21. $5z = \dfrac{1 - 3z}{z - 1}$
22. $3y = \dfrac{y + 2}{y - 1}$
23. $x - 1 = \dfrac{-2x}{x + 1}$
24. $2y + 2 = \dfrac{-y}{y + 2}$

Solve for x in terms of a, b, or c.

25. $x^2 - 2x + c = 0$
26. $x^2 - bx - 1 = 0$
27. $x^2 + bx + c = 0$
28. $ax^2 + bx + c = 0$

PROBLEMS

1. The dimensions of a sheet of paper can be represented by consecutive odd integers. Its area is 99 square inches. Find its dimensions.
2. A chalk board is two feet wider than it is high. Its area is 24 square feet. Find its length and height.
3. A large package of brownie mix made 154 brownies cut into 1-inch squares. If the pan was 3 inches longer than it was wide, find the dimensions of the pan.
4. A bookshelf is 5 times longer than it is deep. If the area of the surface of the bookshelf is 625 square inches, find the dimensions of the surface.
5. Two (square) checkerboards together have an area of 169 square inches. One has a side that is 7 inches longer than the other. Find the side of each.
6. To lay wall-to-wall carpeting in a living room and dining room takes 612 square feet of carpet. The living-room floor is 3 feet longer than it is wide. The dining-room floor is 6 feet wider than the width of the living room, and its length is twice the width of the living room. Find the dimensions of each room.
7. The members of a certain organization decided to donate 2000 dollars to charity, and to share the cost of the contribution equally. If the club had 20 more members, each would have contributed 5 dollars less. How many members now belong to the organization?
8. Each year a certain company divides 7200 dollars equally among its employees. This year the company employs 6 more people than last year, and therefore each employee received 60 dollars less. How much did each employee receive last year?
9. Harry and Bob live 13 miles apart. If Harry drives directly south, and Bob drives directly west, they will meet. If Bob will have driven 7 miles more than Harry when they meet, how far will each have driven?
10. The Boutique Fashion Shop bought some dresses for 240 dollars. The owners kept 2 of the dresses for themselves and sold the rest at a gain of 3 dollars each for a total profit of 30 dollars. How many did they sell?

13–4 Solution by Using the Quadratic Formula

For a given set of coefficients a, b, and c, $a \neq 0$, the quadratic equation $ax^2 + bx + c = 0$ can be transformed to express the variable x directly in terms of a, b, and c by completing the square. Study carefully the following parallel treatment of the standard quadratic equation of a special quadratic equation.

$$ax^2 + bx + c = 0 \qquad\qquad 2x^2 - 5x + 1 = 0$$

$$x^2 + \frac{b}{a}x + \frac{c}{a} = 0 \qquad\qquad x^2 - \tfrac{5}{2}x + \tfrac{1}{2} = 0$$

$$x^2 + \frac{b}{a}x = -\frac{c}{a} \qquad\qquad x^2 - \tfrac{5}{2}x = -\tfrac{1}{2}$$

$$x^2 + \frac{b}{a}x + \left(\frac{b}{2a}\right)^2 = -\frac{c}{a} + \left(\frac{b}{2a}\right)^2 \qquad\qquad x^2 - \tfrac{5}{2}x + (\tfrac{5}{4})^2 = -\tfrac{1}{2} + (\tfrac{5}{4})^2$$

$$\left(x + \frac{b}{2a}\right)^2 = -\frac{c}{a} + \frac{b^2}{4a^2} \qquad\qquad (x - \tfrac{5}{4})^2 = -\tfrac{1}{2} + \tfrac{25}{16}$$

$$\left(x + \frac{b}{2a}\right)^2 = \frac{b^2 - 4ac}{4a^2} \qquad\qquad (x - \tfrac{5}{4})^2 = \tfrac{17}{16}$$

$$x + \frac{b}{2a} = \pm\sqrt{\frac{b^2 - 4ac}{4a^2}}{}^* \qquad\qquad x - \tfrac{5}{4} = \pm\sqrt{\tfrac{17}{16}}$$

$$x = -\frac{b}{2a} \pm \sqrt{\frac{b^2 - 4ac}{4a^2}} \qquad\qquad x = \tfrac{5}{4} \pm \sqrt{\tfrac{17}{16}}$$

$$x = -\frac{b}{2a} \pm \frac{\sqrt{b^2 - 4ac}}{2a} \qquad\qquad x = \frac{5}{4} \pm \frac{\sqrt{17}}{4}$$

$$x = \frac{-b \pm \sqrt{b^2 - 4ac}}{2a} \qquad\qquad x = \frac{5 \pm \sqrt{17}}{4}$$

The last sentence in the preceding proof is actually a disjunction of two linear equations and is called the **quadratic formula**. If the root of either linear equation is taken as the value of x and substituted in the original quadratic equation, the resulting statement is "$0 = 0$." In developing the quadratic formula, notice the assumptions that $a \neq 0$ and that $\sqrt{b^2 - 4ac}$ is a real number ($b^2 - 4ac \geq 0$).

To solve any quadratic equation of the form

$$ax^2 + bx + c = 0,$$

substitute the coefficients in the quadratic formula,

$$x = \frac{-b \pm \sqrt{b^2 - 4ac}}{2a},$$

and evaluate.

*If $b^2 - 4ac \geq 0$.

Example. Solve $2x^2 - 5x + 1 = 0$ by using the quadratic formula.

Solution: $2x^2 - 5x + 1 = 0$

$$x = \frac{-b \pm \sqrt{b^2 - 4ac}}{2a}; \quad a = 2, \quad b = -5, \quad c = 1$$

$$x = \frac{-(-5) \pm \sqrt{(-5)^2 - 4(2)(1)}}{2(2)}$$

$$x = \frac{5 \pm \sqrt{25 - 8}}{4} = \frac{5 \pm \sqrt{17}}{4}$$

Check: Is the sum of the roots $-(-\frac{5}{2})$? Is the product of the roots $\frac{1}{2}$? Checking that each answer is "Yes" is left to you.

∴ the solution set is $\left\{ \dfrac{5 + \sqrt{17}}{4}, \dfrac{5 - \sqrt{17}}{4} \right\}$.

EXERCISES

Use the quadratic formula to solve each equation. Give irrational roots in simplest radical form and also approximate them to the nearest tenth.

1. $t^2 - 9t + 20 = 0$
2. $r^2 + 9r - 10 = 0$
3. $x^2 + 10x - 2 = 0$
4. $y^2 + 6y - 1 = 0$
5. $2n^2 + 4n + 1 = 0$
6. $3p^2 - 7p - 3 = 0$
7. $t^2 - 2t - 5 = 0$
8. $r^2 + 4r + 1 = 0$
9. $3y^2 = 4y + 2$
10. $3s = 1 - 2s^2$
11. $5x^2 + 2x = 2$
12. $5y - 2 = 3y^2$
13. $5t^2 = 5t$
14. $3k^2 = 14$

Factor each polynomial over $\mathcal{R}$.

Example. $x^2 - 2x - 2$

Solution: Use the fact that if r_1 and r_2 are roots of $x^2 - 2x - 2 = 0$, then $(x - r_1)$ and $(x - r_2)$ are factors of $x^2 - 2x - 2$.
Solve $x^2 - 2x - 2 = 0$.
By the quadratic formula,

$$x = \frac{2 \pm \sqrt{4 + 8}}{2} = \frac{2 \pm \sqrt{12}}{2} = \frac{2 \pm 2\sqrt{3}}{2} = 1 \pm \sqrt{3}.$$

∴ $x^2 - 2x - 2 = (x - 1 - \sqrt{3})(x - 1 + \sqrt{3})$, as we can check by multiplication.

15. $x^2 - 2x - 1$
16. $n^2 - 4n - 6$
17. $s^2 - 6s - 1$
18. $r^2 + 4r - 4$

Show that the given equation has no real roots.

19. $x^2 + 2x + 8 = 0$
20. $3x^2 - 2x + 5 = 0$
21. $\sqrt{z} = z + 3$
22. $3\sqrt{z} = z + 5$

PROBLEMS

Give irrational answers to the nearest tenth. Reject inappropriate roots.

1. A one-room apartment of 490 square feet has folding doors such that it can be divided into a square kitchen area and a rectangular living room area (see diagram). The ratio of the length of the living room to its width is $3:2$. Find the dimensions of each division.

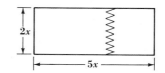

2. A square section of wood floor is 144 square inches in area. The section has 6 rectangular pieces, 4 outside pieces of equal area, and 2 smaller inside pieces of equal area (see diagram). The length of each outside piece is 3 times its width. Find the dimensions of the pieces.

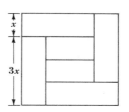

3. A square picture has a two-inch frame. If the area of the picture is $\frac{2}{3}$ of the total area, what are the dimensions of the frame?

4. A quilt is made from 150 small squares. It is possible to make a quilt of the same size using 100 squares 1 inch longer on a side. How long is each small square?

5. The perimeter of a triangle is 3 feet. Two sides form a right angle and are in the ratio of $3:4$. Find the lengths of all 3 sides.

6. A sailboat heads east from a dock at the same time that a motorboat heads north from the dock. The motorboat travels 7 miles per hour faster than the sailboat. The boats are 13 miles apart at the end of one hour. Find the rate of each boat.

7. A jet plane flying 3000 miles cross country ran into a strong tail wind that increased the plane's speed by 100 miles per hour and thus shortened its flying time by 1 hour. What was its ground speed?

8. A book club has a bonus of 16 dollars worth of books per year for each of its 1000 members. The bonus is increased by 1 cent for each new member. How many new members would increase the total bonus money by approximately 900 dollars?

9. A boat takes 1 hour longer to go 36 miles up a river than to return. If the river flows at 3 miles per hour, find the rate at which the boat travels in still water.

10. The average speed of a plane in still air is 425 miles per hour. With a constant wind blowing, the plane flies 900 miles into the wind and returns. The round trip takes $4\frac{1}{4}$ hours of flying time. Find the speed of the wind.

11. A workman earned $180 in a certain number of days. If his daily wage had been $2 less, he would have taken one more day to earn the same amount. Find how many days he worked at the higher rate.

12. One car starts traveling west along a road. At the same time, from the same point, another car starts traveling south at a speed 15 miles per hour faster than that of the first car. After one hour and twenty minutes the cars are 100 miles apart. At what speeds are they traveling?

★ 13. Draw a flow chart of a program to read and then print three positive integers a, b, and c, and to determine whether or not these numbers are the measures of the sides of a right triangle. The output message should be "right triangle" or "not a right triangle" as the case may be. (*Hint:* We have a right triangle if $a^2 + b^2 = c^2$, or $a^2 + c^2 = b^2$, or $b^2 + c^2 = a^2$.)

★ 14. Draw a flow chart of a program to read and print three numbers a, b, and c, and to determine and print the real roots of

$$ax^2 + bx + c = 0.$$

Assume that *not both a* and *b* are zero. (*Hint:* If $a = 0$, the equation is linear and has one root. If $a \neq 0$, the equation is quadratic and has either two or no real roots, depending on the value of $b^2 - 4ac$.)

13–5 The Nature of the Roots of a Quadratic Equation

In Figure 13–1, the parabola at the far left is the graph of the function $x \rightarrow y = x^2 + 12x + 35$, $x \in \mathcal{R}$. The figure also shows the graphs of the functions $x \rightarrow y = -x^2 + 4x - 4$ and $x \rightarrow y = x^2 - 12x + 40$, $x \in \mathcal{R}$. (Recall section 10–10.)

The y-coordinate of every point on the x-axis is 0. Therefore, by replacing y with "0" in the formula for any of the functions and solving the resulting quadratic equation, we can determine the abscissa of any point where the graph intersects the x-axis. Such an abscissa is called an **x-intercept**.

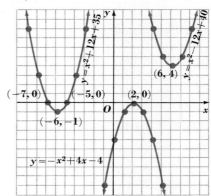

FIGURE 13–1

Case 1
$y = x^2 + 12x + 35$
$0 = x^2 + 12x + 35$
$x = \dfrac{-12 \pm \sqrt{144 - 140}}{2}$
$x = -6 \pm 1$
$x = -5$ or $x = -7$
∴ there are *two* x-intercepts, -5 and -7.

Case 2
$y = -x^2 + 4x - 4$
$0 = -x^2 + 4x - 4$
$0 = x^2 - 4x + 4$
$x = \dfrac{4 \pm \sqrt{16 - 16}}{2}$
$x = 2 \pm 0$
$x = 2$
∴ 2 is a double root (page 186) so that there is *one* x-intercept, 2.

Case 3
$y = x^2 - 12x + 40$
$0 = x^2 - 12x + 40$
$x = \dfrac{12 \pm \sqrt{144 - 160}}{2}$
$x = \dfrac{12 \pm \sqrt{-16}}{2}$
But $\sqrt{-16}$ is not a real number and so neither is $\dfrac{12 \pm \sqrt{-16}}{2}$.
∴ there is *no* x-intercept.

The three cases are analyzed in the following chart:

	Number of points in common with the x-axis	Number of different real roots of the equation	Value of $b^2 - 4ac$
Case 1	2	2	a positive number
Case 2	1	1 (a double root)	zero
Case 3	0	0	a negative number

Notice that the value of $b^2 - 4ac$ is the key to distinguishing these cases. If $b^2 - 4ac > 0$, then $\sqrt{b^2 - 4ac}$ is positive. Thus, "$ax^2 + bx + c = 0$" has two different roots, because

$$\dfrac{-b + \sqrt{b^2 - 4ac}}{2a} \neq \dfrac{-b - \sqrt{b^2 - 4ac}}{2a}.$$

But if $b^2 - 4ac = 0$, then $\sqrt{b^2 - 4ac} = 0$ and

$$\dfrac{-b + 0}{2a} = \dfrac{-b - 0}{2a} = \dfrac{-b}{2a},$$

so that there is only one root. But, for $b^2 - 4ac < 0$, no real root exists, because square roots of negative numbers do not exist in the real number system (page 315).

A quadratic equation with real coefficients can have
1. two different real roots ($b^2 - 4ac > 0$)
2. one (double) real root ($b^2 - 4ac = 0$), or
3. no real roots ($b^2 - 4ac < 0$).

Because the value of $b^2 - 4ac$ distinguishes the three cases, it is called the **discriminant** of the quadratic equation.

EXERCISES

Determine the nature of the roots of each equation (a) graphically and (b) by use of the discriminant.

1. $x^2 - 3x - 4 = 0$
2. $x^2 + 5x - 6 = 0$
3. $3x^2 + 5x + 3 = 0$
4. $-2x^2 - 4x + 6 = 0$
5. $-\frac{1}{3}x^2 - 3x + 4 = 0$
6. $\frac{1}{2}x^2 - 6x + 5 = 0$
7. $-x^2 - 2x - 2 = 0$
8. $-x^2 + 2x - 2 = 0$
9. $2x^2 - 3x + 2 = 0$
10. $3x^2 + 4x + 1 = 0$

Determine whether or not each polynomial can be factored over $\mathcal{R}$. If it can be factored, find the factors.

11. $x^2 + 4 - 5x$
12. $x^2 + 6x - 9$
13. $-3x^2 + 5x - 2$
14. $-2x^2 + 6x + 3$
15. $y^2 - \frac{1}{4}y + \frac{1}{3}$
16. $t^2 - \frac{1}{12}t - \frac{1}{12}$
17. $\frac{1}{9}y - \frac{1}{9}y^2 - \frac{2}{3}y^3$
18. $\frac{1}{6}n + \frac{1}{3}n^2 - \frac{5}{3}n^3$

PROBLEMS

Find the roots in the most efficient way, and reject inappropriate roots. Approximate irrational roots to the nearest tenth. Use the given formulas.

1. The motion of a freely falling body is described approximately by $h = vt - 16t^2$, where h represents the height above the ground, v the initial velocity of the body (positive if the body is propelled away from the earth and negative if propelled toward the earth), and t the elapsed time of fall. In how many seconds will an object thrown upward from the ground with an initial velocity of 64 feet per second first reach a height of 32 feet? When will it again be 32 feet above the ground? When will it strike the ground?

2. If an object is thrown downward with a velocity of 8 feet per second from the top of a building 250 feet high, in how many seconds will it strike the ground? (Use the formula in Problem 1 with $h = -250$.)

3. The sum of the first n positive integers is represented by $\dfrac{n(n+1)}{2}$. How many positive integers yield a sum of 78?

4. How many positive integers yield a sum of 171? (Use the formula in Problem 3.)

5. The formula for the surface area of a cube with edges of length s is $A = 6s^2$. What is the length of each edge of a cube with surface area 600 square inches?

6. The number of straight lines l determined by n points, no three of which lie on the same line, is given by $l = \dfrac{n(n-1)}{2}$. How many such points determine 21 lines?

The distance d between two points (x_1, y_1) and (x_2, y_2) is given by $d^2 = (x_2 - x_1)^2 + (y_2 - y_1)^2$.

7. Find the distance between the points (3, 2) and (5, 1).
8. Find the distance between the points $(-5, 1)$ and $(4, -6)$.
9. Find the abscissas of the two points on the line $y = 3$ that are 13 units from the point $(6, -2)$.
10. Find the ordinate of the two points on the line $x = 6$ that are 10 units from the point $(-2, -2)$.
11. Show that the points $(-5, 0)$, $(-3, 4)$, and $(5, 0)$ are the vertices of a right triangle.
12. Show that the points $(-1, -2)$, $(7, 2)$, and $(1, 4)$ are the vertices of an isosceles triangle.

13-6 Solving Quadratic Inequalities

To solve quadratic equations by factoring, use the zero-product property. To solve a **quadratic inequality**, use:

The Property of the Nonzero Product of Two Real Numbers

A product is greater than zero if and only if both factors are greater than zero or both are less than zero, and a product is less than zero if and only if one factor is greater than zero and the other is less than zero.

Example 1. Graph the solution set of $x^2 + x > 6$.

Solution:
$$x^2 + x > 6$$
$$x^2 + x - 6 > 0$$
$$(x + 3)(x - 2) > 0$$

Both factors are negative,   or both factors are positive.

$x + 3 < 0$ and $x - 2 < 0$	$x + 3 > 0$ and $x - 2 > 0$
$x < -3$ and $x < 2$	$x > -3$ and $x > 2$
The only numbers which satisfy *both* conditions satisfy $x < -3$, whose solution set is the intersection (page 111) of their solution sets:	The only numbers which satisfy *both* conditions satisfy $x > 2$, whose solution set is the intersection of their solution sets:

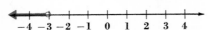   

∴ the graph of the solution set is the union (page 111) of these graphs:

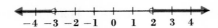

Example 2. Graph the solution set of $x^2 + x \leq 6$.

Solution:
$$x^2 + x - 6 \leq 0$$
$$(x + 3)(x - 2) \leq 0$$

$x + 3 \geq 0$ and $x - 2 \leq 0$ or	$x + 3 \leq 0$ and $x - 2 \geq 0$
$x \geq -3$ and $x \leq 2$	$x \leq -3$ and $x \geq 2$
The numbers satisfying *both* conditions satisfy	The intersection of their solution sets is the empty set.

$$-3 \leq x \leq 2,$$

whose solution set is the intersection of their solution sets:

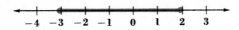

∴ the graph of the solution set is the union of these graphs:

EXERCISES

Graph the solution set of each inequality.

1. $x^2 - 3x - 4 < 0$
2. $x^2 - 3x - 4 > 0$
3. $t^2 + 5t + 6 > 0$
4. $t^2 + 5t + 6 < 0$
5. $x^2 - x < 2$
6. $x^2 - x > 2$
7. $2y^2 - 3y < 2$
8. $3z^2 > 2 - z$
9. $a^2 \leq 4a$
10. $b^2 \geq 5b$
11. $n^2 \geq 4$
12. $m^2 \leq 16$
13. $t^2 - 2t + 1 > 0$
14. $r^2 + 4r + 4 \leq 0$
15. $n^2 + 8n + 16 \leq 0$
16. $x^2 - 6x + 9 \geq 0$
17. $y^2 > 0$
18. $z^2 < 0$

Find the values of x for which each expression represents a real number.

19. $\sqrt{2x^2 + x}$
20. $\sqrt{3x^2 - 27}$
21. $\sqrt{x^2 - 3x - 28}$
22. $\sqrt{x^2 + 12x + 27}$

13-7 Using Graphs of Equations to Solve Inequalities

Figure 13-2 below shows the graph of "$y = x^2 + x - 6$." The abscissas of points at which $y = 0$ form the solution set of "$x^2 + x - 6 = 0$." The values of x for which $y > 0$ give the solution set of "$x^2 + x - 6 > 0$," and the values of x for which $y < 0$ give the solution set of "$x^2 + x - 6 < 0$."

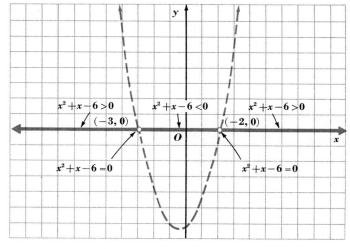

FIGURE 13-2

Notice that we can find the solution sets of these open sentences by determining the values of x for which the graph of the equation is on, above, or below the x-axis.

Compare the solution above with the solutions of Examples 1 and 2 in section 13–6.

Figure 13–3 shows the graph of $y = x^2 - 4x + 4$.

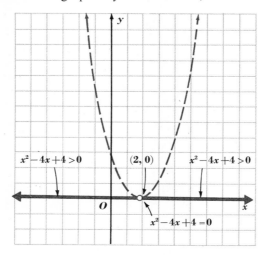

FIGURE 13–3

How many x-intercepts does it have? Does the point $(2, 0)$ satisfy $y = 0$, $y > 0$, or $y < 0$? Which sentence is satisfied by values of x to the left of $(2, 0)$? to the right of $(2, 0)$? Notice that no value of x will make $y < 0$, since the graph does not go below the x-axis.

In general, the values of x for which the graph of "$y = ax^2 + bx + c$" lies *above* the x-axis form the solution set of "$ax^2 + bx + c > 0$," while the x-coordinates of the points on the graph *below* the x-axis form the solution set of "$ax^2 + bx + c < 0$."

EXERCISES

Graph each equation and mark the section of the x-axis it determines as solution sets for $y = 0$, $y > 0$, and $y < 0$.

1. $x^2 - 4 = y$
2. $x^2 - 1 = y$
3. $x^2 + 2x = y$
4. $x^2 - 4x = y$
5. $x^2 - 3x - 4 = y$
6. $x^2 + 3x - 4 = y$
7. $x^2 - 4x + 3 = y$
8. $x^2 + 4x + 3 = y$
9. $4x^2 - 8x + 3 = y$
10. $2x^2 + x - 3 = y$
11. $4 - x^2 = y$
12. $3x - x^2 = y$

13. $x^2 + 2x + 1 = y$
14. $x^2 - 4x + 4 = y$
15. $9 + 6x + x^2 = y$
16. $9 - 6x + x^2 = y$

Find the values of x for which each expression represents a real number.

17. $\sqrt{x^2 - 3x - 4}$
18. $\sqrt{x^2 + 2x}$

chapter summary

1. A quadratic equation containing no linear term can be solved by using the **Property of Square Roots of Equal Numbers**: If r and s are real numbers, $r^2 = s^2$ if and only if $r = s$ or $r = -s$.

2. In a quadratic equation of the form "$x^2 + bx + c = 0$," the **sum of the roots** is equal to $-b$, the opposite of the coefficient of the linear term, and **the product of the roots** is equal to the constant term c. These facts may be used to check the solution set of a quadratic equation.

3. To solve a quadratic equation in one variable by the method of **completing the square**: transform it into an equivalent equation whose quadratic term has coefficient 1; write it in the form "$x^2 + bx = -c$"; add to each member $\left(\dfrac{b}{2}\right)^2$, making the left member a square; apply the property of square roots of equal numbers; solve the resulting linear equations.

4. **A decimal approximation** of an irrational root can be calculated by substituting a decimal approximation for each radical which is accurate to one more place than desired in the root and performing the operations.

5. The roots of the **standard quadratic equation** "$ax^2 + bx + c = 0$" are given by the **quadratic formula**: $x = \dfrac{-b \pm \sqrt{b^2 - 4ac}}{2a}$.

6. To solve a quadratic equation in one variable graphically, write it in the standard form "$ax^2 + bx + c = 0$," and find ordered pairs (x, y) satisfying "$ax^2 + bx + c = y$." Graph the parabola and determine the abscissas of the points of intersection of the parabola with the x-axis.

7. To solve a quadratic inequality in one variable, first transform it into an inequality whose right member is zero, and then factor the left member L. (If $L > 0$, the factors must be both positive or both negative; if $L < 0$, the factors must be opposite in sign.) Solve the resulting linear inequalities.

8. The solution set of "$ax^2 + bx + c > 0$" is the set of values of x for which the graph of "$y = ax^2 + bx + c$" lies above the x-axis. The solution set of "$ax^2 + bx + c < 0$" is the set of values of x for which the graph of "$y = ax^2 + bx + c$" lies below the x-axis.

CHAPTER REVIEW

13-1 Solve, using the property of square roots of equal numbers.

1. $2x^2 - 50 = 0$
2. $3y^2 = \frac{1}{27}$
3. $(x - \frac{3}{5})^2 = \frac{9}{25}$
4. $5(z + 3)^2 = 125$

13-2 Is the given set the solution set of the given equation?

5. $x^2 + 4x - 3$; $\{-2 \pm \sqrt{7}\}$
6. $x^2 + 7 = 0$; $\{\sqrt{7}, -\sqrt{7}\}$

13-3 7. Find the value of c that will make the left member of

$$x^2 - 0.2x + c = 0$$

the square of a binomial, and write an equivalent equation using the square of a binomial.

8. Transform "$y^2 + 5y + 6 = 0$" into an equivalent equation of the form $(y - a)^2 = b$.

Solve by completing the square. Express each root in simplest radical form and also to the nearest tenth.

9. $r^2 + 6r + 4 = 0$
10. $s^2 + 4s = 7$

13-4 11. Use the quadratic formula to solve "$x^2 + 6x + 7 = 0$." Express each root in simplest radical form and also to the nearest tenth.

12. A path of uniform width was constructed around the border of a rectangular flower garden measuring 40 feet by 60 feet. If the path added an area of 438 square feet to the garden, find the width of the path.

13-5 Determine the nature of the roots of the given equation.

13. $y^2 + 4y = 5$
14. $3t^2 - 4t = -2$

13-6 15. Graph the solution set of "$s^2 + s \leq 2$."

13-7 16. Graph "$x^2 - 2x - 8 = y$" and, from the graph, find the solution set of "$x^2 - 2x < 8$."

TABLE 1   FORMULAS

Circle	$A = \pi r^2, C = 2\pi r$	Cube	$V = s^3$
Parallelogram	$A = bh$	Rectangular Box	$V = lwh$
Right Triangle	$A = \frac{1}{2}bh, c^2 = a^2 + b^2$	Cylinder	$V = \pi r^2 h$
Square	$A = s^2$	Pyramid	$V = \frac{1}{3}Bh$
Trapezoid	$A = \frac{1}{2}h(b + b')$	Cone	$V = \frac{1}{3}\pi r^2 h$
Triangle	$A = \frac{1}{2}bh$	Sphere	$V = \frac{4}{3}\pi r^3$
Sphere	$A = 4\pi r^2$		

TABLE 2   WEIGHTS AND MEASURES

AMERICAN SYSTEM OF WEIGHTS AND MEASURES

LENGTH
- 12 inches = 1 foot
- 3 feet = 1 yard
- $5\frac{1}{2}$ yards = 1 rod
- 5280 feet = 1 land mile
- 6076 feet = 1 nautical mile

AREA
- 144 square inches = 1 square foot
- 9 square feet = 1 square yard
- 160 square rods = 1 acre
- 640 acres = 1 square mile

VOLUME
- 1728 cubic inches = 1 cubic foot
- 27 cubic feet = 1 cubic yard

WEIGHT
- 16 ounces = 1 pound
- 2000 pounds = 1 ton
- 2240 pounds = 1 long ton

CAPACITY

Dry Measure
- 2 pints = 1 quart
- 8 quarts = 1 peck
- 4 pecks = 1 bushel

Liquid Measure
- 16 fluid ounces = 1 pint
- 2 pints = 1 quart
- 4 quarts = 1 gallon
- 231 cubic inches = 1 gallon

METRIC SYSTEM OF WEIGHTS AND MEASURES

LENGTH	10 millimeters (mm) = 1 centimeter (cm)	$\doteq$ 0.3937	inch
	100 centimeters = 1 meter (m)	$\doteq$ 39.37	inches
	1000 meters = 1 kilometer (km)	$\doteq$ 0.6	mile
CAPACITY	1000 milliliters (ml) = 1 liter (l)	$\doteq$ 1.1	quart
	1000 liters (l) = 1 kiloliter (kl)	$\doteq$ 264.2	gallons
WEIGHT	1000 milligrams (mg) = 1 gram (g)	$\doteq$ 0.035	ounce
	1000 grams = 1 kilogram (kg)	$\doteq$ 2.2	pounds

TABLE 3   SQUARES OF INTEGERS FROM 1 TO 100

Number	Square	Number	Square	Number	Square	Number	Square
1	1	26	676	51	2601	76	5776
2	4	27	729	52	2704	77	5929
3	9	28	784	53	2809	78	6084
4	16	29	841	54	2916	79	6241
5	25	30	900	55	3025	80	6400
6	36	31	961	56	3136	81	6561
7	49	32	1024	57	3249	82	6724
8	64	33	1089	58	3364	83	6889
9	81	34	1156	59	3481	84	7056
10	100	35	1225	60	3600	85	7225
11	121	36	1296	61	3721	86	7396
12	144	37	1369	62	3844	87	7569
13	169	38	1444	63	3969	88	7744
14	196	39	1521	64	4096	89	7921
15	225	40	1600	65	4225	90	8100
16	256	41	1681	66	4356	91	8281
17	289	42	1764	67	4489	92	8464
18	324	43	1849	68	4624	93	8649
19	361	44	1936	69	4761	94	8836
20	400	45	2025	70	4900	95	9025
21	441	46	2116	71	5041	96	9216
22	484	47	2209	72	5184	97	9409
23	529	48	2304	73	5329	98	9604
24	576	49	2401	74	5476	99	9801
25	625	50	2500	75	5625	100	10,000

TABLE 4 SQUARE ROOTS OF INTEGERS FROM 1 TO 100

Exact square roots are shown in color. For the others, rational approximations are given correct to three decimal places.

Number N	Positive Square Root $\sqrt{N}$	Number N	Positive Square Root $\sqrt{N}$	Number N	Positive Square Root $\sqrt{N}$	Number N	Positive Square Root $\sqrt{N}$
1	1	26	5.099	51	7.141	76	8.718
2	1.414	27	5.196	52	7.211	77	8.775
3	1.732	28	5.292	53	7.280	78	8.832
4	2	29	5.385	54	7.348	79	8.888
5	2.236	30	5.477	55	7.416	80	8.944
6	2.449	31	5.568	56	7.483	81	9
7	2.646	32	5.657	57	7.550	82	9.055
8	2.828	33	5.745	58	7.616	83	9.110
9	3	34	5.831	59	7.681	84	9.165
10	3.162	35	5.916	60	7.746	85	9.220
11	3.317	36	6	61	7.810	86	9.274
12	3.464	37	6.083	62	7.874	87	9.327
13	3.606	38	6.164	63	7.937	88	9.381
14	3.742	39	6.245	64	8	89	9.434
15	3.873	40	6.325	65	8.062	90	9.487
16	4	41	6.403	66	8.124	91	9.539
17	4.123	42	6.481	67	8.185	92	9.592
18	4.243	43	6.557	68	8.246	93	9.644
19	4.359	44	6.633	69	8.307	94	9.695
20	4.472	45	6.708	70	8.367	95	9.747
21	4.583	46	6.782	71	8.426	96	9.798
22	4.690	47	6.856	72	8.485	97	9.849
23	4.796	48	6.928	73	8.544	98	9.899
24	4.899	49	7	74	8.602	99	9.950
25	5	50	7.071	75	8.660	100	10

answers for odd-numbered exercises

Page 2 1. = 3. ≠ 5. = 7. = 9. = 11. = 13. 7 15. Any num. except 4 17. 0 19. 0 21. 7 23. 8 25. 14 27. $\frac{1}{2}$ 29. 0.6 31. 1 33. 0

Page 5 1. 29 3. 0 5. 65 7. 11 9. $\frac{6}{5}$ 11. 8 13. = 15. = 17. ≠ 19. = 21. ≠ 23. ≠ 25. = 27. = 29. =

Page 7 1–7.
9. 8 11. 0 13. ⁻3 15. $\frac{1}{2}$ 17. 0 19. $3\frac{1}{2}$ 21. 0.5 23. ⁻5 25. ⁻$1\frac{1}{2}$ 27. 1 29. 1

Page 10 1. < 3. > 5. < 7. > 9. = 11. < 13. > 15. < 17. < 19. < 21. 0 23. 0 25. Any no. < 13 27. 2 29. Any no. > 5 31. Any no. 33. $\frac{8}{15}$ 35. Any no. < $\frac{9}{2}$ 37. 12 39. Any no. > 4 and < 6 41. 1 × 1 = 1 43. Any no. > $\frac{25}{8}$ and < $\frac{27}{8}$ 45. Any no. > $\frac{2}{5}$ and < $\frac{3}{5}$ 47. Any no. > 1 and < 1.2 49. Any no. > 3.14159 and < 3.15159 51. Any no. > 0 and < 1

Page 13 1. {Truman, Johnson} 3. {January, February, March, April, May} 5. {Dolciani, Sorgenfrey} 7. {The four "corner" states of the continental U.S.} 9. {the two most populated cities in Texas} 11. Answers will vary.

13.
15.

17.
19.
21.
23.
25.
27.
29.

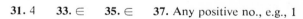

31. 4 33. ∈ 35. ∈ 37. Any positive no., e.g., 1

Page 17 1. {6, 5, 4, 3, 2, 1}; finite 3. {13, 14, 15, ...}; infinite 5. {7, 9, 11 ... 39}; finite 7. ∅; finite 9. {the integers between 21 and 29, inclusive} 11. {the positive odd integers less than 10} 13. {the nonnegative integers that are multiples of 5}

15.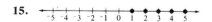
17.

19. {3, 5, 7} 21. {1} 23. {4, 6, 8, 9, 10, 11, ...}

Page 18 1. a. ≠ b. = 3. 12 5. a. < b. > c. < 7. {The three Pacific coast states of the continental U.S.} 9. a. equal b. not equal

Page 21 1. 6 3. 6 5. 0 7. 18 9. 36 11. 72 13. 0 15. 24 17. 3 19. 6 21. 24 23. 8

Page 22 (Answers to Problems involving π are approximations.) **1.** 5040 mi. **3.** 744 ft. **5.** 24 in. **7.** 93.75 sq. cm. **9.** 254 yd. **11.** 616 sq. in. **13.** 22.28 ft. **15.** 515.625 cu. in. **17.** 94.20 ft. per min. **19.** 3518 **21.** 8.7 cm.

Page 25 **1.** x^3 **3.** z^2 **5.** $10xy^2$ **7.** $23u^3(w+3)$ **9.** $(x-2)^3$ **11.** $(x+8)^3$ **13.** $5(y+z)^3$ **15.** $(a+b)^4$ **17.** 36 **19.** 54 **21.** $\frac{2}{9}$ **23.** 36 **25.** 27 **27.** 25 **29.** $\frac{7}{5}$ **31.** 3 **33.** 37 **35.** 0

Page 26 (Answers to Problems involving π or g are approximations.) **1.** 54 cu. ft. **3.** 2552 sq. in. **5.** 616 sq. cm. **7.** 753.60 cu. in. **9.** 373.66 sq. ft. **11.** 12.8 watts **13.** 907,500 ft-lb. **15.** 3.27 ft.

Page 29 **1.** 3 **3.** 18 **5.** 9 **7.** 2 **9.** 1 **11.** 9 **13.** 600 **15.** 6 **17.** 1 **19.** 1 **21.** 3 **23.** 6 **25.** 6 **27.** $^-2$ **29.** 25 **31.** 11 **33.** 3 **35.** 1 **37.** 3 **39.** $\frac{42}{13}$ **41.** 0

Page 31 **1.** True only for 4; {4} **3.** True only for 5; {5} **5.** True for 2, 3; {2, 3} **7.** True only for 3; {3} **9.** True for 1, 2, 3; {1, 2, 3} **11.** True for 1, 2; {1, 2} **13.** {6} **15.** {3} **17.** {6} **19.** {5} **21.** {2}

23.

25.

27.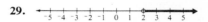

29.

31. {the real nos.}   **33.**

35. {2} **37.** True for $3+4=7$; $4+3=7$ **39.** True for $5+2(3)=11$ **41.** True for $3 < 2(2)$ **43.** True for $1+(2)(1)=1(2+1)$ **45.** True for $3(5)-2(3)=9$ **47.** True for $3(5)+2(6)>26$; $3(6)+2(5)>26$; $3(6)+2(6)>26$ **49.** True for all except $0+3(0) \neq 2(0)+0$

Page 34 **1.** $t=3$ **3.** Any integer y such that $y > 2$ **5.** Any integer x such that $x \neq 1$ **7.** $m=0$ **9.** $b=5a$

Page 36 **1.** $k+8=53$ **3.** $g+(g+1)=19$ **5.** $s+s+(s-8)=35$ **7.** $t+2=2(t-7)$ **9.** $10d+5(d+11)>250$

Page 39 **1.** if $y=5$; if $y \neq 5$ **3.** 6

Page 40 **1.** 12 **3.** $8\frac{1}{2}$ **5.** $\{t, 1\}$ **7.** $\{x, y, \frac{1}{2}, \frac{1}{2}x, \frac{1}{2}y, xy, 1\}$ **9.** $\frac{1}{8}$ **11.** $\frac{9}{2}$ **13.** 16 **15.** 8 **17.** 4 **19.** {2} **21.** F **23.** F **25.** $l=4w+2$

Page 43 **1.** Not closed under add. or subt.; closed under mult. and div. **3.** Closed under add., subt., and mult.; not defined for div. **5.** Not closed under add. or subt.; closed under mult. and div. **7.** Not closed under add., subt., mult., or div. **9.** Closed under add. and mult.; not closed under subt. or div. **11.** Closed under add., mult., and div.; not closed under subt. **13.** Not closed under subt. or div.; closed under add. and mult. **15.** symmetric prop. **17.** closure for add. **19.** reflexive prop. **21.** transitive prop. **23.** transitive prop.

Page 46 **1.** 540 **3.** 3400 **5.** 15 **7.** 11 **9.** 24 **11.** {the real nos.} **13.** {6} **15.** ∅ **17. a.** 8 **b.** closed **c.** * is comm. and assoc. **19. a.** $^-3$ **b.** not closed **c.** * is neither assoc. nor comm.

ANSWERS FOR ODD-NUMBERED EXERCISES

Page 49 1. a. $^-8$ b. $^-4 + (3 + ^-7) = ^-4 + ^-4 = ^-8$ 3. a. $^-6$ b. $12 + (^-24 + 6) = 12 + ^-18 = ^-6$ 5. a. 60 b. $38 + ^-8 + 30 = 30 + 30 = 60$ 7. a. 0.6 b. $(^-2.3 + 1.3) + 1.6 = ^-1 + 1.6 = 0.6$ 9. a. 3 b. $\frac{7}{2} + (^-\frac{1}{2} + 0) = \frac{7}{2} + ^-\frac{1}{2} = 3$ 11. a. $^-\frac{2}{3}$ b. $^-2 + (5 + ^-3\frac{2}{3}) = ^-2 + 1\frac{1}{3} = ^-\frac{2}{3}$ 13. a. $^-12$ b. $^-2 + [(3 + ^-5) + ^-8] = ^-2 + [^-2 + ^-8] = ^-2 + ^-10 = ^-12$ 15. $\{^-2\}$ 17. $\{2\}$ 19. $\{0\}$ 21. $\{7\}$ 23. $\{12\}$ 25. $\{^-24\}$ 27. $\{^-6\}$ 29. $\{0\}$ 31. $\{^-2, ^-1, 0, 1, 2\}$; A is not closed under add. because 2 and $^-2$ are not elements of A.

Page 50 1. a. $42 + ^-57$ b. $^-15$ c. 15 mi. south of A 3. a. $27,000 + ^-8000 + 3500$ b. 22,500 c. 22,500 ft. 5. a. $^-2.14 + ^-3.05 + ^-3.40 + ^-2.85 + 2.25 + 2.85 + 3.15 + 2.75$ b. $^-0.45$ c. He lost 45¢. 7. a. $650 + ^-75 + ^-50 + 350 + ^-125$ b. 750 c. 750 shares 9. a. $142 + 17 + ^-22 + 8$ b. 145 c. 3 points higher

Page 53 1. 10 3. -5 5. -14 7. 1 9. -7.3 11. 3 13. -2 15. $-\frac{1}{2}$ 17. a. True for -2 only b. $\{-2\}$ 19. a. True for 0 only b. $\{0\}$ 21. a. All false b. $\emptyset$ 23. a. True for -2 only b. $\{-2\}$ 25. a. True for -2 only b. $\{-2\}$ 27. a. All false b. $\emptyset$ 29. a. True for -2 b. $\{-2\}$ 31. a. True for -1 and 0 b. $\{-1, 0\}$ 33. a. True for 1 b. $\{1\}$ 35. $3\frac{1}{4}$ 37. $-1\frac{3}{4}$ 39. $\frac{1}{2}$

Page 55 1. 30 3. 12 5. 1 7. -2 9. 13 11. -8 13. $\{2, -2\}$ 15. $\{5, -5\}$ 17. $\{5, -5\}$

19.

21.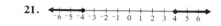

23.

25.

27. The statement is true. If a is a positive no. or zero, then $|a| = a$. If a is negative no., then $|a|$ equals the opposite of a or $-a$.

Page 58 1. 8 3. 1 5. -110 7. -5 9. -7 11. 20 13. -25 15. 6.7 17. 11 19. -3 21. -10 23. -3 25. $-\frac{7}{5}$

Page 59 1. On the second floor 3. 2 yd. 5. 10 fewer patients 7. 39 ft. below sea level 9. $-2, -4, 2, -4, 8$

Page 62 1. $47x$ 3. $82b$ 5. $52z$ 7. $24a^2 + 14$ 9. $7m + 9k$ 11. $4m + 24$ 13. $4r + 5s$ 15. $7c + 9d$ 17. $9h^2 + 2h + 6$ 19. $9ab + 5a$ 21. $20x + 17y$ 23. $15a + 39$ 25. $15x^2 + 3x + 24$ 27. $13a + 6b + 27$ 29. $52t^2 + 45t + 20$ 31. $9a + 11b$ 33. $y^2 + y + 3$ 35. $45y + 44$ 37. $120p + 164q + 234$ 39. 5400 41. 65 43. 35 45. 628,628 47. 370

Page 62 49.

Start → Read a, b, c → Compute the value of $ab+ac$ → Write the value of $ab+ac$ → End

Page 66 1. -36 3. -9 5. 0 7. -50 9. 21 11. 0 13. -350 15. -17 17. -18 19. -1.7 21. $x + 3y$ 23. $5m - 5n$ 25. $2t - 4$ 27. $-3xy + 5yz$ 29. $-4x - 2$ 31. $2t^2 + 3t$ 33. $0.6x + 0.4y$ 35. $8u^2v - 5uv^2$ 37. $-4u + v$ 39. $a - 9b$ 41. $y + 7z$ 43. $2x - 26y$ 45. 4 47. 0 49. -11 51. 2 53. 0.624 55. -0.648 57. 0.4 59. -0.512

Page 68 1. 7 3. -17 5. 3 7. -3 9. $-ab$ 11. $3b^2$ 13. $-3k^3$ 15. $5y$ 17. -5 19. $5x + 4$ 21. $5c - 14d$ 23. $9mk + k$ 25. $-a^2 + b^2$ 27. $-a + 4b$ 29. $2s$ 31. $-8a^2$ 33. a

Page 71 **1.** 3, 6, 18 **3.** 650, $\frac{1}{2}$, 325 **5.** 50 **7.** 8, 11 **9.** 2 **11.** $ax + 1 = b$ has no positive integer root. **13.** c **15.** b **17.** c

19. **21.** **23.**

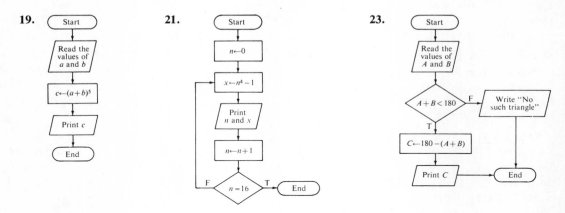

Page 74 **1.** Ax. of closure for mult. **3.** Symmetric prop. of equality **5.** Comm. ax. of add.

7. **9.** $-8\frac{3}{4}$ **11.** -20 **13.** 1 **15.** 27 **17.** 2; origin **19.** -2

21. $19y$ **23.** identity **25.** 1 **27.** -15 **29.** reciprocal

Page 79 **1.** {37} **3.** {72} **5.** {0} **7.** {50} **9.** {-17} **11.** {50} **13.** {6.4} **15.** {$\frac{16}{5}$}
17. {$\frac{1}{2}$} **19.** {-21} **21.** {0} **23.** {-5} **25.** {-1} **27.** {9} **29.** {-22} **31.** {10}
33. {$-\frac{1}{2}$} **35.** {5, -5} **37.** {3, -3}

Page 82 **1.** -6 **3.** 247 **5.** 10 **7.** {89} **9.** {-41} **11.** {0} **Page 82** **11.** 4; 3; 1
13. {-3} **15.** {7} **17.** {-1} **19.** $-3 - 8 = -11$ **21.** $x - (x - 5) = 5$ **23.** $(3\pi - 2) - (3\pi + 2) = -4$ **25.** $(-8 + 17) - (-6) = 15$ **27.** not closed; $7 - 10 = -3$ **29.** closed
Problems **1.** 1014 ft. **3.** 4 hr. 12 min. **5.** 69 yr. **7.** 67° **9.** 227 yd.

Page 85 **1.** {136} **3.** {-135} **5.** {32} **7.** {-4} **9.** {-65}
11. {-40.7} **13.** {-3} **15.** {-12} **17.** {1} **19.** {13}
21. {-23} **23.** {0} **25.** {30} **27.** {1} **29.** {33} **31.** $\emptyset$

Page 88 **1.** -5 **3.** 0 **5.** 51 **7.** -0.3 **9.** -20 **11.** -0.29
13. 0 **15.** $-21x$ **17.** {20} **19.** {$13\frac{1}{2}$} **21.** {-6} **23.** {-1.5}
25. -10 **27.** 0 **29.** 1 **31.** -3 **33.** -9 **35.** 9 **37.** 5 **39.** 1 **41.** 5
43. 1 **45.** closed **47.** closed **49.** not closed

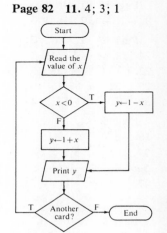

Page 91 **1.** {5} **3.** {18} **5.** {−9} **7.** {−1} **9.** {2} **11.** {1} **13.** {8} **15.** {20} **17.** {10} **19.** {9} **21.** {8} **23.** {−4} **25.** {−3} **27.** {−3} **29.** {2} **31.** {3} **33.** {−16} **35.** {−$\frac{3}{2}$} **37.** {−13} **39.** {1} **41.** {39.6} **43.** {−3} **45.** {$\frac{1}{3}$} **47.** {5} **49.** {8, −8}

Page 93 **1.** 13 yd. **3.** 2549.5 ft. **5.** *News*, 20 cents; *Times*, 50 cents **7.** 353 people **9.** 25 g. **11.** 42 ft.; 51 ft. **13.** 12 **15.** 18 **17.** 38 **19.** 36 ft. **21.** 15 ten-cent stamps, 45 four-cent stamps, 60 six-cents stamps **23.** I, 8000; II, 16,000; III, 64,000

25.

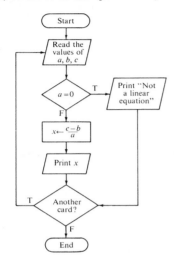

Page 99 25.

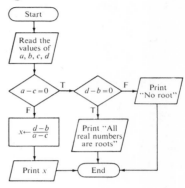

Page 97 **1.** {7} **3.** {6} **5.** {−6} **7.** {7} **9.** {14} **11.** {−65} **13.** {−6} **15.** {15} **17.** {−3} **19.** {3} **21.** {7} **23.** {8} **25.** {5$\frac{3}{5}$} **27.** {$\frac{8}{3}$} **29.** $w = \frac{P-1}{2}$ **31.** $u = 3w - 2t$ **33.** $x = \frac{1-3y}{2}$ **35.** $y = 3M - x - z$ **37.** {2} **39.** {−5} **41.** $h = \frac{T - 2\pi r^2}{2\pi r}$ **43.** $r = 1 - \frac{a}{S}$ **45.** {−5} **47.** identity **49.** ∅ **51.** {−64} **53.** $x = \frac{7b}{4}$ **55.** {1} **57.** {6} **59.** {−$\frac{3}{2}$} **61.** {−17$\frac{1}{2}$}

Page 99 **1.** 4 **3.** −15 **5.** 4; 13 **7.** −5; 20 **9.** 9 yr.; 36 yr. **11.** silver nitrate, 3; sucrose, 11 **13.** 5 m; 7 m **15.** 11 in.; 6 in. **17.** 90 mg. **19.** Monument, 565 ft.; arch, 650 ft. **21.** 78 ft.; 27 ft. **23.** 170 mi. **25.** See right, above

Page 103 **1.** $f: x \rightarrow 4x - 3$ **3.** {2, 3, 4} **5.** {−1, 5, 7} **7.** {5, 3, 9} **9.** {4, 11, 20} **11.** {2, 3, 4} **13.** {0, 35, 80} **15.** {51, 102, 123} **17.** 4 **19.** 16 **21.** 4 **23.** 96 **25.** 17 **27.** 7 **29.** −3; 3; 3 **31.** −1; 0; 1 **33.** 3 **35.** 5

Page 105 **1.** {39} **3.** {2} **5.** {−48} **7.** {3} **9.** {−7} **11.** {$\frac{1}{18}$} **13.** {3} **15.** {−24} **17.** {−3} **19.** {−3} **21.** $1.35; $2.04 **23.** {−1} **25.** $R = \{4, 1, 0\}$

Page 109

1. $x < -60$
3. $z \geq -7$;
5. $z > 6$;
7. $x \leq -10$;
9. $n < 5$;
11. $y > 0$;
13. $y \leq -3$;
15. $a < 5$;
17. $x \geq 1$;
19. $z \leq 1$;
21. $n \geq 4$;
23. $t < 7$ 25. $x \geq 33$ 27. $z \leq 5$ 29. $\emptyset$ 31. $\Re$ 33. $x < \dfrac{11a}{6}$ 35. $t \leq -44c$

Page 112   1. $\{-5\}$; $\{0\}$; $\{10\}$ 3. $\emptyset$ 5. $\{3, 5\}$; $\{-1, 3, 4, 5, 7\}$ 7. $\{-3, 4\}$; $\{-3, 0, 3, 4\}$
9. $\emptyset$, disjoint; $\{0, 1, 2, 3, 4, 5\}$ 11. $\{2\}$; $\{1, 2, 3, 4, 6, 8, \ldots\}$ 13. $\emptyset$, disjoint; {the real nos. except 0}

15. a. b. c. $A \cap B$ d. $A \cup B$

17. a. b. c. $A \cap B$ d. $A \cup B$

19. a. b. c. $A \cap B$ d. $A \cup B$

21. a. b. c. $A \cap B$ d. $A \cup B$

23. a. b. c. $A \cap B$ d. $A \cup B$

25. a. b. c. $A \cap B$ d. $A \cup B$

27. 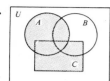   29. 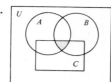   31. 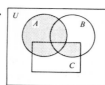   33.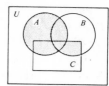

35. $S = \{3, 6\}$ 37. $S = \{3, 6\}$

Page 115

1. $-5 \leq x < 1$;

3. $-3 \leq a \leq 2$;

5. $0 < t < 4$;
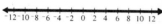

7. $y > 3$ or $y < -7$;
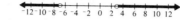

9. $r \leq 0$ or $r \geq 1$;

11. $-1 \leq p \leq 3$;

13. $c \leq -2$ or $c > 3$;

15. $1 < z \leq 4$;

17. $\Re$;
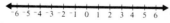

19. $\Re$;

21. $m < -1$;
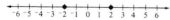

23. $z \geq -6$;

25. $\emptyset$; no graph

27. $-4 < x < 1$;

Page 117

1. $\{-2, 6\}$;

3. $a \leq -3$ or $a \geq 3$;

5. $1 < t < 11$;

7. $1 \leq y \leq 3$;

9. $\{-\frac{1}{3}, 1\}$;

11. $4 \leq v \leq 5$;

13. $\Re$;

15. $0 < p < 7$;

17. $\{2, -2\}$;

19. $\{6, -4\}$;

21. $-1 < y < 2$;

23. $s \geq 3$ or $s \leq -3$;

25. $\Re$;

27. $\emptyset$; no graph

29. $t > 2$ or $t < \frac{1}{4}$;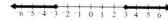

Page 119 **1.** 22, 23 **3.** $-12, -11, -10$ **5.** $-27, -26, -25, -24$ **7.** 72, 73, 74 **9.** 9, 11, **11.** 8, 10 **13.** 6, 7, 8, 9 **15.** 4, 6 **17.** 8, 9 **19.** $\{-28, -24, -20\}$; $\{-24, -20, -16\}$; $\{-20, -16, -12\}$

Page 123 **1.** 8 hr. **3.** 2 hr. **5.** 240 mi. **7.** 2 mi. **9.** 1000 mi. **11.** $1\frac{1}{2}$ hr. **13.** 50 m.p.h.; 60 m.p.h. **15.** 300 mi. **17.** 140 m.p.h. **19.** 5:00; bus

Page 126 **1.** 6 coins of each type **3.** 12 first class passengers **5.** 20 lb. **7.** 6 bags of peanuts **9.** 39 quarters; 27 dimes **11.** inconsistent **13.** insufficient information **15.** 20 nickels; 40 dimes; 60 quarters **17.** insufficient information **19.** insufficient information **21.** $5200 **23.** inconsistent **25.** insufficient information

Page 119 21. 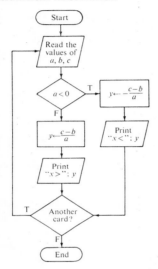 **Page 126 27.**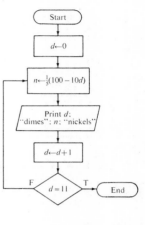

Page 128 **1.** $x < -4\frac{3}{5}$;

3. $y > 2$ or $y < -\frac{1}{2}$; **5.** $x < -1$ or $x > 6$ **7.** 10:00 A.M.

Page 132 **1.** $7m + n$ **3.** $-x - 7y$ **5.** $2x^2 + x$ **7.** $6rs - 20s$ **9.** $6.6y + 5.2x$ **11.** $-\frac{1}{7}t + \frac{3}{13}s$ **13.** $4x^2 + x + 5$ **15.** $3x^3 - 2x^2 + x + 7$ **17.** Correct **19.** Correct **21.** Correct **23.** $3n + 8$ **25.** $\frac{3}{4}z - 1$ **27.** $4.3x^2 - 2.2$ **29.** $4z^3 - z^2 - 4z$ **31.** $5x^3 + x^2 - 2x$ **33.** $3a^4 - a^3 - 2a^2 - 3$ **35.** $3t^6 + t^5 + t - 2$

Page 134 **1.** $2a + 3$ **3.** $2y + 8$ **5.** $7x - 3y$ **7.** $-x + 3$ **9.** $a^2 - 3a + 7$ **11.** $3ax$ **13.** $2r + s$ **15.** $-u + 2v$ **17.** $2z^2 - z$ **19.** $x^4 - x^2 + x - 1$ **21.** $\{-7\}$ **23.** $\{-6\}$ **25.** $\{6\}$ **27.** $\{21\}$ **29.** $\{2\}$ **31.** $\{-3\}$ **33.** $\{-10\}$ **35.** $\{10\}$ **37.** $\{8\}$ **39.** $\{17\}$ **41.** $8x^2 - 4x + 6$ **43.** $6 - 3x - x^2$ **45.** $4x$

Page 136 **1.** $6y^3z$ **3.** $-2m^3n^4$ **5.** $-2x^3y^2$ **7.** $-8y^3$ **9.** $30a^3b^6$ **11.** $-x^4y^2z^3$ **13.** $\frac{1}{15}m^3n^6$ **15.** u^7v^8 **17.** $-27a^6b^6c^3$ **19.** x^{n+1} **21.** $2z^{n+2}$ **23.** $5r^6$ **25.** $-9h^3k^4$ **27.** $9m^4n^3p^2$ **29.** $5y^5 - 2y^4$

ANSWERS FOR ODD-NUMBERED EXERCISES

Page 137 1. $16a^4$ 3. $16t^4$ 5. $12z^3$ 7. $-2s^3t^2$ 9. x^7y^8 11. $4a^9b^8$ 13. $4r^8s^6$ 15. $-3c^{12}k^5$ 17. $x^{n+2}z^{n+1}$ 19. $32a^3b^6$ 21. $5r^7s^2t$ 23. 0 25. $-x^2y^2 + 6x^2y - 6xy^2$

Page 139 1. $3a^2 + 6ab + 6b^2$ 3. $-12 + 8x + 12x^2$ 5. $2x^3 + 2x^3y - 6x^2y^2$ 7. $12x^2y + 6x^2y^2 - 3x^3y^3$ 9. $-12p^3q + 8p^2q^2 - 6pq^3$ 11. $21xz^2 + 15x^2z^2 - 9xz^3 + 6x^4z^3$ 13. $-8c^5d^2 - 8c^3d^4 + 32c^4d^2 + 40c^3d^3$ 15. $\{9\}$ 17. $\{-8\}$ 19. $\{15\}$ 21. $\{6\}$ 23. $\{\frac{37}{10}\}$ 25. $\{1\}$ 27. $\{43\}$ 29. $\{6\}$ **Problems** 1. $w^2 + 5w$ (sq. in.) 3. $\frac{1}{2}b^2 - \frac{3}{2}b$ (sq. ft.) 5. $120 + x$ (mi.) 7. $520 + 7x$ 9. $4v + 30$ (mi.) 11. $4x^2 + 128x + 960$ (sq. ft.)

Page 141 1. $a^2 + 3a + 2$ 3. $x^2 - 4$ 5. $z^2 + 3z - 18$ 7. $2a^2 + 5a + 2$ 9. $2n^2 - 7n - 15$ 11. $3x^2 - 18x + 15$ 13. $6r^2 + 13r + 6$ 15. $18d^2 + 3d - 10$ 17. $2x^2 + xy - y^2$ 19. $4x^2 - 49z^2$ 21. $4x^2 - 12xt + 9t^2$ 23. $0.28z^2 + z - 2$ 25. $x^4 - y^2$ 27. $a^4 - 2a^2b^2 + b^4$ 29. $x^3 + 4x^2 + x - 2$ 31. $2x^3 + 5x^2 + 7x - 5$ 33. $x^3 - y^3$ 35. $5x^2 + x - 4$ 37. $4a^2 + 6ab - 4b^2$ 39. $x^4 + x^3 + 7x - 3$ 41. $2r^2 + 3rs - rt - 2s^2 + 3st - t^2$ 43. $t^5 - 1$

Page 143 1. 6 in.; 12 in. 3. 11 in.; 14 in. 5. 10 in.; 12 in. 7. 12 in. by 12 in. and 9 in. by 16 in. 9. 4 ft.; 7 ft. 11. 20 in.

Page 145 1. $x^2 + 2x + 1$ 3. $a^2 - 6a + 9$ 5. $a^2 + 2ab + b^2$ 7. $4x^2 - 4xz + z^2$ 9. $a^3 + 3a^2b + 3ab^2 + b^3$ 11. $8x^3 + 12x^2z + 6xz^2 + z^3$ 13. $x^2 + x + \frac{1}{4}$ 15. $a^2 - 0.6a + 0.09$ 17. $2x^3 + 20x^2 + 50x$ 19. $x^3 + x^2y - xy^2 - y^3$ 21. $5x^2 + 4x + 4$ 23. $x^2 + 2xy + y^2 + 2xz + 2yz + z^2$ 25. $x^4 + 4x^3y + 6x^2y^2 + 4xy^3 + y^4$ 27. $16a^2b^2$ 29. $6c^2d + 2d^3$ 31. $x^{3n} + 9x^{2n} + 27x^n + 27$

Page 146 1. 7 in.; 10 in. 3. 13, 14 5. 9, 11 7. 15, 16 9. 33 in.; 15 in. 11. 2 in.; 4 in.

Page 149 1. t^7 3. 3 5. $-4x^2y$ 7. $\frac{n}{2}$ 9. $\frac{8}{p^5q^3}$ 11. $\frac{3}{r^{12}}$ 13. $\frac{3x}{2y}$ 15. $-\frac{2m^4}{n^2}$ 17. $-\frac{1}{3x^3y^5}$ 19. $7t$ 21. $4c^2$ 23. $-4x^4y^2$ 25. cd 27. $6a^2$ 29. $-2m^4n^2$

Page 151 1. $\frac{x}{y^3}$ 3. $\frac{1}{r^2s}$ 5. $\frac{u^2}{z^2}$ 7. 1 9. $\frac{r^6}{s^3}$ 11. $\frac{z}{2}$ 13. a^2b^{-2} 15. $x^{-4}y^3$ 17. $a^{-3}b^2c^{-1}d^4$ 19. $4y^2$ 21. 0 23. $5x^3y^3$ 25. $-2x$ 27. 2×10^9 29. $\frac{41}{9}$ 31. $1 + x - y$

Page 153 1. $2a + 4$ 3. $2 - z$ 5. $z + 2$ 7. $4p - 1$ 9. $3 + \frac{5}{t}$ 11. $2 + \frac{1}{2k}$ 13. $\frac{3m}{2n} - \frac{1}{2}$ 15. $\frac{4}{y} + \frac{2}{x}$ 17. $-4x^2 + 3x - 2$ 19. $x + 2y - 3xy$ 21. $-z + 4 - \frac{2}{z}$ 23. $10x - 14$ 25. $5k - 4t + 3$ 27. $2x^2y + 4x - xy$ 29. 1.6 31. $\emptyset$ 33. $\{1\}$

Page 156 1. $y + 2$ 3. $x - 3$ 5. $a - 5$ 7. $x + 1 + \frac{2}{x+1}$ 9. $x + 4$ 11. $m - 2$ 13. $x + 2$ 15. $3p + 3 + \frac{11}{2p-3}$ 17. $x^2 + 2xy + 4y^2$ 19. $2t + 3s$ 21. $16n^2 + 20mn + 25m^2$ 23. $x^2 - 2x + 3$ 25. $3r^2 - 4r + 5 + \frac{2}{2r+3}$ 27. $x^2 + 5x + 2$ 29. $x^6 + x^4 + x^2 + 1$ 31. $x^2 - 2x + 3$ 33. no 35. -21

Page 158 1. $38y + 70$ 3. $3a + b - c$ 5. $\{4\}$ 7. $3x^6y^3z^3$ 9. $-27r^6s^3t^9$ 11. $-15s^2 + 18s$ 13. $t^2 - 3t - 4$ 15. 8 in.; 11 in. 17. $-\dfrac{4xy}{z^2}$ 19. $\dfrac{x^5z^2}{y^2}$ 21. $-2x^2 + 3x - 4$ 23. $x^2 - 2x + 4$

Page 162 1. $2 \cdot 17$ 3. $3^2 \cdot 7$ 5. $2^4 \cdot 3^2$ 7. prime 9. 1, 3, 5, 15 11. 1, 2, 3, 4, 6, 8, 12, 24 13. 1, 2, 3, 5, 6, 10, 15, 30 15. 1, 2, 3, 4, 6, 12, 37, 74, 111, 148, 222, 444 17. 12 19. 35 21. 63 23. $22ab$ 25. $21x^2yz^2$ 27. $16s^2$ 29. ab 31. $3x$ 33. $-7uv$ 35. $4s$ 37. z^4 39. $(2t)^3$ 41. $(5xy)^2$ 43. $(4a^3b^2)^3$

Page 164 1. $4(x^2 - 2)$ 3. $a(4a + 5)$ 5. $5z^2(2z - 1)$ 7. $3ab(2a + b)$ 9. $3(p^2 + p - 3)$ 11. $5(n^2 - 3n + 4)$ 13. $ax(a^2x^2 + ax - 1)$ 15. $4b^2(y^2 - 2y + 6)$ 17. $5xy(3x - 6 + 7y)$ 19. $3x^3y^2(3x^2 - 2xy + y^2)$ 21. $(n + 3)(n - 1)$ 23. $(3a - b)(b + a)$ 25. $(t^2 - 5)(y + 5)$ 27. $(n^2 + 1)(2n + 1)$ 29. $(a^2 + b)(a - b)$ 31. $(5c - 1)(a^3 + b)$ 33. $(n + p)(n + 2)$ 35. $(3a - b)(b + a)$ 37. $n(m + 1)(n + 2)$

Page 165 1. $A = r^2(4 - \pi)$ 3. $A = 2r^2(6 - \pi)$ 5. $A = \pi(R^2 - r^2)$ 7. $A = \pi(R^2 - 3r^2)$

Page 167 1. 96 3. 391 5. 375 7. 1599 9. 1564 11. 8096 13. 9975 15. $35\tfrac{8}{9}$ 17. 999,600 19. 0.91 21. $y^2 - 4$ 23. $x^2 - y^2$ 25. $t^2 - 36$ 27. $4a^2 - 1$ 29. $y^4 - 25$ 31. $9r^2 - \tfrac{1}{4}$

Page 168 1. $(t + 3)(t - 3)$ 3. $(2m + 1)(2m - 1)$ 5. $(t + u)(t - u)$ 7. $(5v + 7)(5v - 7)$ 9. $(15a + b)(15a - b)$ 11. $(ab^2 + c)(ab^2 - c)$ 13. $(2y + 1)(2y - 1)$ 15. $(5x + 6y)(5x - 6y)$ 17. $4(r + 4)(r - 4)$ 19. $(x + 13)(x - 13)$ 21. $(9t + 10s)(9t - 10s)$ 23. $3(t + 3) \times (t - 3)$ 25. $x^2(x + 5)(x - 5)$ 27. $(c + b)(c - b)$ 29. $3(7z + 5)(7z - 5)$ 31. $(t^2 + k^2) \times (t + k)(t - k)$ 33. $(1 + n^n)(1 - n^n)$ 35. $(x^n + 3)(x^n - 3)$ 37. $(k + a^n)(k - a^n)$ 39. $(l^3 + w^2h)(l^3 - w^2h)$ 41. $(x^{2n} + y^{2n})(x^n + y^n)(x^n - y^n)$ 43. $3(2x + 3)$ 45. $4x$ 47. a. $(n + 1)^2 - n^2 = 2n + 1 = n + (n + 1)$ b. $(2n + 3)^2 - (2n + 1)^2 = 2(4n + 4) = 2[(2n + 3) + (2n + 1)]$

Page 170 1. $p^2 + 14p + 49$ 3. $4x^2 - 4x + 1$ 5. $16t^2 + 24t + 9$ 7. $36r^2 + 60r + 25$ 9. $4x^2 - 12xy + 9y^2$ 11. $x^2y^2 - 2xy + 1$ 13. $x^4 + 4x^2 + 4$ 15. $u^4v^4 + 14u^2v^2 + 49$ 17. $x^2 - \tfrac{4}{3}x + \tfrac{4}{9}$ 19. $\tfrac{1}{4} - m + m^2$ 21. $n^2 - 0.6n + 0.09$ 23. $\tfrac{9}{16}y^2 + 2y + \tfrac{16}{9}$ 25. $0.36x^2 + 1.8x + 2.25$ 27. $x^6 + 2x^3y^2z + y^4z^2$

Page 172 1. $(n - 1)^2$ 3. $(k - 3)^2$ 5. $(r + 5)^2$ 7. $(2p - 1)^2$ 9. $(4c + 1)^2$ 11. $(2x - y)^2$ 13. $(1 - 2t)^2$ 15. $(8y - z)^2$ 17. $(3 + 2p)^2$ 19. $(2xy - 3z)^2$ 21. $(x^2 + 1)^2$ 23. $(5y^2 - x)^2$ 25. $7x(x + 1)^2$ 27. $x(x - 5)^2$ 29. $6x(2 + x)^2$ 31. $4(y + x)^2(y - x)^2$ 33. $(x + y + 1) \times (x - y + 1)$ 35. $(a + b - 1)(a - b + 1)$ 37. 9 39. 4

Page 174 1. $n^2 + 9n + 18$ 3. $x^2 + 5x - 50$ 5. $2x^2 + 5x + 2$ 7. $3t^2 + 11t - 4$ 9. $6x^2 - 13x + 6$ 11. $6r^2 - 7r - 5$ 13. $6 + x - x^2$ 15. $3 - 2s - 8s^2$ 17. $-x^2 + 5x - 6$ 19. $2z^2 + 5z + 3$ 21. $x^3 + x^2 - 2x$ 23. $6y^3 + 28y^2 - 10y$ 25. $\{2\}$ 27. $\{1\}$ 29. $x^2 + \tfrac{1}{3}x - \tfrac{2}{9}$ 31. $y^2 + \tfrac{2}{5}y - \tfrac{8}{25}$ 33. $a^2 - 2.9 - 0.62$ 35. $6.3z^2 - 2.97z - 0.34$ 37. $\{2\}$ 39. $\{-4\}$

Page 177 1. $(n + 7)(n + 1)$ 3. $(y + 4)(y + 1)$ 5. $(k - 6)(k - 1)$ 7. $(a - 8)(a - 1)$ 9. $(x + 5)(x + 2)$ 11. $(n - 13)(n - 2)$ 13. $(7 + k)(2 + k)$ 15. $(7 - u)(6 - u)$ 17. $(x + 7y)(x + y)$ 19. $(m - 7n)(m - 4n)$ 21. $(r - 4t)(r - 19t)$ 23. $(a - 24b)(a - 2b)$ 25. $11, 7, -11, -7$ 27. $21, 12, 9, -21, -12, -9$ 29. 2 31. $5, 8, 9$ 33. $(x + y - 3) \times (x + y - 1)$ 35. $(x + 1)(x - 1)$ 37. $5 = 5 \cdot 1, (-5)(-1); 5 + 1 \neq 3, -5 + (-1) \neq 3$ 39. $1 = 1 \cdot 1, (-1)(-1); 1 + 1 \neq 0, -1 + (-1) \neq 0$

Page 179 1. $(a + 3)(a - 2)$ 3. $(x - 3)(x + 1)$ 5. $(c - 5)(c + 2)$ 7. $(u + 9)(u - 2)$ 9. $(z - 7)(z + 3)$ 11. $(x + 8)(x - 7)$ 13. $(a + 5)(a - 4)$ 15. $(x - 10)(x + 6)$ 17. $(a - 4b) \times (a + 2b)$ 19. $(p - 8q)(p + 3q)$ 21. $(r - 12s)(r + 2s)$ 23. $(x - 12y)(x + 3y)$ 25. $-9, -3, 9, 3$ 27. $-11, -4, -1, 1, 4, 11$ 29. $-23, -10, -5, -2, 2, 5, 10, 23$ 31. $-2, -6$ 33. $-4, -10$ 35. $-3, -8$ 37. $(s + t - 11)(s + t + 6)$ 39. $(y + 7)(y - 7)$

Page 181 1. $(2x + 1)(x + 1)$ 3. $(3t + 1)(t + 2)$ 5. $(5r - 2)(r - 1)$ 7. $(3y - 2)(y + 3)$ 9. $(t - 3)(4t + 1)$ 11. $(5n + 2)(n - 1)$ 13. $(7x - 3)(x - 1)$ 15. $(5k - 7)(k + 1)$ 17. $(2y - 3x)(y + 2x)$ 19. $(3x + 5)(4x - 3)$ 21. $(3t + 2)(2t + 7)$ 23. $(3u + 4)(2u - 3)$ 25. $(10y - 9)(y + 2)$ 27. $[2(x + y) + 5][4(x + y) - 3]$ 29. $(4x - 7)(6x - 5)$

Page 182 1. $4(x + 1)(x - 1)$ 3. $2(y + 2)(y + 1)$ 5. $(7x - 1)(x + 2)$ 7. $-3a(a^2 + b^2)$ 9. $a(r - 4)(r + 1)$ 11. $(6t - 5)(t - 1)$ 13. $2(5r - 1)^2$ 15. $(3y - 5)(y + 1)$ 17. $(12y + 7x^2) \times (12y - 7x^2)$ 19. $(2n - 3)(3n + 5)$ 21. $(7 + s)(3 - s)$ 23. $(4x + 9)(7x + 6)$ 25. $2ab(3a + 2b)(a - 5b)$ 27. $6n(7m + 3n)(m - n)$ 29. $s^2(3s - 2)(2s - 5)$ 31. $(z^2 + r^2) \times (z + r)(z - r)$ 33. $(z + 1)(z - 1)(z + 3)(z - 3)$ 35. $-3a(7 + z)(2 + z)$ 37. $(u - 3) \times (u - 12)(u + 2)$ 39. $(x + 5)(2x + 3)(x - 3)$ 41. $(x + 3)(x - 3)(2x - 5)(2x - 3)$ 43. $(3x - 2)(x + 1)(3x + 2)(x - 1)$ 45. $6b(z + w)(z - w)$ 47. $(-5x + 2y)(4x - 7y)$ 49. $(x^2 - 6)(x^2 + 7)$ 51. $(n + 1)^2(n - 1)$ 53. -18 55. 40

Page 184 1. $\{0, 5\}$ 3. $\{6, 8\}$ 5. $\{-2, 7\}$ 7. $\{-4, \frac{3}{2}\}$ 9. $\{\frac{1}{2}, \frac{7}{3}\}$ 11. $\{-\frac{7}{2}, -\frac{5}{2}\}$ 13. $\{0, 1, -3\}$ 15. $\mathcal{R}$ 17. $\{6\}$ 19. $\{\frac{2}{3}, -\frac{1}{4}\}$ 21. $\{14, \frac{8}{3}\}$ 23. $\{\frac{35}{3}, -\frac{6}{5}\}$

Page 187 1. $\{-3, 2\}$ 3. $\{-3, 1\}$ 5. $\{4, -4\}$ 7. $\{0, 5\}$ 9. $\{7, -1\}$ 11. $\{-8, 7\}$ 13. $\{-3, -5\}$ 15. $\{-\frac{1}{2}\}$ 17. $\{-\frac{3}{2}, 2\}$ 19. $\{-\frac{5}{2}, -2\}$ 21. $\{-6, 11\}$ 23. $\{\frac{2}{3}, 5\}$ 25. $\{\frac{2}{3}, 5\}$ 27. $\{\frac{3}{2}, -5\}$ 29. $\{0, 4\}$ 31. $\{-1, 1, -2, 2\}$ 33. $\{0, \frac{5}{2}, -2\}$ 35. $\{0, \frac{3}{2}\}$ 37. $\{2, -1\}$ 39. $\{-\frac{1}{2}, 2\}$ 41. $x^2 - 9x + 18 = 0$ 43. $p^3 - 7p^2 + 10p = 0$ 45. $p^3 - 3p^2 - p + 3 = 0$

Page 190 1. $7, 8$ 3. $-8, -7$ 5. 4 ft.; 7 ft. 7. $3\frac{1}{2}$ sec. 9. 40 ft. 11. 23 in.; 7 in. 13. 5 sec. 15. 3 sec. 17. 6 in. 19. -3 21. 7 25. 4 in.

Page 193 1. 30 3. $15t^2(3t - 1)$ 5. $(x^2 + 5)(x - 3)$ 7. $x^2y^2 - 9$ 9. $(z + 11)(z - 11)$ 11. $16a^2 - 24ab + 9b^2$ 13. -1 15. $(x - 7y)(x - y)$ 17. $(p + 6)(p - 2)$ 19. $(3a - 4b) \times (2a + 3b)$ 21. $\{-3, 9\}$ 23. $\{-11, 8\}$ 25. 3 ft.; 8 ft.

Page 195 1. $\dfrac{6}{x}$; $x \neq 0$ 3. $\dfrac{23}{100}$ 5. $\dfrac{y}{2}$ 7. $\dfrac{x - 1}{x}$; $x \neq 0$ 9. $\dfrac{h}{3h - 6}$; $h \neq 2$ 11. $\dfrac{7z^2 + 2}{5}$ 13. $\dfrac{p - 3}{7p + 14}$; $p \neq -2$ 15. $\dfrac{1}{z(z - 7)}$; $z \neq 0, z \neq 7$ 17. $\{3, 5\}$ 19. $\{7, -4\}$ 21. $\{-\frac{1}{3}, 2\}$ 23. $\{7, -7\}$ 25. $c \neq 0, c \neq d$ 27. $x \neq y$ 29. $p \neq \dfrac{r}{2}, p \neq -r$

Page 198 1. $5a$ 3. $\dfrac{2}{3x}$; $x \neq 0$ 5. $-a$; $a \neq 0, b \neq 0$ 7. $\tfrac{2}{5}$; $c \neq -d$ 9. $\dfrac{7}{a+b}$; $a \neq -b$ 11. $-3 - k$; $k \neq 3$ 13. $\dfrac{3p}{q + 2p}$; $p \neq 0, q \neq 0, q \neq -2p$ 15. $\dfrac{x-3}{x+3}$; $x \neq -3$ 17. $\dfrac{p}{q}$; $q \neq 0, q = \dfrac{1}{p}$ 19. $\dfrac{z^2+2}{z+2}$; $z \neq -2$ 21. $\dfrac{x}{x+1}$; $x \neq 3, x \neq -1$ 23. $\dfrac{x+3}{x-2}$; $x \neq 2, x \neq -2$ 25. $\dfrac{x+3y}{x-3y}$; $x \neq 3y, x \neq -3y$ 27. $\dfrac{y+2}{y}$; $y \neq 0, y \neq 4$ 29. $\dfrac{n-3}{n-2}$; $n \neq 2$ 31. $\dfrac{n+3}{n-5}$; $n \neq 5, n \neq 2$ 33. y is not a factor of num. or denom. 35. t is not a factor of num. or denom. 37. $\dfrac{2y-5}{y-4}$; $y \neq 4, y \neq -2$ 39. $\dfrac{r-1}{r+5}$; $r \neq -5, r \neq -\tfrac{3}{2}$ 41. $\dfrac{3(2x+3)}{4x-5}$; $x \neq 0, x \neq \tfrac{4}{3}$, $x \neq \tfrac{5}{4}$ 43. $\dfrac{x-3}{x+a}$; $x \neq -a, x \neq -2$

Page 200 1. $\tfrac{2}{1}$ 3. $\tfrac{9}{25}$ 5. $\tfrac{1}{16}$ 7. $\tfrac{62}{1}$ 9. $\tfrac{15}{16}$ 11. $\tfrac{4}{3}$ 13. $\tfrac{3}{2}$ 15. $\tfrac{9}{5}$ 17. $\tfrac{1}{2}$ 19. $\tfrac{3}{4}$ 21. $-\tfrac{3}{8}$ 23. $\tfrac{5}{3}$ 25. $\tfrac{1}{1}$

Page 202 1. 35 3. 200 men 5. 91 mi. 7. $12,000 9. 38 ft. 11. 16 oz. at 72¢ 13. 646,926 trees 15. cement, 110 lb.; sand, 440 lb.; gravel, 550 lb. 17. first person, $960; second person, $600; third person, $240

Page 204 1. 21.6 3. 7.77 5. 4 7. 30 9. 4 11. 4.92 13. 30 15. 85 17. 218 19. 615 21. 75% 23. 300% 25. 0.5% 27. 1600% **Problems** 1. 2304 persons 3. 9 min. 5. $3100 7. $37,200 9. 4.5% 11. 7% 13. 5% 15. $2840 17. $422.22

19. a. 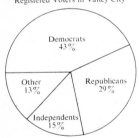 b. 4154 21.

Page 207 1. $\tfrac{18}{55}$ 3. $\tfrac{2}{15}$ 5. $\tfrac{1}{2}$ 7. $\tfrac{1}{3}$ 9. $4a$ 11. $\dfrac{3z^2}{2}$ 13. $-2x^2y^2$ 15. $\dfrac{3(x+5)}{2(x+10)}$ 17. $\tfrac{1}{4}$ 19. $\dfrac{a-b}{a-4}$ 21. $\dfrac{2(z-3)}{z}$ 23. $\dfrac{x(x+2)}{2}$ 25. $\dfrac{r+6}{r-2}$ 27. $\dfrac{x-y}{x^2}$ 29. $\dfrac{n-6}{n-3}$ 31. 1 33. a 35. 1 37. $n+1$ 39. $\dfrac{y-3}{6-y}$ 41. -1

Page 209 1. $\tfrac{4}{7}$ 3. $\dfrac{1}{xy}$ 5. $\dfrac{2q^2}{p^2}$ 7. $\dfrac{s}{8t}$ 9. $-12z$ 11. $\tfrac{1}{6}$ 13. $\dfrac{y-2}{2y}$ 15. $\dfrac{z+2}{z+3}$ 17. 4 19. $\dfrac{3}{2(a-z)}$ 21. $\dfrac{n(n+m)}{n-4}$ 23. $\dfrac{(a+b)(a+3)}{3(a-3)}$ 25. $\dfrac{t(1-2t)}{2(t-2)}$ 27. $\dfrac{(n+3)(n-3)(n+1)}{2(2n-3)(n+7)(n+2)}$ 29. -1 31. $(a+2)(b-3)$ 33. $\dfrac{(a-b)(x-y)}{3ax(x+y)}$

ANSWERS FOR ODD-NUMBERED EXERCISES

Page 211 1. $\dfrac{3}{xy}$ 3. $\dfrac{s^2}{r}$ 5. 2 7. $\dfrac{2st}{9}$ 9. p 11. $\dfrac{x+1}{x(x-3)}$ 13. $\dfrac{t}{2s(st-2)}$ 15. $\dfrac{2(x+1)}{x-1}$ 17. $\dfrac{2(2n+m)}{9(2n-m)}$ 19. $2y - c$

Page 213 1. $\tfrac{11}{17}$ 3. $\tfrac{6}{7}$ 5. $\dfrac{1}{3x}$ 7. $\dfrac{3(x+1)}{2}$ 9. 4 11. $x + y$ 13. $\dfrac{3}{b}$ 15. $a + b$ 17. $\dfrac{k-1}{k-3}$ 19. $\dfrac{1}{z+3}$ 21. $\dfrac{a}{a^2+9}$

Page 215 1. $\dfrac{6+a}{3a}$ 3. $\dfrac{1}{2a}$ 5. $\dfrac{x+6}{6}$ 7. $\dfrac{5z+4}{9}$ 9. $\dfrac{17t+10}{6}$ 11. $\dfrac{2x^2+3x-1}{x^3}$ 13. $\dfrac{5(c+2)}{6c}$ 15. $\dfrac{4x+7}{ax}$ 17. $\dfrac{6x^2-5xy+6y^2}{12xy}$ 19. $\dfrac{2-3c+3d}{c^2-d^2}$ 21. $-\dfrac{2}{3(r+1)}$ 23. $-\dfrac{4y}{(y+2)(y-2)}$ 25. $\dfrac{5t+12}{t^2+5t+6}$ 27. $-\dfrac{1}{y^2+5y+6}$ 29. $\dfrac{5x-2}{(x-3)(x-1)}$ 31. $\dfrac{2x^2+10x-3}{16x^2}$ 33. $-\dfrac{2z-7}{(z-2)(z+1)^2}$ 35. $\dfrac{6z+9}{z^2-4}$ 37. $\dfrac{-2a^3+12a^2+50a+84}{(a^2-9)(a^2-1)(a^2+2a+3)}$ 39. $\dfrac{t+1}{(t-3)(t+2)(t-2)}$

Page 217 1. $\dfrac{(a+1)(a+2)}{a+3}$ 3. $\dfrac{3x}{x-y}$ 5. $\dfrac{y^2-3}{y+2}$ 7. $\dfrac{n+6}{n+2}$ 9. $\dfrac{(a+b)^2}{ab}$ 11. $\dfrac{y^2+3y-7}{y-2}$ 13. $6\tfrac{3}{4}$ 15. $\dfrac{2}{z} + 3z^2$ 17. $2p - \dfrac{3}{7p^2}$ 19. $xy + \tfrac{3}{4}$ 21. $y - 6 + \dfrac{20}{y+3}$ 23. $3t + \dfrac{5}{3t+2}$

Page 219 1. $\tfrac{2}{3}$ 3. 1 5. $\dfrac{2x}{3}$ 7. $\dfrac{xy+y^2}{x^2-xy}$ 9. $\dfrac{y-3}{y}$ 11. $\tfrac{1}{2}$ 13. $\dfrac{x+3}{3}$ 15. $\dfrac{a+2b}{b-a}$ 17. $\dfrac{x-y}{2(x+y)}$ 19. $\dfrac{x+6}{x-1}$ 21. $-\dfrac{x^2+y^2}{4xy}$ 23. $-\dfrac{1}{x-3}$ 25. $\{-2\}$

Page 220 1. $-1, 1$ 3. $\dfrac{n}{n-1}$ 5. $\tfrac{9}{2}$ 7. $3.15 9. $\dfrac{2}{3z}$ 11. 1 13. $\dfrac{3p^2+8p-3}{3p+1}$ 15. $\tfrac{13}{18}$ 17. $\dfrac{1}{2(r-5)}$ 19. $2c + 3 - \dfrac{4}{c}$ 21. $\dfrac{t+1}{2}$

Page 223 1. $\{10\}$ 3. $z < 1$; 5. $\{-4\}$ 7. $b \geq 10$; 9. $\{1\}$ 11. $x > 12$; 13. $\{1\}$ 15. $\{12\}$ 17. $x \leq 84$; 19. $\{\tfrac{5}{2}\}$ 21. $\{-4\}$ 23. $\{6000\}$ 25. $\{3\}$ 27. $\{-3\}$ 29. $\{\tfrac{21}{16}\}$

Page 225 1. 15 lb. 3. 6.25 lb. 5. 24 lb. 7. 6.66 lb. 9. 32 lb. 11. 35% silver alloy, 16 lb.; 65% silver alloy, 4 lb. 13. 16 qt.

Page 226 1. $56 3. $2000 5. $763 7. $4500 9. $1200 at 4%; $2400 at 5% 11. $800 13. $2000 at 4%; $6000 at 5% 15. $9000 at 5%; $12,000 at 4% 17. $4200 at 7%; $3150 at 6%; $3000 at 5% 19. $60,000

Page 229 **1.** {2} **3.** {5} **5.** {3} **7.** {−3, 2} **9.** {−1, 9} **11.** {−2, 1} **13.** {−2} **15.** {10, 20} **17.** {5, 10} **19.** {−1, 12} **21.** {2, 8} **23.** {1, 4}

Page 231 **1.** $34\frac{2}{7}$ sec. **3.** $1\frac{7}{8}$ days **5.** 8 min. **7.** 12 min.; 24 min. **9.** 30 hr. **11.** $10\frac{1}{5}$ hr. **13.** first spillway, $37\frac{1}{2}$ days; second spillway, 25 days **15.** 9 hr.

Page 234 **1.** 15 m.p.h. **3.** 3 m.p.h. **5.** $\frac{1}{2}$ m.p.h. **7.** 5 m.p.h. **9.** 60 m.p.h. **11.** James, 6 m.p.h.; Mike, 12 m.p.h. **13.** 5 yd.

Page 239

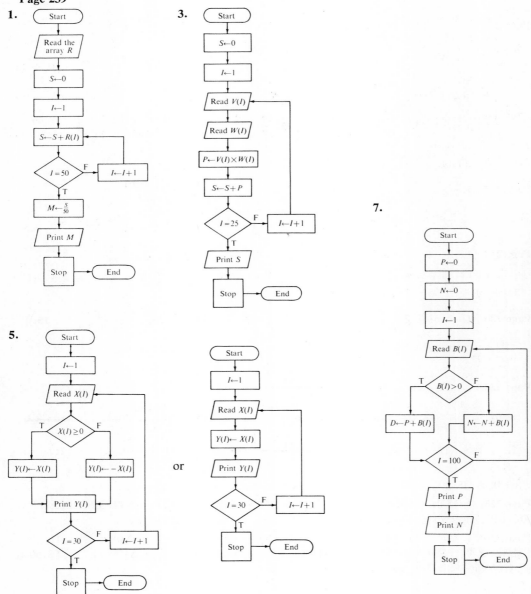

ANSWERS FOR ODD-NUMBERED EXERCISES 375

Page 240 1. $x \leq \frac{3}{2}$; 3. $5000 at 4%; $5000 at 6%
5. $13\frac{1}{3}$ hr.

Page 244
1. domain = {Telstar, Tiros 5, Discoverer 18, Mariner 4}; range = {170, 280, 300, 575} (see right)

3. domain = {1930, 1940, 1950, 1960, 1970}; range = {30,000, 32,000, 35,000, 46,000, 52,000} (see right)

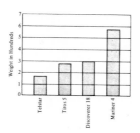

Ex. 1

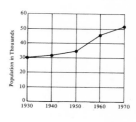

Ex. 3

5.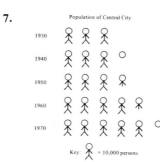

7.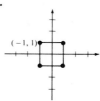

9. $a = 6; b = 3$ 11. $a = \frac{3}{2}; b = 2$ 13. $a = 0; b = 18$ 15. $a = -\frac{1}{2}; b = 4$ or $b = -4$ 17. $a = 1; b = 0$ or $b = 1$ 19. no solution

Page 247
1–11.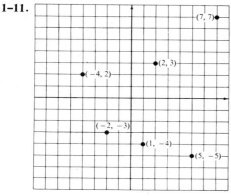

13.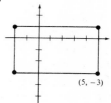

15.

17.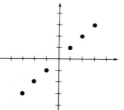

19.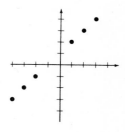

21.

Page 248

1.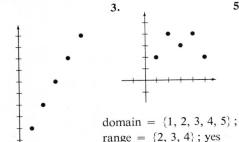

domain = {1, 2, 3, 4, 5};
range = {2, 4, 6, 8, 10}; yes

3.

domain = {1, 2, 3, 4, 5};
range = {2, 3, 4}; yes

5.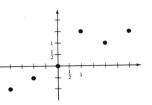

domain = {−2, −1, 0, 1, 2, 3}; range = {−1, −0.5, 0, 1, 1.5}; yes

7.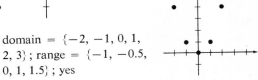

domain = {−2, −1, 0, 1, 2, 3};
range = {0, 1, 4, 9}; yes

9.

domain = {1, 2, 3}
range = {0, 2, 3, 4}; no

11.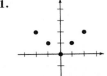

domain = {−2, −1, 0, 1, 2};
range = {0, 1, 2}; yes

13.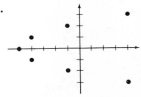

domain = {−5, −4, −1, 4};
range = {−3, −2, −1, 0, 1, 2, 3}; no

15. domain = {all real nos. ≥ 0 and ≤ 3}; range = {all real nos. ≥ 0 and ≤ 3}; yes **17.** domain = {all real nos. ≥ 0 and ≤ 2}; range = {all real nos. ≤ 2 and ≥ −2}; no **19.** domain = {all real nos. ≥ −3 and ≤ 0}; range = {all real nos. ≥ 0 and ≤ 3}; yes **21.** domain = {all real nos. ≥ −2 and ≤ 2}; range = {all real nos. ≥ −2 and ≤ 0}; yes **23.** domain = {all real nos.}; range = {2, −2}; no **25.** domain = {all real nos. ≥ −1 and ≤ 1}; range = {all real nos. ≥ 0 and ≤ 2}; no

Page 251

1.

{(−2, −2), (−1, −1), (0, 0), (1, 1), (2, 2)}

3.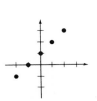

{(−2, −1), (−1, 0), (0, 1), (1, 2), (2, 3)}

5.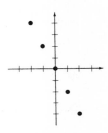

{(−2, 4), (−1, 2), (0, 0), (1, −2), (2, −4)}

7.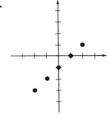
{(−2, −3), (−1, −2), (0, −1), (1, 0), (2, 1)}

9.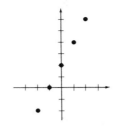
{(−2, −2), (−1, 0), (0, 2), (1, 4), (2, 6)}

11.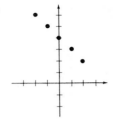
{(−2, 6), (−1, 5), (0, 4), (1, 3), (2, 2)}

13.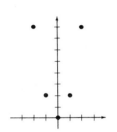
{(−2, 8), (−1, 2), (0, 0), (1, 2), (2, 8)}

15.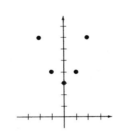
{(−2, 7), (−1, 4), (0, 3), (1, 4), (2, 7)}

17.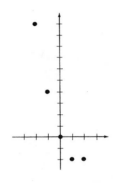
{(−2, 10), (−1, 4), (0, 0), (1, −2), (2, −2)}

19.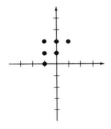
{(−1, 0), (−1, 1), (−1, 2), (0, 1), (0, 2), (1, 2)}

21.
{(0, −1), (0, −2), (1, 1), (1, 0), (1, −1), (1, −2)}

23.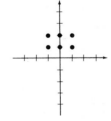
{(−1, 1), (−1, 2), (0, 1), (0, 2), (1, 1), (1, 2)}

25. domain = {1, 2, 3, 4, 5}; range = {4, 5, 6, 7, 8}; $y = x + 3$ **27.** domain = {1, 2, 3, 4, 5}; range = {1, 2, 3, 4, 5}; $y = x$ **29.** {(−3, 1), (0, 4)}   **31.** {(3, 0), (1, 0), (2, −1), (0, 3), (4, 3)}   **33.** {(5, 12), (10, 9), (15, 6), (20, 3)}

Page 254

1.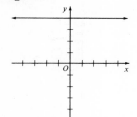

3.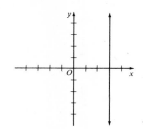

5.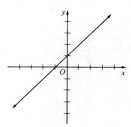

7.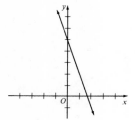

9.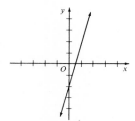

11.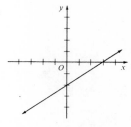

13.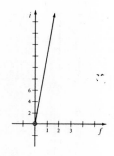

$i = 12f$

15.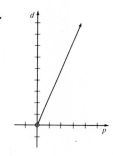

$d = 2.4p$

17.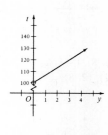

$t = 7y + 100$

19.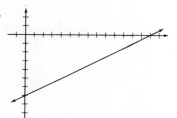

21.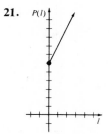

23.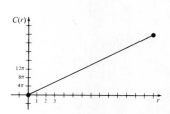

25. a. (5, 0) b. (0, 7) 27. a. (10, 0) b. (0, 5) 29. a. (0, 0) b. (0, 0)

ANSWERS FOR ODD-NUMBERED EXERCISES 379

33.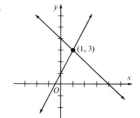

35.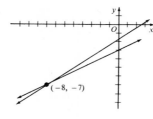

37. no;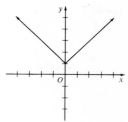

39. no;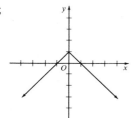

41. no;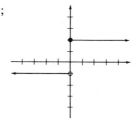

43. no;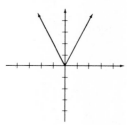

Page 257

1.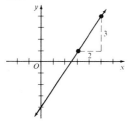

$m = \dfrac{4 - 1}{5 - 3} = \dfrac{3}{2}$

3.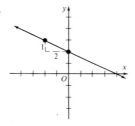

$m = \dfrac{2 - 3}{0 - (-2)} = -\dfrac{1}{2}$

5.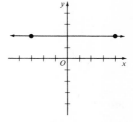

$m = \dfrac{2 - 2}{-3 - 4} = 0$

7.

9.

11.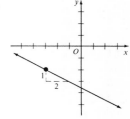

13. $b = 1$;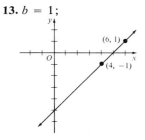

15. $b = 4$;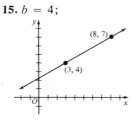

17. b is any real no.

Page 259 **1.** $-2x + y = 5$ **3.** $2x + y = -3$ **5.** $-x + 2y = 14$ **7.** $y = 2$ **9.** $4x + 6y = -1$

11. 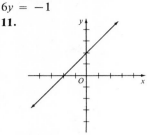   **13.** 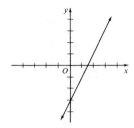   **15.**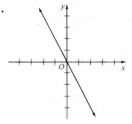

17.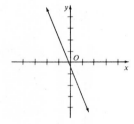

Page 261 **1.** $y = 2x$ **3.** $y = -2x$ **5.** $3y = 2x - 4$ **7.** $y = -\frac{3}{4}x$ **9.** $y = x + 3$ **11.** $x = 2$ **13.** $2y = -x + 1$ **15.** $y = \frac{1}{2}x$ **17.** $a = -1$ **19.** $a = -3$ **21.** $a = 2$ **23.** $y = -x + 3$ **25.** $y = 2x - 2$ **27.** $(-\frac{10}{7}, 0)$ **29.** $(3, -3)$

Page 261 **31.**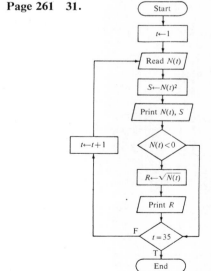

Page 264 **1. a.** $y = 7$ **b.** $y = 140$ **c.** $y = -14$ **3.** $y_2 = 24$ **5.** $m_2 = -25$ **7.** $s_2 = 6$ **9.** $x = 18$ **11.** $w = 10$ **13.** $y = 4$ **15.** $z = 8$ **17.** $x = 8$ or $x = -8$ **19.** $x = 3$ **21.** $y = 11$ **23.** $x = 1$ **25.** $t = kw$ **27.** $d = kr$ **29.** $P = kT$

Page 266 **1.** 58 lb.   **3.** $3\frac{1}{9}$ gal.   **5.** 11,880 votes   **7.** 0.294 ohm   **9.** $3\frac{3}{5}$ in.   **11.** 37 divisions   **13.** Cube of gold is heavier by 4.5384 g.   **15.** $3.78   **17.** 111,059 sq. mi.

ANSWERS FOR ODD-NUMBERED EXERCISES

Page 269

1.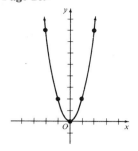
3.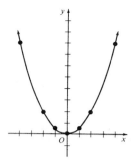
5.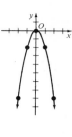

7.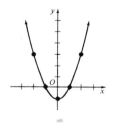
9.
11.

13.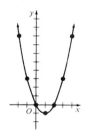
15.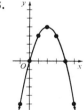
17.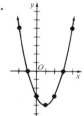
19.

Page 270 1. 25:4   3. 550 ft.   5. 8.75 lb. per sq. ft.   7. 9π sq. in.   9. 9:4   11. 14,400 ft.

Page 273 1. 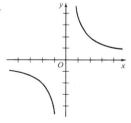   3. 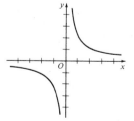   19.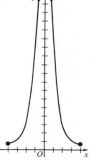

5. $y_2 = 234$ 7. $p_1 = 2.10$ 9. t is reduced to $\frac{1}{3}t$.
11. m is reduced to $\frac{1}{9}m$.   13. $t = 5$ 15. 2 ft.   17. $Q = 40$

Page 274   **1.** 6 hr.   **3.** 4 ft.   **5.** 5%   **7.** 16 in.   **21.**
9. 35 cu. ft.   **11.** 360 r.p.m.   **13.** 5.5 ft.   **15.** 352 lb.
17. Wavelength of higher note is half that of lower note.
19. 1.25 ft.

Page 278   **1.** $r = k\dfrac{d}{t}$ **3.** $t = k\dfrac{w}{r}$ **5.** $T = kVP$
7. $V = khr^2$ **9.** $F = k\dfrac{q_1 q_2}{d^2}$ **Problems**   **1.** H is
multiplied by 36.   **3. a.** $k = 27$ **b.** $W = \dfrac{27xy}{z^2}$
c. $W = 72$ **5.** 57 boys   **7.** 72 lb.   **9.** 9300 g.
11. 18 m.p.h.

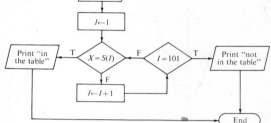

Page 281
1.
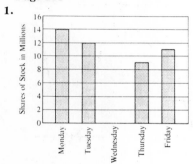

3. $a = 8; b = 3$ **5.**
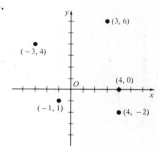

7.
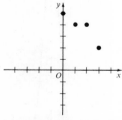

domain = $\{0, 1, 2, 3\}$;
range = $\{5, 4, 2\}$; yes

9. $3; -1$ **11.**
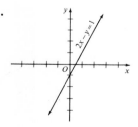

13. $\tfrac{2}{3}$ **15.** $a = -2$

17. $2x + y = 3$ **19.** $2x + y = 4$ **21.** 265 women   **23.** $y_1 = 4\tfrac{1}{2}$ **25.** I is $\tfrac{3}{8}$ as large.

ANSWERS FOR ODD-NUMBERED EXERCISES **383**

Page 285
1. {(1, 1)};
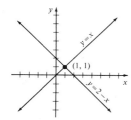

3. {(2, 1)};
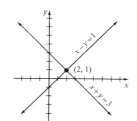

5. {(−2, 2)};
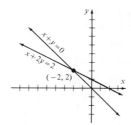

7. ∅;
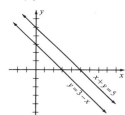

9. {(−2, 0)}
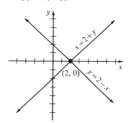

11. {(−1, −1)};
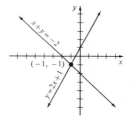

13. {(5.7, 0.7)};
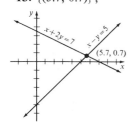

15. {(−3.6, 1.8)};
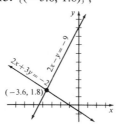

17. {(1, 1)};
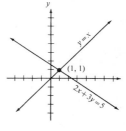

19. 25;
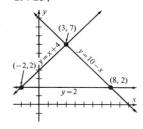

Page 288 **1.** {(8, −1)} **3.** {(5, −3)} **5.** {(8, 2)} **7.** {(7, −2)}
9. {(−3, 2)} **11.** {(−2, −1)} **13.** {(0, −3)} **15.** {(10, 2)}
17. {(5, −2)} **19.** {(6, 2)} **21.** {(3, 1)} **23.** {(45, 29)}
25. {(6, 8)}

Page 289 **1.** 200 ft.; 150 ft. **3.** 10, 7 **5.** 225 points; 200 points **7.** 3 large boxes; 2 small boxes **9.** hamburgers, 45¢; hot dogs, 20¢ **11.** 6 nickels; 9 dimes

384 ANSWERS FOR ODD-NUMBERED EXERCISES

Page 291   **1.** $\{(4, -1)\}$ **3.** $\{(3, 4)\}$ **5.** $\{(1, -1)\}$ **7.** $\{(-3, 2)\}$ **9.** $\{(2, 3)\}$ **11.** $\{(3, -2)\}$ **13.** $\{(\frac{1}{5}, \frac{1}{2})\}$ **15.** $\{(-5, 3)\}$ **17.** $\{(3, 5)\}$

Page 292   **1.** $140   **3.** 15 yr.   **5.** 150 ft.   **7.** potato chips, 59¢; pretzels, 39¢   **9.** stock A, $2000; stock B, $3000

Page 293   **1.** $\{(1, 2)\}$ **3.** $\{(-3, -2)\}$ **5.** $\{(1, 1)\}$ **7.** $\{(5, -1)\}$ **9.** $\{(1, 0)\}$ **11.** $\{(3, 1)\}$ **13.** $\{(6, 4)\}$ **15.** $\{(5, 2)\}$ **17.** $\{(5, 2)\}$

Page 294   **1.** 6, 9   **3.** Lynn, 5 blocks; Janet, 3 blocks   **5.** 7 men   **7.** $\frac{7}{12}, \frac{1}{3}$ **9.** $2000 in special account; $500 in regular account

Page 294   **11.**

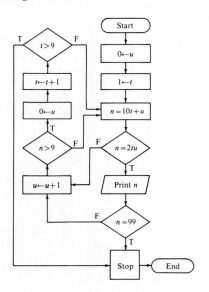

Page 298   **11.**

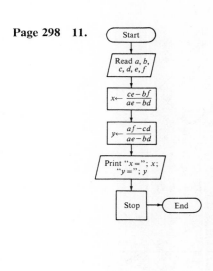

Page 296   **1.** 38   **3.** 39   **5.** 39   **7.** 52   **9.** 79¢   **11.** 632   **13.** 49   **15.** 0.36   **17.** 0.25

Page 298   **1.** 15 m.p.h.   **3.** 5 m.p.h.   **5.** $1\frac{3}{4}$ m.p.h.   **7.** 3 m.p.h.; $2\frac{2}{3}$ m.p.h.   **9.** approx. 144 m.p.h.; 29 m.p.h.

Page 299   **1.** 2 yr.   **3.** 18 yr.   **5.** Ann, 18 yr.; Judy, 24 yr.   **7.** 18 yr.   **9.** $m = \frac{3}{2}l$ **11.** Mary, 16 yr.; Jane, 12 yr.

Page 301   **1.** $\frac{3}{7}$ **3.** $\frac{7}{8}$ **5.** $\frac{12}{18}$ **7.** $\frac{18}{81}$ **9.** $\frac{37}{73}$ **11.** $\{\frac{21}{12}, \frac{42}{24}, \frac{63}{36}, \frac{84}{48}\}$

ANSWERS FOR ODD-NUMBERED EXERCISES 385

Page 304

1.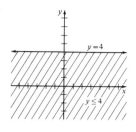

3.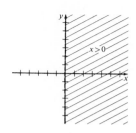

5.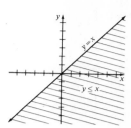

7.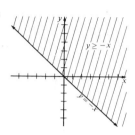

9.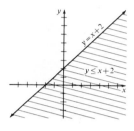

11.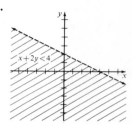

13.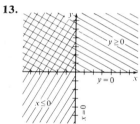

15.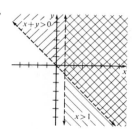

17.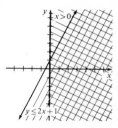

19.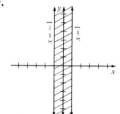

21.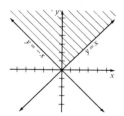

23.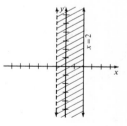

Page 305

1.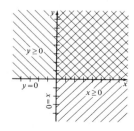

3.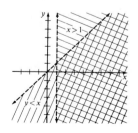

5.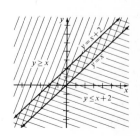

7.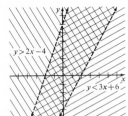

9.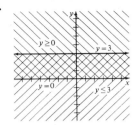

11.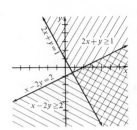

13.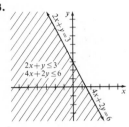

15.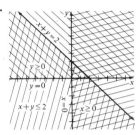

17.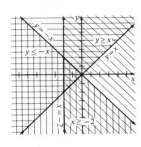

19.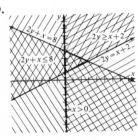

21. a.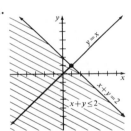

b. $\{y = x;\ x \leq 1;\ y \leq 1\}$

23. a.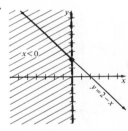

b. $\{y = 2 - x;\ x < 0;\ y > 2\}$

Page 307

1.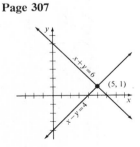

15.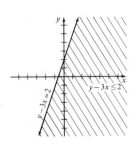

17.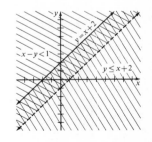

3. $\{(-1, 5)\}$ **5.** 23 lb.; 19 lb. **7.** $\{(2, 1)\}$ **9.** $\{(2, -2)\}$ **11.** 24 **13.** Maria, 28 yr.; Carl, 12 yr. **15.** See above **17.** See above

Page 311 **1.** $>$ **3.** $>$ **5.** $<$ **7.** $<$ **9.** $\{-\frac{3}{4}, \frac{5}{9}, \frac{2}{3}\}$ **11.** $\{-2.1, -2.0, -\frac{4}{3}\}$ **13.** $\{\frac{1}{4}, \frac{4}{15}, \frac{7}{24}, \frac{5}{16}\}$ **15.** $\frac{19}{30}$ **17.** $-\frac{11}{200}$ **19.** $4\frac{31}{48}$ **21.** $\frac{15}{16}$ **27. a.** F **b.** T **c.** T **d.** F

ANSWERS FOR ODD-NUMBERED EXERCISES   387

Page 314   **1.** 0.14   **3.** 0.09375   **5.** $-0.\overline{81}$ **7.** $-0.\overline{428571}$ **9.** 2.05   **11.** -0.1875 **13.** $\frac{33}{100}$
15. $\frac{107}{900}$ **17.** $\frac{13}{110}$ **19.** $\frac{7}{33}$ **21.** $0.00\overline{18}$; e.g., 0.1818   **23.** $0.0\overline{8}$; e.g., 0.88   **25.** 0.001;
e.g., 0.1255   **27.** $\{0.\overline{142857}, 0.\overline{285714}, 0.\overline{428571}, 0.\overline{571428}, 0.\overline{714285}, 0.\overline{857142}\}$ **29.** $\{0.\overline{09}, 0.\overline{18},$
$0.\overline{27}, 0.\overline{36}, 0.\overline{45}, 0.\overline{54}, 0.\overline{63}, 0.\overline{72}, 0.\overline{81}, 0.\overline{90}\}$

Page 317   **1.** 15   **3.** 26   **5.** -36 **7.** $-\frac{16}{7}$ **9.** $\pm\frac{1}{16}$ **11.** $\pm\frac{5}{22}$ **13.** $-7|a|$ **15.** $9a^4|b|$
17. $\pm\frac{k^2}{4}$ **19.** $\{4, -4\}$ **21.** $\{\frac{2}{3}, -\frac{2}{3}\}$ **23.** $\{4, -4\}$ **25.** 5   **27.** 31   **29.** 4   **31.** $\{-3\}$

Page 321   **1.** 1.7   **3.** 3.6   **5.** 43   **7.** -17.6 **9.** 1.73   **11.** 6.38   **13.** -16.53 **15.** 13.78
17. $\{29, -29\}$ **19.** $\{2.4, -2.4\}$ **21.** $\{11.0, -11.0\}$ **23.** 6.4   **25.** $\{2.9, -2.9\}$ **27.** $\{1.8, -1.8\}$
Problems   **1.** 8.2 in.   **3.** 4.89 cm.   **5.** 1.3 m.   **7.** 26.4 m.; 6.6 m.

Page 324   **1.** yes   **3.** yes   **5.** 10 ft.   **7.** $12\frac{1}{2}$ yd.   **9.** 12 mi.   **Problems**   **1.** 8 ft.   **3.** 58.31 ft.
5. 8 in.; 15 in.; 17 in.   **7.** 16 cm.; 30 cm.   **9.** 9.56 ft.

Page 326   **1.** 30   **3.** 10   **5.** $2\sqrt{105}$ **7.** 1   **9.** $\frac{\sqrt{5}}{2}$ **11.** 3   **13.** $4\sqrt{3}$ **15.** $5\sqrt{6}$ **17.** $18\sqrt{2}$
19. 4   **21.** $4\sqrt{2}$ **23.** 1   **25.** $10\sqrt{6}$ **27.** $\frac{4\sqrt{6}}{5}$ **29.** $\frac{\sqrt{5}}{3}$ **31.** $3a\sqrt{6}$ **33.** $-8x^2y$ **35.** $a +$
$2\sqrt{a}$ **37.** -48 **39.** $14x\sqrt{2}$ **41.** $18x^3$ **43.** $12\sqrt{3} + 3\sqrt{6}$ **45.** $-\sqrt{21}$ **47.** $-\frac{3}{5}\sqrt{30}$
49. $2 + 3\sqrt{2}$ **51.** $2 - \sqrt{2}$ **53.** $5x + 1$

Page 327   **1.** 0.9; 4.3   **3.** 6.9 in.; 5.2 in.   **5.** 9.0 sq. in.   **7.** 3.7 ft.   **9.** no; diagonal = 8.5 in.;
diameter = 8.3 in.   **11.** 0.9 ft.

Page 329   **1.** $4\sqrt{3}$ **3.** $\sqrt{7}$ **5.** $5\sqrt{2}$ **7.** 0   **9.** $\frac{4}{3}\sqrt{3}$ **11.** $8\sqrt{6}$ **13.** $\frac{6}{5}\sqrt{10}$ **15.** $4\sqrt{2}$
17. $-\sqrt{2}$ **19.** $2\sqrt{10}$ **21.** $15\sqrt{5}$ **23.** $6\sqrt{2y}$ **25.** $\frac{5}{12}x$ **27.** $\{\frac{5}{2}\}$ **29.** $\{25\}$

Page 330   **1.** 7   **3.** 3   **5.** $13 - 5\sqrt{5}$ **7.** $30 + 10\sqrt{5}$ **9.** $21 - 12\sqrt{3}$ **11.** $15 + 30\sqrt{2}$
13. $28 - \sqrt{10}$ **15.** $-2 + 4\sqrt{3}$ **17.** $\sqrt{2} - 1$ **19.** $-3 - 2\sqrt{3}$ **21.** $-\frac{5 + 3\sqrt{3}}{2}$ **23.** $\frac{10\sqrt{7} - 15}{19}$
25. $3 - 4\sqrt{2}$ **27.** 0   **31.** $a - b^2$ **33.** $6a^2b + 5ac\sqrt{b} - 4c^2$

Page 333   **1.** $\{12\}$ **3.** $\{\frac{1}{28}\}$ **5.** $\{1\}$ **7.** $\{\frac{121}{9}\}$ **9.** $\{3\}$ **11.** $\{14\}$ **13.** $\{2\}$ **15.** $\emptyset$
17. $\{\frac{2}{3}\}$ **19.** $\{48\}$ **21.** $\{37\}$ **23.** $\{18\}$ **25.** $\{4, -4\}$ **27.** $m = \frac{2E}{V^2}$ **29.** $\{\frac{1}{2}\}$ **31.** $\{0\}$
33. $\{5\}$ **35.** $\{0\}$ **37.** $\{4\}$ **39.** $\emptyset$

Page 334   **1.** 121   **3.** 315   **5.** $38\frac{1}{2}$ sq. in.   **7.** 8   **9.** $2023   **11.** 85.8°   **13.** $w = 6(5t)^2 =$
600 (lb.)   **15.** 52

Page 336   **1.** $\frac{7}{25}, \frac{4}{23}, \frac{3}{20}$ **3.** $\frac{47}{48}$ **5.** 0.075   **7.** $-0.\overline{45}$ **9.** $\frac{2}{11}$ **11.** $3\frac{13}{99}$ **13.** 40   **15.** $\{50, -50\}$
17. 1.78   **19.** 12   **21.** 66   **23.** $\frac{1}{2}\sqrt{2}$ **25.** $-2\sqrt{7}$ **27.** $\sqrt{3}$ **29.** $-1 - \sqrt{3}$ **31.** $\frac{-3 - \sqrt{5}}{2}$
33. $\{9\}$

Page 338 **1.** {11, −11} **3.** {$\frac{9}{2}$, −$\frac{9}{2}$} **5.** {$2\sqrt{3}$, −$2\sqrt{3}$} **7.** {$\frac{1}{5}$, −$\frac{1}{5}$} **9.** {$\frac{1}{12}$, −$\frac{1}{12}$}
11. {18, −18} **13.** {1, −5} **15.** {6, −4} **17.** {0, −6} **19.** {1, −$\frac{1}{5}$} **21.** {2, −4}
23. {7, 3} **25.** {$4 + \sqrt{3}$, $4 − \sqrt{3}$} **27.** {$−5 + \sqrt{7}$, $−5 − \sqrt{7}$} **29.** $\left\{\frac{−1 + \sqrt{2}}{2}, \frac{−1 − \sqrt{2}}{2}\right\}$
31. {0, 1, −1} **33.** {0, −3, 3} **35.** {$\sqrt{5}$, 0, −$\sqrt{5}$} **37.** {$\sqrt{5}$} **39.** {−$\sqrt{6}$}

Page 340 **1.** yes **3.** yes **5.** yes **7.** yes **9.** yes **11.** no **13.** $x^2 − 4x + 3 = 0$
15. $x^2 + x − 6 = 0$ **17.** $x^2 − 5x = 0$ **19.** $x^2 − 2x\sqrt{11} + 11 = 0$ **21.** $x^2 − 6 = 0$
23. $x^2 − 4x + 1 = 0$

Page 343 **1.** {4, −6} **3.** {$4 + 2\sqrt{5}$, 8.5; $4 − 2\sqrt{5}$, −0.5} **5.** {$−1 + \sqrt{2}$, 0.4; $−1 − \sqrt{2}$, −2.4} **7.** {5, 0} **9.** $\left\{\frac{1 + \sqrt{13}}{2}, 2.3; \frac{1 − \sqrt{13}}{2}, −1.3\right\}$ **11.** $\left\{\frac{3 + \sqrt{19}}{2}, 3.7; \frac{3 − \sqrt{19}}{2}, −0.7\right\}$
13. $\left\{\frac{1 + \sqrt{17}}{4}, 1.3; \frac{1 − \sqrt{17}}{4}, −0.8\right\}$ **15.** $\left\{\frac{−1 + \sqrt{13}}{6}, 0.4; \frac{−1 − \sqrt{13}}{6}, −0.8\right\}$ **17.** $\left\{\frac{2 + \sqrt{13}}{3}, 1.9; \frac{2 − \sqrt{13}}{3}, −0.5\right\}$ **19.** $\left\{\frac{−1 + \sqrt{7}}{6}, 0.3; \frac{−1 − \sqrt{7}}{6}, −0.6\right\}$ **21.** $\left\{\frac{1 + \sqrt{6}}{5}, 0.7; \frac{1 − \sqrt{6}}{5}, −0.3\right\}$ **23.** {$−1 + \sqrt{2}$, 0.4; $−1 − \sqrt{2}$, −2.4} **25.** $x = 1 \pm \sqrt{1 − c}$ **27.** $x = \frac{−b \pm \sqrt{b^2 − 4c}}{2}$

Page 344 **1.** 9 in.; 11 in. **3.** 11 in.; 14 in. **5.** 5 in.; 12 in. **7.** 80 members **9.** Harry, 5 mi.; Bob, 12 mi.

Page 346 **1.** {5, 4} **3.** {$−5 + 3\sqrt{3}$, 0.2; $−5 − 3\sqrt{3}$, −10.2} **5.** $\left\{\frac{−2 + \sqrt{2}}{2}, −0.3; \frac{−2 − \sqrt{2}}{2}, −1.7\right\}$ **7.** {$1 + \sqrt{6}$, 3.4; $1 − \sqrt{6}$, −1.4} **Page 347** **13.**
9. $\left\{\frac{2 + \sqrt{10}}{3}, 1.7; \frac{2 − \sqrt{10}}{3}, −0.4\right\}$ **11.** $\left\{\frac{−1 + \sqrt{11}}{5}, 0.5; \frac{−1 − \sqrt{11}}{5}, −0.9\right\}$ **13.** {0, 1} **15.** $(x − 1 − \sqrt{2}) \times (x − 1 + \sqrt{2})$ **17.** $(s − 3 − \sqrt{10})(s − 3 + \sqrt{10})$
19. $x = \frac{−2 \pm \sqrt{−28}}{2}$ **21.** $z = \frac{−5 \pm \sqrt{−11}}{2}$

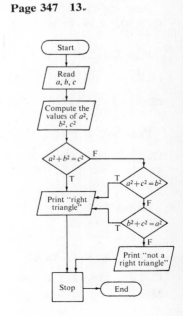

Page 347 **1.** living room, 14 ft. by 21 ft.; kitchen, 14 ft. by 14 ft. **3.** approx. 21.8 in. by 21.8 in. **5.** 9 in.; 12 in.; 15 in. **7.** 500 m.p.h. **9.** 15 m.p.h. **11.** 9 days

ANSWERS FOR ODD-NUMBERED EXERCISES   389

Page 350 **1. a.** Two different real roots
b.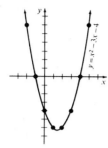

3. a. No real roots
b.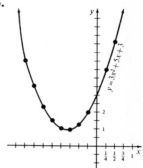

5. a. Two different real roots
b.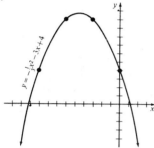

7. a. No real roots
b.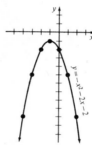

9. a. No real roots
b.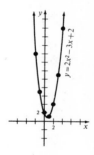

11. $(x - 4)(x - 1)$ **13.** $(-3x + 2)(x - 1)$ **15.** Not factorable over $\Re$ **17.** $\frac{1}{9}y(1 + 2y) \times$
$(1 - 3y)$ **Problems**   **1.** 0.6 sec.; 3.4 sec.; 4 sec. after object is thrown   **3.** 12   **5.** 10 in.   **7.** 2.2
9. $-6; 18$

Page 353

1. $-1 < x < 4$
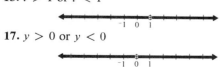

3. $t < -3$ or $t > -2$

5. $-1 < x < 2$

7. $-\frac{1}{2} < y < 2$

9. $0 \leq a \leq 4$

11. $n \leq -2$ or $n \geq 2$

13. $t > 1$ or $t < 1$

15. $\{-4\}$

17. $y > 0$ or $y < 0$

19. $x \geq 0$ or $x \leq -\frac{1}{2}$

21. $x \geq 7$ or $x \leq -4$

Page 354

1.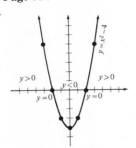

3.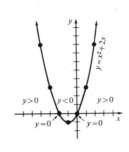

5.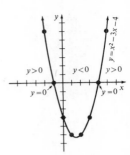

7.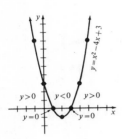

9.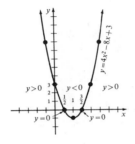

11.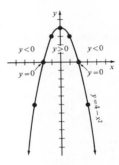

17. $x \geq 4$ or $x \leq -1$

13.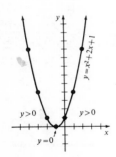

15.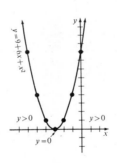

Page 356 **1.** $\{5, -5\}$ **3.** $\{0, \frac{6}{5}\}$ **5.** yes **7.** 0.01; $(x - 0.1)^2 = 0$ **9.** $\{-3 + \sqrt{5}, -0.8; -3 - \sqrt{5}, -5.2\}$ **11.** $\{-3 + \sqrt{2}, -1.6; -3 - \sqrt{2}, -4.4\}$ **13.** Two different real roots

15. $-2 \leq s \leq 1$

index

Abscissa, 246
Absolute value, 54–55
  in open sentences, 116–117
Addition
  axioms for, *see* Axioms
  of fractions, 212–215
  identity element for, 49
  on the number line, 47–49
  of polynomials, 130–131
  of radical expressions, 328–329
  rules for, 57
Additive inverse, 52
Array, 237
Associative axioms, 45
Assumption, 76
Axiom(s), 42
  addition, of order, 107
  additive, of zero, 49
  associative, of addition, 45
  associative, of multiplication, 45
  closure, for addition, 42
  closure, for multiplication, 42
  commutative, of addition, 44
  commutative, of multiplication, 44
  of comparison, 106
  distributive, 60–62
  of equality, 43
  multiplication, of order, 108
  multiplicative, of one, 63
  of opposites, 52
  of reciprocals, 67
  reflexive property, 43
  symmetric property, 43
  transitive, of order, 106
  transitive property, 43

Bar graph, 242
Base
  of percentage, 203
  of a power, 24
Between, 10
Binary operations, 45

Binomial(s), 130, 166–178
  factoring products of, 175–178
  multiplying, at sight, 173–174
  multiplying sum and difference, 166
  squaring, 169–170
Boundary line of a half-plane, 302
Braces, 3
Brackets, 3
Broken-line graph, 243

Circle graph, 205
Closed half-plane, 302
Closure axioms, 42
Coefficient, 24
Commutative axioms, 42
Comparing rational numbers, 309
Comparison, axiom of, 106
Completeness, property of, 319
Completing a trinomial square, 341–343
Complex fractions, 218
Components, of ordered pairs, 243
Compound interest, 235
Conclusion, 76
Conjugates, 330
Conjunction
  of inequalities, 113–114
  of statements, 10
Consecutive integers, 117–119
Consistent system of equations, 284
Constant, 20
  of proportionality (variation), 262
Constant term, 174
Converse of the Pythagorean theorem, 324
Coordinate(s)
  in a plane, 245–246
  of points on a line, 6
Coordinate axes, 246
Coordinate plane, 246
Correspondence, one-to-one, 15
Counting numbers, 15
Cubic equation, 185

Decimal(s)
  approximation of a square root, 320
  and common fractions, 312–313
  rounding off, 314
  terminating, 312
Decimal forms of rational numbers, 312–314
Degree
  of a monomial, 130
  of a polynomial, 130–131
  of a polynomial equation, 185
Denominator, 195
  excluded values, 195
  least common, 214
  rationalizing the, 326, 330
Density, property of, 310
Difference, 79
Difference of two squares, 167–168
Direct proof, 76
Direct variation, 262–264
  linear, 262
  quadratic, 268
Directed number, 7
Discriminant, 350
Disjoint sets, 111
Disjunction, 115
Displacements on the number line, 47
Distance between two points, 54, 351
Distributive axiom, 60–62
Division, 85–88
  of fractions, 209–210
  of polynomials, 152–156
  of radical expressions, 325–326, 330
  rules of exponents, 148
  transformation of an equation by, 84, 87–88
  by zero, 86–87
Domain
  of a function, 101, 241
  of a relation, 247
  of a variable, 20
Double root, 186

Element of a set, 12
Empty set, 13
  and subsets, 110
Equality, 1–2
  addition property of, 76
  axioms of, 43
  multiplication property of, 83
  of ordered pairs, 243
  of sets, 15
Equation(s), 11
  consistent, 284

cubic, 185
degree of, 185
equivalent, 77
fractional, 228–229
and functions, 101–102
graphs of, 252–260
identity, 96
inconsistent, 284
of a line, 260–261
linear, 185, 252–253
member, 77
polynomial, 185–187
pure quadratic, 337
quadratic, 185, 337–350
radical, 331–333
roots of, 30, 186, 249, 339–340
simultaneous, 284
slope-intercept form, 258–259
standard form of polynomial, 185
standard form of quadratic, 337
systems of linear, 283–301
transforming, 76–102
See also Solving equations
Equivalent equations, 77
Equivalent expressions, 61
Equivalent inequalities, 108
Equivalent systems of open sentences, 285
Evaluating an expression, 21
Expanded form of a power, 24
Expanding an expression, 145
Exponent(s), 24
  negative, 150
  positive, 135, 137, 148
  rule of, for division, 148
  rule of, for multiplication, 135
  rule of, for a power of a power, 137
  rule of, for a power of a product, 137
  zero, 150
Exponential form of a power, 24
Expression(s),
  equivalent, 61
  evaluating, 21
  expanding, 145
  mathematical, 21, 35–36
  mixed, 217
  numerical, 1
  open, 21
  radical, 315, 325–330
  rational, 200
  simplifying, 61
  value of, 1
  variable, 20
Extracting a root, 315
Extremes of a proportion, 263

Factor(s), 24, 160–163
  greatest common, 161
  monomial, 163
  polynomial, 161
  prime, 160
Factoring, 160–162
  difference of two squares, 167–168
  polynomials, 160–165, 167–168, 172–173
  problem solving, 188–190
  products of binomial sums and differences, 175–177, 178
  quadratic trinomials, 179–181
  solving equations by, 185–187
  trinomial squares, 172
  trinomials, 175–176
Finite decimal, 312
Finite set, 16
Flow chart(s), 36–37, 69–70, 235–238
  arrays, 237
  loops in, 235–238
  open sentences in, 36–37
Formula(s), 22–23, 26–27, 96, 242, 357
  quadratic, 344–346
  See also Problems
Fraction(s), 195
  adding, 212–215
  complex, 218
  decimal form of, 313
  denominator, 195
  dividing, 208–209, 210
  excluded values, 195
  lowest terms, 197
  multiplication property of, 196
  multiplying, 206–207, 210
  numerator, 195
  reducing, to lowest terms, 196–198
  rule for multiplying, 206
  subtracting, 212–215
Fractional equations, 228–229
Function(s), 101, 241, 248
  arrow notation, 101
  described by bar graphs, 242
  described by tables, 241
  domain, 101, 241
  graph of, 242–243, 245–246, 252–271
  linear, 254
  as a list of ordered pairs, 241
  quadratic, 267–269
  range, 101
  value of, 102

Graph(s)
  and absolute value, 54–55
  bar, 242
  broken-line, 243
  circle, 205
  of equations, 252–257
  of functions, 242–243, 245–246, 252–271
  hyperbola, 272
  of inequalities, 54–55, 108–109, 302–303, 353–354
  of linear equations, 252–254
  on the number line, 6, 13
  of open sentences, 30
  of ordered pairs, 245–246
  parabola, 269
  pictorial, 242
  of quadratic functions, 267–269
  of relations, 248
  of sets, 12–16
  of systems of inequalities, 304
  of systems of simultaneous equations, 283–285
Greatest Common Factor, 161
Greatest Monomial Factor, 163
Grouping symbols, 3–4, 21, 28

Half-planes, 302
Heron's formula for area of a triangle, 327
Horizontal axis, 245
Hyperbola, 272
Hypotenuse, 322
Hypothesis, 76

Identity, 96
Identity element,
  for addition, 49
  for multiplication, 63
Inclusion symbols, 3–4, 28
Inconsistent equations, 284
Index, root, 315
Inequalities, 2
  axioms, 106–109
  combining, 113
  comparing numbers, 9–10
  equivalent, 108
  with fractions, 222
  graphs of, 54–55, 302–303
  graphs of systems of linear, 304
  quadratic, 351–352
  solving, 106–126, 302–303, 353–354
  system of, 303–304
  transformations of, 108
Infinite set, 15
Integers, 16, 308
  consecutive, 117–119
Interest,
  compound, 235
  simple, 226
Intersection of sets, 110–111

Inverse operations, 89
Inverse variation, 271
Inverses
    additive, 52
    multiplicative, 67
Irrational numbers, 319
Irreducible polynomial, 177

Joint variation, 276–277

Least common denominator, 214
Lever, law of, 242
Like terms, 62
Line(s),
    equation of, 260–261
    number, 6–7
    parallel, 283
    slope of, 255
    y-intercept of, 258
Linear direct variation, 262
Linear equation(s), 185–187, 252–264
    graphs of, 252–254
    slope-intercept form of, 258–259
    systems of, 283–285
Linear function(s), 254
    graphs of, 252–254, 262–264
Linear term, 174
Loops in flow charts, 235–237
Lowest terms, fractions in, 197

Mathematical expression, 20
Means, in a proportion, 263
Member of an equation, 77
Member of a set, 12
Minuend, 133
Mixed expression, 217
Mixed numeral, 216
Monomial(s), 130
    degree of, 130
    division by, 152
    as factors, 163
    multiplication by, 135–136, 138
Multiple, of a real number, 118
Multiple root, 186
Multiplication
    axioms for, *see* Axioms
    of binomials, 140–141, 145, 166, 169–170, 173–174
    of fractions, 206–207, 210
    identity element for, 63
    of monomials, 135–138
    of polynomials, 135–141, 145, 166, 169–170, 173–174
    of radical expressions, 315, 325–330
    rules for, 63–65, 196
    See also, Products

Multiplicative axiom of one, 63
Multiplicative inverse, 67
Multiplicative property of -1, 64
Multiplicative property of zero, 63–64

Natural numbers, 15
    system of, 308
Negative exponents, 150
Negative of a number, 52
Negative numbers, 6
Nested loop, 237
Nonterminating decimal, 312
Null set, 13
Number(s), 1
    comparing, 9–10
    counting, 15
    directed, 7
    graphing, 6, 13
    natural, 15
    negative, 6
    negative of a, 52
    and numerals, 1
    opposite of a, 52
    positive, 6
    prime, 160
    rational, 200, 308–314
    real, 7, 16, 51–53, 308
    whole, 16
Number line, 6–7
    addition on, 47–49
    comparing numbers on, 9–10
    coordinate of a point on, 6
    displacements on, 47
    origin of, 6
Number scale, *see* Number line
Number systems, 308
Numeral, 1
    mixed, 216
Numerator, 195
Numerical coefficient, 24
Numerical expression, 1

One, multiplicative axiom of, 63
One-to-one correspondence, 15
Open expression, 20
Open half-plane, 302
Open sentence(s), 30
    absolute value in, 116–117
    graph of, 30
    root, 30, 249
    solution (truth) set of, 249
    in two variables, 249–250
    See also Equation(s), Inequalities

Operation(s)
  associative, 45
  binary, 45
  closure under, 42
  commutative, 45
  inverse, 89
  order of, 28
  rational, 308
Opposite(s), 80
  axiom of, 52
  of a number, 52
  in products, property of, 65
  of a sum, property of, 56
Order axioms for real numbers, 106–108
Order of operations, 3, 28
Order of real numbers, 106
Ordered pairs, 241
  graphing, 245–246
Ordinate, 246
Origin
  of coordinate plane, 245
  of number line, 6

Parabola, 269
Parallel lines, 283
Parenthesis, 3
Percent, 203, 223–226
Percentage, 203
Perfect squares, 318
Periodic decimals, 312
Pictograph, 242
Plane rectangular coordinate system, 246
Point(s)
  distance between two, 54
  of intersection, 283
  plotting, 245
Polynomial(s), 130
  addition, 130–131
  degree of, 130–131
  division of, 152–156
  expanding powers, 145
  factoring, 160–165, 167–168, 172, 175–190
  irreducible, 177
  multiplication, 138–141, 166–174
  powers of, 145
  prime, 177
  quadratic, 174
  simple form, 130
  subtraction of, 133
Polynomial equation(s), 185
Positive numbers, 6
Postulate, 42
Power(s), 24
  of polynomials, 145
  of a power, 137
  of a product, 136–137

  product of, 135–136
  quotient of, 147–149
Prime factor, 160
Prime number, 160
Prime polynomial, 177
Principal square root, 315
Problem solving, 91–92, 117–128, 288–289
  plan, 92
  using factoring, 188–190
Problems,
  age, 298–299
  area, 142–143
  digit, 295
  about fractions, 300–301
  integers, 117–120
  investment, 225–226
  mixture, 125–126, 223
  motion, 232, 297
  percent, 203–206, 223–225
  rate-of-work, 230–231
  with two variables, 288–290
  uniform-motion, 120–125
Product(s)
  of binomials, 166, 169–170, 173–174
  power of a, 136–137
  of powers, 135–136
  property of opposites in, 65
  property of reciprocal of a, 68
  property of square roots, 316
Program, 36–37
Proof, 76
Property(ies)
  addition, of equality, 76
  of completeness, 319
  density, 310
  multiplication, of fractions, 196
  multiplicative, of equality, 83
  multiplicative, of -1, 64
  multiplicative, of zero, 63–64
  of nonzero product of real numbers, 351
  of opposite of a sum, 56
  of opposites in products, 65
  of pairs of divisors, 319
  product, of square roots, 316
  quotient, of square roots, 316
  of quotients, 147
  of the reciprocal of a product, 68
  reflexive, 43
  square roots of equal numbers, 337
  substitution, 4
  of the sum and product of the roots of
    a quadratic equation, 339
  symmetric, 43
  transitive, 43
  zero-product, 183–184
  See also Axioms

Proportion, 262–264
Proportionality, constant of, 262
Pythagorean Theorem, 322–324
    converse of, 324

Quadrants of a coordinate plane, 245
Quadratic direct variation, 268
Quadratic equation(s), 185
    parabolas, 269
    pure, 337
    roots of, 348–350
    solving, by completing the square, 341–343
    solving, by factoring, 185–187
    solving, by using the quadratic formula, 344–346
    standard form of, 337
    sum and product of roots, 339–340
Quadratic formula, 344–346
Quadratic functions, 267–269
Quadratic inequalities, 351–354
Quadratic polynomial, 174
Quadratic term, 174
Quadratic trinomials, 179–181
Quantifiers, 33–34
Quotient(s), 85
    complete, 154
    of powers, 147–149
    property of, 147

$\mathcal{R}$, 78
Radical equations, 331–333
Radical expression(s), 315
    operations with, 325–330
    rationalizing the denominator, 326, 330
    in simplest form, 325
Radicand, 315
Raising to a power, 315
Range
    of a function, 101
    of a relation, 247–248
Rate of percentage, 203
Ratio, 199–200
Rational expressions, 200
Rational number(s), 200, 308–310
    comparing, 309
    decimal forms for, 312–314
    density property of, 310
    system of, 308
Rational operations, 308
Rationalizing denominators, 326, 330
Real number(s), 7
    and the number line, 7, 12

    opposite of, 51–53
    set of, 12, 16
    system of, 308
Reciprocals, 67–68
Reducing fractions, 196–198
Reflexive property of equality, 43
Relation(s), 247–248
Repeating decimal, 312
Replacement set
    of open sentence, 30
    of a variable, 20
Rise, 255
Root(s)
    double, 186
    of equations, 30, 186, 249, 339–340
    multiple, 186
    nth, 315
    of numbers, 315–316
    of open sentences in two variables, 249
    of polynomial equations, 185–187
    of quadratic equation, 339–340
Root index, 315
Roster
    of a relation, 248
    of a set, 12, 15
Rounding decimals, 314
Rule(s)
    for adding and subtracting fractions, 212
    for addition of real numbers, 57
    for division, 86–88
    for division of fractions, 209
    of exponents for division, 148
    of exponents for multiplication, 135
    of exponents for a power of a power, 137
    of exponents for a power of a product, 137
    for multiplication, 63
    for multiplication of fractions, 206
    for negative exponents, 150
    for order of operations, 28
    for a set, 12
    for subtraction, 80
Run, 255

Set(s), 12–16
    comparing, 15–17
    disjoint, 111
    element of, 12
    empty, 13
    equal, 15
    finite, 16
    graph of, 12–13
    incomplete roster, 15

infinite, 15
intersection of, 110–111
member of, 12
null, 13
replacement, 20
roster of, 12, 15
solution, 30, 249
specifying, 12–13
subset, 110
truth, 30
union of, 111
universal, 110
Similar terms, 62
Simple form
of a polynomial, 130
of a radical expression, 325
Simple interest, 226
Simplifying expressions, 4, 61
Simplifying polynomials, 138
Simultaneous equations, 284
Slope of a line, 255–257
Slope-intercept form of linear equations, 258–259
Solution set
of an open sentence, 30, 249
of a system of linear equations, 284
Solving equations, 30, 76–102
fractional equations, 228–229
by inspection, 90
plan for, 90
quadratic, by completing the trinomial square, 341–343
quadratic, by factoring, 185–187
quadratic, by the quadratic formula, 344–346
systems of linear, 283–293
by transformations, 76–90, 95–97, 222
Solving inequalities, 106–126
quadratic, 351–354
systems of linear, 304
Square(s)
of a binomial, 169–170
factoring the difference of two, 167–168
factoring a trinomial, 172
of a number, 24
perfect, 318
trinomial, 169–170, 341–343
Square root(s)
decimal approximation of, 318–321
geometric interpretation, 322
principle, 315
product properties of, 316
quotient properties of, 316
simplifying, 325–330

Standard form
of a polynomial equation, 185
of a quadratic equation, 337
Statement(s), 1
Subscript, in an array, 237–238
Subset, 110
Substitution, transformation by, 78
Substitution Principle, 4
Subtraction
of fractions, 212–215
of polynomials, 133
of radical expressions, 328–329
rule for, 80
transformation by, 76–81
Subtrahend, 133
Sum, property of opposite of a, 56
Symbols of inclusion, 3–4
Symmetric property of equality, 43
System of linear equations, 283–301
equivalent, 285
solution of, 284
solving by the addition-subtraction method, 286–287, 290–291
solving by the graphic method, 283–285
solving by the substitution method, 293
System of linear inequalities, 303–304

Term(s), 21
constant, 174
like, 62
linear, 174
quadratic, 174
similar, 62
Theorem(s), 76
converse, 324
Pythagorean, 322–324
Transformation of equations
by addition, 78
by division, 85–88
with fractions, 222
by multiplication, 83–84
steps in, 89–90
by substitution, 78
by subtraction, 76–81
Transformation of inequalities, 108
Transitive axiom of order, 106
Transitive property of equality, 43
Trinomial(s), 130
factoring, 172, 175–176, 179–181
quadratic, 169–181
Trinomial square(s)
completing, 341–343
factoring, 172
Truth set, 30

Updating a loop, 236
Unending decimal, 312
Union of sets, 111
Universal set, 110
Universe, 110

Value
 absolute, 54–55, 116–117
 of an expression, 21
 of a function, 102
 of a variable, 20
Variable(s), 20, 36
 domain of, 20
 and quantifiers, 33–34
 replacement set of, 20
 subscripted, in a loop, 238
 value of, 20
Variable expression, 20
Variation
 combined, 277
 constant of, 262
 direct, 262–264
 inverse, 271
 joint, 276–277
 quadratic direct, 268
Venn diagrams, 110
Vertical axis, 245

Whole numbers, 16
Work equation, 230

x-axis, 252
x-intercept, 348

y-axis, 252
y-intercept, 258

Zero
 additive axiom of, 49
 division by, 86–87
 as an exponent, 150
 multiplication by, in transforming equations, 83–85
 multiplicative property of, 63–64
Zero-product property of real numbers, 183–184

821.9208

3 8002 02129 966 6

COVENTRY LIBRARIES

2014

PS130553 DISK 4

Please return this book on or before
the last date stamped below.

checked
20/10/18

Central

To renew this book take it to any of
the City Libraries before
the date due for return

Coventry City Council

CLASSIFICATION: POETRY

This book is sold under the condition that it shall not, by way of trade or otherwise, be lent, resold, hired out or otherwise circulated without the publisher's prior consent in any form of binding or cover other than that in which it is published and without a similar condition including this condition being imposed on the subsequent purchaser.

A CIP catalogue record for this book is available from the British Library.

Printed and bound in Great Britain.

Paper used in the production of books published by United Press comes only from sustainable forests.

ISBN 978-0-85781-322-0

First published in Great Britain in 2013 by
United Press Ltd
Admail 3735
London
EC1B 1JB
Tel: 0844 800 9177
Fax: 0844 800 9178
All Rights Reserved

© Copyright contributors 2013

www.unitedpress.co.uk

National Poetry Anthology 2013

This anthology features all the winning entries from an annual competition which is free to enter. Every winner receives a free copy of the anthology and votes for one overall winner who receives £1,000 and a magnificent trophy to keep for life. If you would like to enter for next year's anthology, send a loose second class stamp and up to three poems (25 lines and 160 words maximum each), to United Press, Admail 3735, London, EC1B 1JB by the annual closing date of June 30th. You can also call us on 0844 800 9177 or visit our website on www.unitedpress.co.uk

■ **Nico Russell receives his trophy and £1,000 cheque from Bebington library manager Claire Oxley, with Peter Quinn (left).**

Youngest Ever National Poetry Anthology Champion

Nico Russell has become the youngest ever winner of the National Poetry Anthology at the age of 30. Nico, from the Wirral, was truly shocked to discover that he had won this - the biggest free-to-enter annual poetry competition in the UK.

"It was a real surprise when I was told I'd won," said Nico. "I never imagined it. This success is almost like a life-changing experience." Nico, who works in creative media, submitted his poem *A British Heart* for the National Poetry Anthology competition, which is open to all UK residents and attracts many thousands of entries.